Radar Evaluation Handbook

Radar Evaluation Handbook

David K. Barton, Charles E. Cook,
and Paul Hamilton, Editors,
and the staff of ANRO Engineering, Inc.

This work was produced under a contract with the
Department of the Army,
Contract No. DAADO5-88-C-0098

Artech House
Boston • London

Library of Congress Cataloging-in-Publication Data

Radar evaluation handbook / David K. Barton, Charles E. Cook, and Paul
  Hamilton, editors, and the staff of ANRO Engineering, Inc.
       p.   cm.
    Includes bibliographical references.
    ISBN 0-89006-488-1
    1. Radar--Evaluation--Handbook, manuals, etc. I. Barton, David
Knox, 1927-   . II. Cook, Charles E. (Charles Emerson), 1926-   .
III. Hamilton, Paul (Paul Charles). IV. ANRO Engineering, Inc.
TK6580.R315   1991   90-23303
621.3848--dc20   CIP

This work was produced under a contract with the Department
of the Army, Contract No. DAADO5-88-C-0098.

Published by
Artech House, Inc.
685 Canton Street
Norwood, MA 02062

International Standard Book Number: 0-89006-488-1
Library of Congress Catalog Card Number: 90-23303

10   9   8   7   6   5   4   3   2   1

Table of Contents

Chapter Six

Chapter Seven

Chapter Eight

Chapter Nine

Chapter Ten

Chapter Eleven

Chapter Thirteen

Appendices

Foreword

The *Radar Evaluation Handbook* was initiated as a Small Business Innovation Research project by the U. S. Army Test and Evaluation Command, and a Phase I contract was awarded to ANRO Engineering in 1987. After this successful planning and outline phase, the Phase II Contract DAAH05-88-C-0098 was awarded to ANRO in September 1988, calling for the preparation of this handbook and the accompanying *Radar Evaluation Software*.

The general outline of this handbook follows that of the engineering text *Modern Radar System Analysis* by David K. Barton. Material from that text has been simplified and organized for use by those technical personnel who are not designers or specialists in the radar field. Mathematical and statistical background, beyond that normally associated with work in technical and electronic fields, is not a prerequisite for understanding and using this book. Material specifically addressed to radar evaluation problems and practices has been added, and appendices have been included to define terms and symbols, to discuss certain problems in greater depth, and to illustrate the system parameters and performance of typical radar systems.

Chapters 1, 2, 8, 9, 10, and 11 and Appendices A, B, C, D, and H were prepared by David K. Barton. Chapters 3, 4, 5, and 6 and Appendices E and F were prepared by Paul Hamilton. Charles E. Cook prepared the portion of Chapter 7 dealing with pulse compression and Appendix G. The discussions of MTI and doppler radar were prepared by Paul Hamilton with assistance from Leonard Hopkins and William Hardenburg. The final section of Chapter 7, on impulse radar, was prepared by Dr. Gerald F. Ross. Robert N. Maglathlin prepared Chapters 12 and 13 and Appendices I, J, and K. All material was reviewed by Barton, Hamilton, and Cook, as editors, by Dr. Walter K. Kahn, and by a panel of government engineers.

The editors would like to express their appreciation to Mr. Joseph E. Knox, Headquarters Test and Evaluation Command, U. S. Army, for his management of the project and to the following government reviewers: Mr. Walter E. Wagner, Jr., Mr. Richard A. McGee, and Mr. Troy L. West.

The camera-ready page copy was typed by Kathleen Chiungos and prepared for publication by Will Klump.

Chapter 1

INTRODUCTION AND RADAR FUNDAMENTALS

1.1 Fundamentals of Radar

1.1.1 Basic Principles

We know that a charged particle in motion, such as an electron, has, in addition to an electric field, a magnetic field associated with it. One does not exist without the other, and the combination is called an *electromagnetic field*.[1] An electric current is said to exist when there is a flow of charge from one place to another. If this electric current is unidirectional, i.e., always flows around the electrical circuit in the same direction, we have *direct current*, or DC, and the electromagnetic field is essentially constant. We can detect and measure the strength of this field, which falls off as the inverse of the distance squared. Direct current is useful in many applications; it is the type of current that we get from batteries and is used in some types of powerful electric motors used for winches, elevators, and the like. Because the electromagnetic field produced by DC is static, there is no radiation resulting, and DC is of no use in sending signals through space.

If the electric charges oscillate, or change direction in a regular manner, then the flow of electrons will move first in one direction and then in the opposite direction, as a function of time. The result is an *alternating current*, or AC, which behaves quite differently from direct current. Because the electrons that make up the current change their direction of motion periodically, an *oscillating electric field* (E-field) results. This, in turn, causes an *oscillating magnetic field* (H-field) to occur. The process keeps being repeated, and the two fields move away from the source in a wave at the speed of light. The result is *electromagnetic (EM) radiation*, upon which the principle of radar is based. Figure 1.1 depicts a representation of a typical electromagnetic wave at one point in time. Note that the spatial orientation of the two fields is orthogonal, or 90° apart. The orientation of the electric or E-field defines the *polarization* of the EM radiation; the polarization shown here is horizontal. The direction of propagation of the wave is perpendicular to both the electric and the magnetic fields.

---

[1] Terms printed in bold italics are defined in the Glossary of Radar Terms, Appendix A of this handbook.

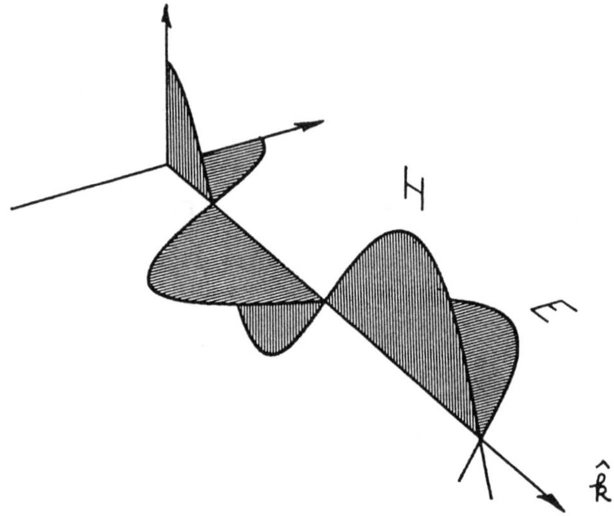

Figure 1.1 Electromagnetic wave in space [1.1].
©Artech House, reprinted by permission.

The principal ways in which EM radiation is described are by its *frequency* (f),[2] *wavelength* (λ), and *intensity* (or field strength). Field strength will be discussed later. The frequency of a particular EM wave (or signal, in radar and in radio communications), is simply the number of oscillations in the wave during one second. Figure 1.2 shows an example of a *waveform* typically found in nature, and one which is of immense importance to radar, the *sine wave*. In this example, the waveform completes one oscillation, or one *cycle*, in one second, i.e., it moves through one complete oscillation, returning to its original value after one second. The term *hertz* (abbreviated Hz) has replaced the term cycles per second previously used, in honor of the scientist who first confirmed the existence of EM radiation.

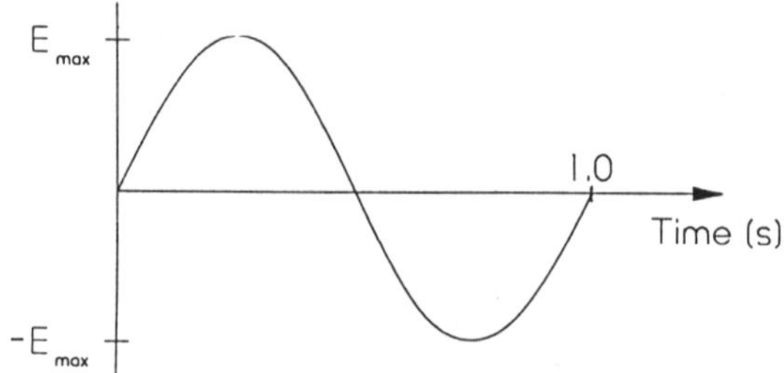

Figure 1.2 Electromagnetic wave in time.

---

[2] The symbols used in this handbook are listed and defined in Appendix B.

What is changing with time in Figure 1.2 is the magnitude of the electric field *E*, which is measured in volts. The waveform is called a sine wave because the amplitude of the electric field at any point in time is given by the expression:

$$E = E_{max}\sin(2\pi f t) \tag{1.1}$$

where *f* is the frequency of the signal and *t* is time. The wavelength is simply the distance through which the wave propagates in one complete cycle. The wavelength is related to frequency by the equation:

$$\lambda = c/f \tag{1.2}$$

where *c* is equal to the speed of light (and of all EM radiation) through space, approximately 300,000 km/s. Thus, the wavelength of a signal having a frequency of 10^9 Hz = 1000 MHz would be

$$\lambda = \frac{3 \times 10^8}{1 \times 10^9} \frac{\text{meters/sec}}{\text{Hz}} = 3 \times 10^{-1} \text{ meters, or 0.3 meters}$$

In nature, any object whose temperature is above absolute zero (-273° Celsius) radiates energy. If radiation is arranged in order of frequency, the *electromagnetic spectrum* can be represented on a bar chart (Figure 1.3). If there were a zero frequency on this chart, it would represent the limiting case, DC. Frequencies from about 15 Hz to 20,000 Hz are commonly referred to as audio frequencies, because the human ear can detect oscillations in the air pressure at these frequencies. Radiation which occurs in nature can be represented as sine (or cosine) waves, at frequencies all the way from a few Hz up to and beyond 10^{20} Hz. Visible light occupies a very narrow portion of the spectrum, with frequencies in the order of 10^{15} Hz, and with corresponding wavelengths in the order of 10^{-7} meters. Radio communication typically takes place at frequencies in the 10 kHz to 3000 MHz = 3 GHz regime, but while radio and radar frequencies can overlap, most radars designed after World War II and up to the present have operated in the 1 GHz to 40 GHz region of the spectrum.

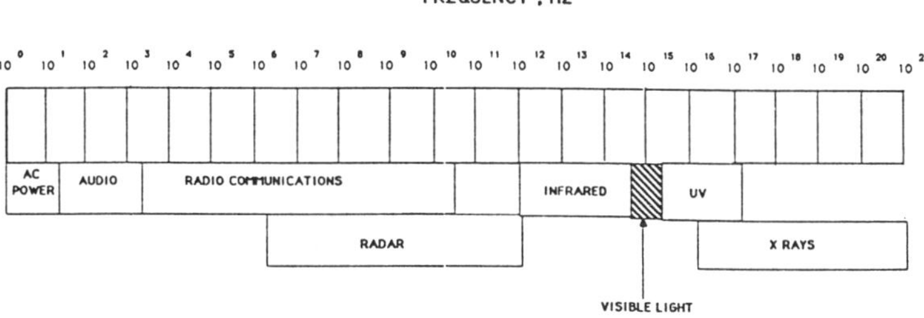

Figure 1.3 The electromagnetic spectrum.

Some characteristics of EM radiation are a function of its frequency. When radiation is caused by atomic interactions (e.g., the dropping of an electron from one energy state to another), the change in energy is directly related to the frequency of the radiation. Similarly, when the radiation is absorbed by atoms, the energy delivered by the EM wave in changing the atomic state is proportional to the frequency of the wave. Long-wave radiation delivers its energy through thermal agitation of entire molecules, and can be sensed as heat, but at the higher end of the spectrum, the extremely short wave (and extremely high frequency) radiations deliver the highest energy per interaction. The example of high-energy X-rays interacting with matter, allowing us to see through materials opaque to ordinary light, is a common one, and the most energetic X-rays can destroy human tissue and transform chemical compounds.

What determines the frequencies that are usable in radar? First, the EM radiation must pass through the atmosphere without too great a transfer of energy to the air molecules and the water molecules contained in clouds and precipitation particles. As the wavelength of the radiation decreases (and the frequency increases), approaching the dimensions of particles in the environment, more of the radiation is scattered and absorbed by those particles. Thus, we observe that radiation at the frequency of light is severely attenuated by humid air, fog, smoke, and rain. Secondly, there is a relationship between the size D of an antenna, the wavelength λ, and the width θ of the radar beam:

$$\theta \approx \lambda/D \quad \text{radians, where 1 radian} = 57.3° \tag{1.3}$$

This means that at very short (mm) wavelengths the beamwidths generated by the large antennas needed for long-range radar are so small that it would require an inordinate amount of time to scan even a modest volume of space in a search function. The situation here is similar to finding an object with a high-magnification telescope: a second, broad-beam telescope is required to help the observer position the narrow beam of the main instrument. Thirdly, the radiation used must be modulated with signals which permit the target to be separated from surrounding objects, so that it can be detected and its range (and often its radial velocity) can be measured. Such modulations can be applied to light radiated by lasers, but in most cases the use of longer wavelengths provides better control of the modulation and better supports the target detection and resolution functions.

The frequency bands used by radar are actually spread widely over the EM spectrum, as shown in Table 1.1. In the U.S. and NATO countries, almost all radars will operate within one of the specific frequency ranges listed as International Telecommunications Union (ITU) assignments for radar. For convenience, the letter band designations shown are often used in place of specific frequency ranges. This gives sufficiently precise information to determine the radar and atmospheric conditions for a particular radar, without disclosing details of tuning range which might be considered classified.

Table 1.1 Standard Radar-Frequency Letter Band Nomenclature
(IEEE Standard 521-1984).

Band Designation	Nominal Frequency Range	Specific Frequency Ranges for Radar Based on ITU Assignments for Region 2
HF	3 MHz–30 MHz	
VHF	30 MHz–300 MHz	138 MHz–144 MHz
		216 MHz–225 MHz
UHF	300 MHz–1000 MHz	420 MHz–450 MHz
		890 MHz–942 MHz
L	1000 MHz–2000 MHz	1215 MHz–1400 MHz
S	2000 MHz–4000 MHz	2300 MHz–2500 MHz
		2700 MHz–3700 MHz
C	4000 MHz–8000 MHz	5250 MHz–5925 MHz
X	8000 MHz–12 000 MHz	8500 MHz–10 680 MHz
K_u	12.0 GHz–18 GHz	13.4 GHz–14.0 GHz
		15.7 GHz–17.7 GHz
K	18 GHz–27 GHz	24.05 GHz–24.25 GHz
K_a	27 GHz–40 GHz	33.4 GHz–36.0 GHz
V	40 GHz–75 GHz	59 GHz–64 GHz
W	75 GHz–110 GHz	76 GHz–81 GHz
		92 GHz–100 GHz
mm	110 GHz–300 GHz	126 GHz–142 GHz
		144 GHz–149 GHz
		231 GHz–235 GHz
		238 GHz–248 GHz

1.1.2 The Radar Concept

The term *radar* is derived from the original name given to this technique by its British inventors during World War II, which was *Ra*dio *D*etection *A*nd *R*anging. This reflects the basic functions that the earliest radar systems performed. The fundamental components of a radar system are shown in Figure 1.4. Pulses of EM energy are generated by the transmitter, sent to the antenna which focuses the energy into a beam directed at the target, and receives the energy returned from the target. The return signal is then routed to the receiver and signal processor, where it is detected and processed, and then to the display, where the target data are used by the operator.

In general, the radar is designed to search for and detect the target of interest, and to determine certain target parameters. These may include target position in range and angle, and its heading relative to the radar location. Typical ground-based air-search radars are required to estimate additional parameters, such as the number of targets within the beam and their radial velocities with respect to the radar, i.e., determine if the target is approaching or receding from the radar, and at what rate.

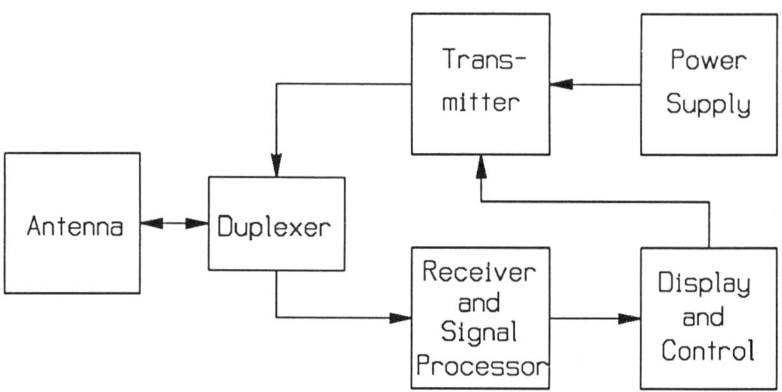

Figure 1.4 Basic components of a radar.

Target range is determined in a pulsed radar by the simple mechanism of measuring the time it takes for a pulse of radiated energy to travel to the target and back to the radar. Because the speed at which a pulse moves is known (the speed of light), this round-trip delay time t_d yields the range. Since range = speed × time, and the round-trip distance is twice the range,

$$2R = c \times t_d$$

$$R = ct_d/2 \tag{1.4}$$

If, for example, the round-trip delay for one pulse was equal to 100 μs, the solution would be

$$R = 3 \times 10^8 \, (\text{m/s}) \times (100 \times 10^{-6} \, \text{s}) \, (1/2) = 15,000 \, \text{m}$$

The target's position in angle is determined by measuring the position of the radar antenna at the time it detected the target. If this angle is a measure only of azimuth, then other means must be used to determine target elevation, e.g., from a separate height-finding antenna. Radars which scan in both azimuth and elevation will measure both these angles simultaneously, and from trigonometric relationships will be able to determine altitude, as shown in Figure 1.5. Observation of the target over time, either by continuous tracking or by processing target data on successive scans, will yield estimates of target course and speed, as well as rates of change in these parameters. A more extensive discussion of the target data obtained by radar is contained in Chapter 10.

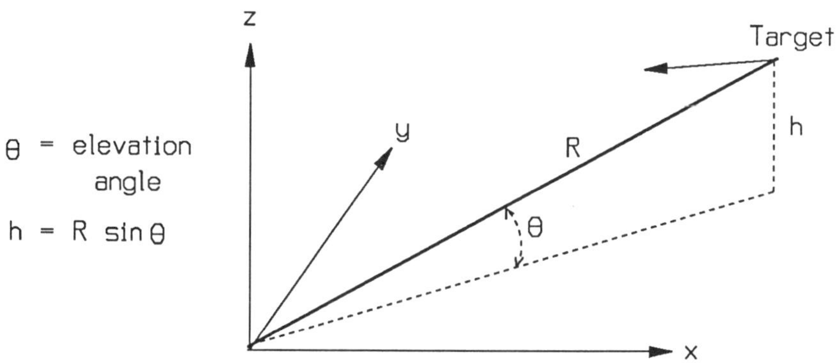

Figure 1.5 Measurement of target height.

The pulse of energy transmitted by the radar described thus far represents an additional component, or ***modulation***, that is superimposed on the ***carrier*** wave. The carrier is so named because it can be thought of as the vehicle which carries the additional information contained in the modulation. The carrier is simply the original signal generated by the transmitter at the radar frequency, or *RF*, discussed previously in the context of the radar spectrum. The desired modulation can be imposed on the carrier in any of several forms by the modulator, shown in the more detailed radar block diagram of Figure 1.6 as part of the transmitter. This modulation is referred to as the radar waveform, and can take on a wide range of characteristics, depending on the radar requirements and the technique employed to encode, or modulate, the desired waveform.

The topic of waveforms is discussed in Chapter 7, and only a summary of the concept is given here. As described previously, the simplest periodic signal to generate in an oscillator is a sine wave. If that signal is transmitted at RF continuously, then it is known as a ***continuous-wave (CW)*** carrier. Radars which employ CW waveforms are useful in land-based radars that have as their overriding requirement the detection and tracking of moving targets in a strong clutter environment, such as that which exists when targets fly at low altitude. They are also useful in measuring the speed of targets such as automobiles, when range and angle data are of lesser importance. CW waveforms do not provide target range information by direct measurement of time delay, as in the case of pulsed radar. Hence the CW carrier must be modulated in some manner, frequency modulation (FM) or phase modulation (PM) being the most common, and the returned echo must be processed to extract any needed range data. If the CW signal is not transmitted continuously, but in discrete bursts of a periodic nature, then the signal is referred to as *interrupted CW*. Interrupted CW waveforms can have application to special functions, such as missile guidance or fuzing, but in general are inferior to other alternative forms for most radar applications.

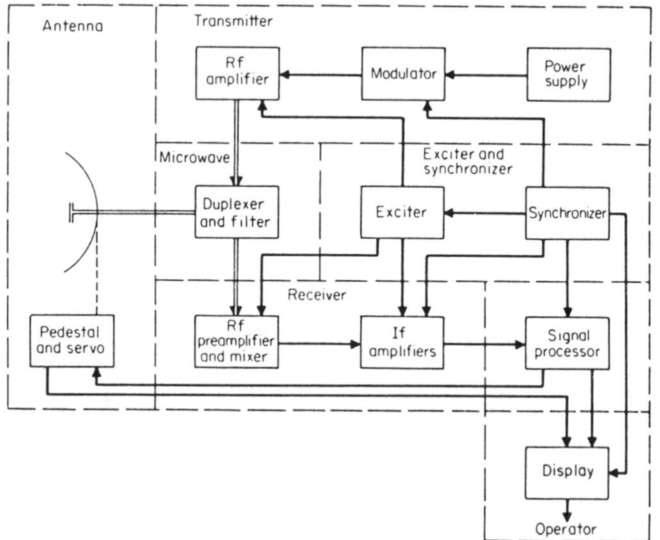

Figure 1.6 Block diagram of typical radar [1.2],
©Artech House, reprinted by permission.

1.1.3 Pulsed Radar

Pulsed systems are chiefly characterized by their pulse width τ, and the *pulse repetition interval (PRI)* t_r, which is the time between leading edges of successive pulses. When the PRI is uniform, its reciprocal gives the number of pulses per unit time, or *pulse repetition frequency (PRF)*, denoted by f_r. These relationships are illustrated in Figure 1.7(a) and (b).

The pulse width τ establishes the minimum range at which targets can be detected efficiently. Because the receiver will be effectively disconnected from the antenna to protect it from the high-energy pulse during transmission, it cannot receive pulses which travel to the target and back in less than τ seconds. Thus, the minimum range $R_{min} = \tau c/2$. Short pulses are therefore advantageous in detecting nearby targets, and also in resolving closely spaced targets and in forming images of targets. Since long pulses contain more energy by definition, their use can extend the range at which targets can be detected.

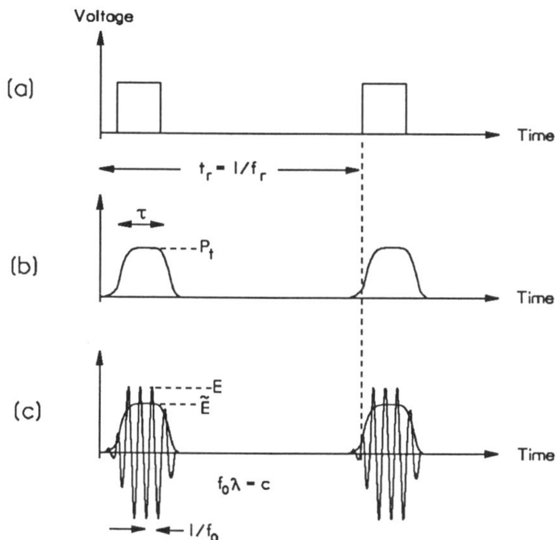

Figure 1.7 Pulsed radar waveform. (a) Idealized rectangular pulses.
(b) Actual pulse envelopes. (c) RF carrier within pulse envelopes.

The PRI is usually chosen by first determining the maximum range at which
the radar will have to detect targets. The PRI will need to be greater than the time
it takes a radar pulse to travel to the target and back:

$$t_r > 2R_{max}/c \tag{1.5}$$

With the PRI determined, the ***maximum unambiguous range*** R_u for the waveform
is established:

$$R_u = ct_r/2 \tag{1.6}$$

If the radar were capable of detecting targets at ranges greater than R_u, it would not
be possible, without special techniques such as the use of different PRIs, to
determine if the target were actually within the unambiguous range or in range
intervals covering integral multiples of that range.

The ratio of the pulse width τ to the pulse repetition interval t_r defines the duty
factor of the radar waveform, and as seen in Figure 1.7, this is simply the ratio of
the radar waveform on-time to off-time.

The radar transmitter puts out pulses each having ***peak power*** P_t, but as shown
in Figure 1.7 the time between pulses is much longer than the width of the pulses,
so this peak power does not describe the radar's average power over time. The
average power P_{av} is given by the product of peak power and duty factor:

$$P_{av} = P_t D_u = P_t \tau/t_r = P_t \tau f_r \qquad\qquad (1.7)$$

It is the average power of any radar that is important in establishing its maximum detection range capability. In most pulsed radars, the pulse width τ is quite short, in order to allow detection of targets whose range is a small fraction of the unambiguous range. The PRI, having been established by the maximum range requirement, is very long by comparison, with the result that the waveform duty factor D_u is very small. Because a reasonable level of average power is required to detect targets, the peak power P_t requirement for typical pulsed radars is very high.

Modern radars often combine the benefits of high pulse energy (and average power) associated with long pulsewidths, with the high resolution inherent in short pulsewidths, through the use of a technique known as *pulse compression*. In pulse compression (see Chapter 7), a wideband frequency (or phase) modulation is superimposed on the carrier during the pulse transmission time. One form of pulse compression uses a linear change in frequency over the pulse. When the return signal is received, it is processed through a linear time-delay filter that delays the output in proportion to frequency. Figure 1.8 shows a swept frequency modulation imposed on the carrier. This modulation starts out at the high end of the sweep range at time t_0, and sweeps down in frequency, ending at the lower limit of its range, f_1, when the transmission ends at t_1. Because the receiving filter is designed to delay the high-frequency part of the return pulse more than the lower-frequency part, the output pulse is much shorter than the original. The pulse has, in effect, been compressed, and the resolution of the radar is improved by the ratio of the transmitted pulse width to the compressed pulse width at the filter output. Use of pulse compression permits a high duty factor to be used, in a waveform having long unambiguous range and high resolution, and hence increases the average power available for a given peak power. In pulse compression systems, the minimum range is still set by the actual transmitted pulse length τ. Additionally, radars which utilize this technique must employ amplifier-type transmitters (see Chapter 7), rather than the simpler magnetron or pulsed oscillator types of transmitter.

1.1.4 Doppler Radars

A major consideration in waveform selection is the ability of radars to select targets and reject clutter based upon the *doppler shift* in the return echo. A doppler shift is the difference in the frequency of the received signal compared to the transmitted signal, caused by the radial velocity of the target relative to the radar. The doppler effect exists for all forms of EM radiation, but a more familiar example is the change in the sound of a locomotive's whistle as it first approaches and then recedes from the listener. It is possible to develop a useful approximation for determination of the doppler frequency of a target.

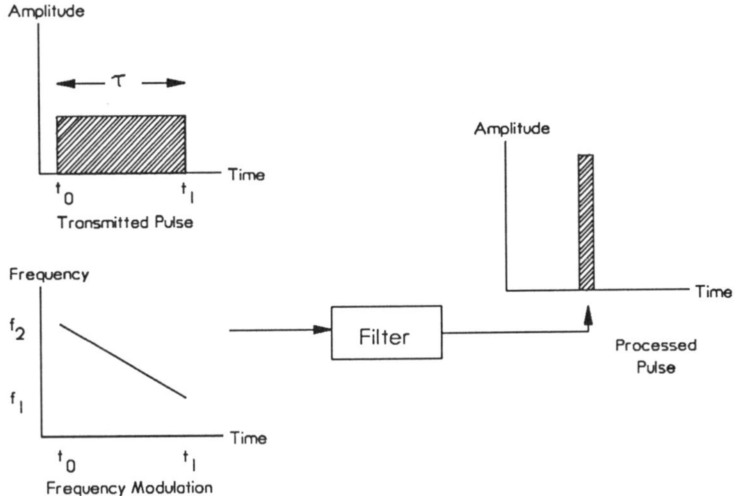

Figure 1.8 Principle of pulse compression.

At any instant of time, the total number of wavelengths contained in the two-way path from the radar to the target and back is equal to $2R/\lambda$. Since one wavelength corresponds to a phase excursion of the sine wave carrier through 2π radians, the total phase excursion Ω made by the EM wave during the two-way travel is

$$\Omega = 2\pi(2R/\lambda) = 4\pi R/\lambda \quad \text{(radians)} \tag{1.8}$$

Because the target is moving, R and Ω are changing with time. The phase change with time, $d\Omega/dt$, is equivalent to a frequency, i.e.,

$$\omega_d = 2\pi f_d = d\Omega/dt = 4\pi v_r/\lambda \quad \text{(radians/s)} \tag{1.9}$$

$$f_d = 2v_r/\lambda \quad \text{(hertz)} \tag{1.10}$$

As an example of using equation (1.11), suppose that an aircraft is approaching an X-band radar with a radial speed of 300 m/s. The wavelength of the X-band wave is

$$\lambda = c/f_t = 3 \times 10^8/10^{10} = \quad 0.03 \text{ m}$$

and the doppler shift is

$$f_d = 2 \times 300/0.03 = 20,000 \text{ Hz} = 20 \text{ kHz}$$

Figure 1.9 shows a target at range R from a fixed radar, and moving toward the radar with a radial component v_r. The radar is transmitting with frequency f_t. The signal received by the radar will, however, be shifted by the amount of the doppler shift f_d. Since the target is moving toward the radar, this shift will add to f_t, making the received frequency f_s higher by this amount:

$$f_s = f_t + f_d \tag{1.11}$$

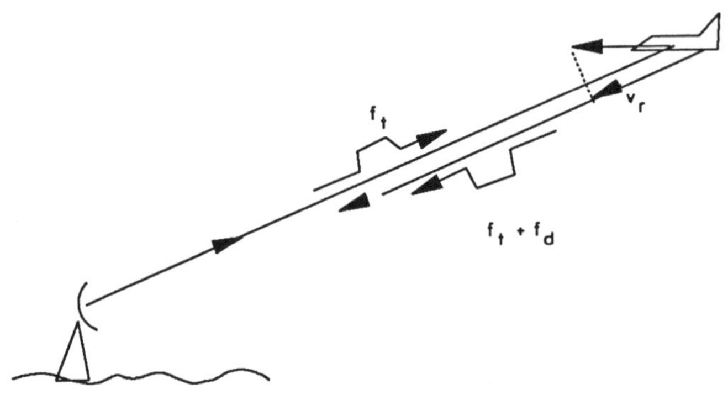

Figure 1.9 Target with doppler shift.

Figure 1.10 shows that the frequency spectrum of a pulse of width τ will extend over a bandwidth $B = 1/\tau$ on each side of the carrier frequency. In most cases, the small value of f_d will not produce a resolvable, or even a noticeable, shift in this spectrum. If strong clutter is present within the radar beam, its echo spectrum will obscure that of the target. For pulse width $\tau = 1$ µs, the bandwidth $B = 1$ MHz, and the 20 kHz doppler shift will move the spectrum by only 2% of its mainlobe width.

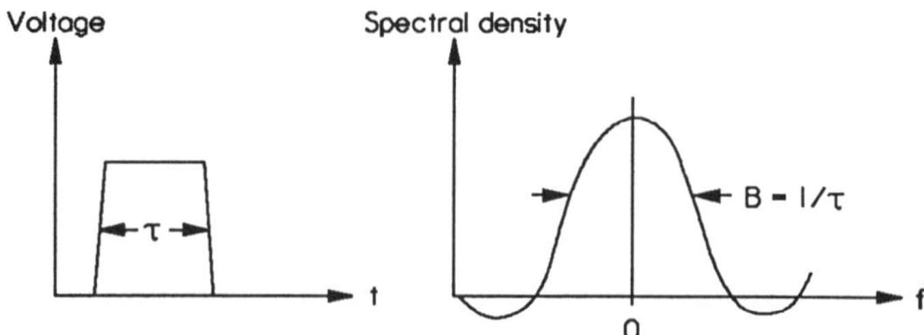

Figure 1.10 Spectrum of a rectangular pulse.

The solution to this problem is either to extend the pulse width in order to narrow the bandwidth B (approaching, as a limit, a CW waveform), or to transmit a coherent train of narrow pulses over some observation interval t_o. The spectrum of this waveform will consist of narrow lines, of width $B_f = 1/t_o$, spaced at intervals

$1/t_r = f_r$ across the original pulse spectrum (Figure 1.11). The term *coherent* means that the phase of the RF carrier in the received pulses remains consistent from pulse to pulse over the duration of the pulse train. This type of transmission restricts the received spectrum of fixed clutter to narrow lines at frequencies f_t, $f_t + f_r$, $f_t + 2f_r$, ... , while the target lines appear at $f_t + f_d$, $f_t + f_r + f_d$, $f_t + 2f_r + f_d$, ... In principle, this makes it possible to resolve and detect the moving target signal regardless of how strong the clutter return may be.

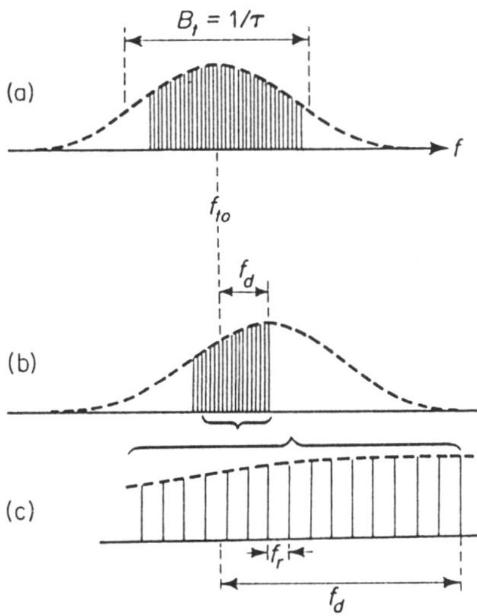

Figure 1.11 Spectra for coherent pulse train. (a) Transmitted spectrum.
(b) Received spectrum. (c) Detail of received spectrum [1.2].
©Artech House, reprinted by permission.

The use of pulse waveforms to reject clutter on the basis of doppler shift leads to **blind speeds** in the radar response. Blind speeds are those radial velocities which cause the target doppler return to fall into the clutter rejection notch (Figure 1.12). For a given f_r, these speeds are

$$v_{bi} = if_r\lambda/2, \qquad i = 0, \pm1, \pm2, \ldots \tag{1.12}$$

The basic blind speed is simply $v_b = f_r\lambda/2$ for $i = 1$. The rejection notch is formed either by use of an *MTI canceler* (Section 7.2) or by generating a bank of narrow doppler filters, omitting those at zero frequency and at multiples of the PRF.

It is apparent from this discussion that the selection of a waveform will depend on the wavelength λ of the RF carrier, the spatial and velocity regions in which targets and clutter are expected, the time interval t_o over which the target signal can be observed, received, and processed, and the strengths of the target and clutter echoes. The properties of the radar antenna and its motion will have a direct effect on the latter issues.

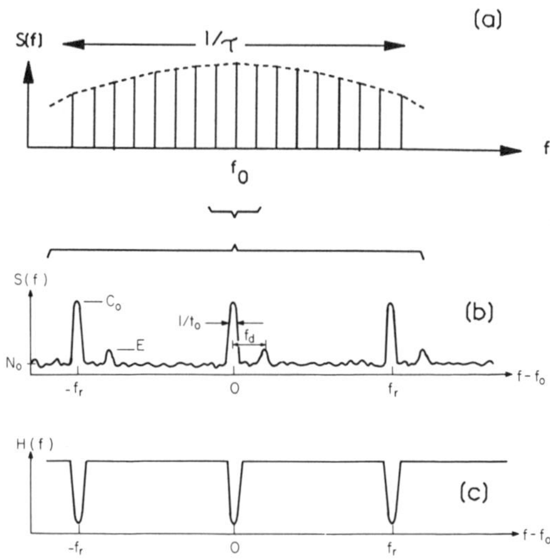

Figure 1.12 Response of doppler processor with clutter
rejection notches at multiples of the PRF.

1.1.5 Antennas

The radar antenna focuses the radar's transmitted and received energy into
beams which, when directed toward the target, produce a large enough echo signal
for detection, while discriminating against unwanted targets, clutter, and jamming.
The narrower we make the antenna beam, the greater will be its response for objects
within the beam, and the less the volume containing clutter and potential jamming
sources. On the other hand, a narrow beam must scan over many positions if it is
to cover an assigned search volume, acquire a target designated to the radar by an
external source, or map a ground region. Choice of radar beamwidth will depend
first on the compromise between the volume requiring search, or *scan field*, and
the time allowed to search over this volume, or *scan time*, and second on the physical
or electrical limitations of the antenna *aperture*. The antenna aperture is an area
from which the antenna radiates into space and receives signals from space. For
a reflector antenna, it is easily identified as the projected area of the reflector,
viewed from the direction of the beam. For antennas composed of dipoles and
similar wire structures, it is an area in front of the physical antenna from which the
antenna can capture energy from a received wave (or equivalent area for the
transmission mode).

Basic electromagnetic theory dictates that the beamwidth of an antenna will be proportional to wavelength λ and inversely proportional to the aperture width w:

$$\theta_3 = K_\theta \lambda/w \quad \text{(radian)} \tag{1.13}$$

where θ_3 denotes the width of the beam between its one-way, half-power points, and K_θ is a constant of proportionality near unity. The constant is 0.88 for antennas which radiate (or receive) with equal intensity over the entire aperture, and increases to about 1.3 for antennas using tapered intensity for sidelobe reduction. It will be shown in Chapter 6 that the diameter D of circular antennas may be substituted for w, with little change in the beamwidth constant. A useful form of this expression, applicable over a wide range of conditions, is

$$\theta_3 \approx 1.22\lambda/w \quad \text{(radian)} \tag{1.14}$$

corresponding to $K_\theta = 1.22$. While there are theoretical ways in which narrower beams than those resulting from these equations can be generated (*superresolution*, or *supergain* antennas), these theories have little application to radar problems (see Chapter 12 in regard to ECCM considerations).

The ***directive gain*** of an antenna is simply a measure of the antenna's ability to focus energy. It is equal to the ratio of the maximum radiation intensity at a given range, on the axis of the beam, to the intensity that would have been produced at that same range by an antenna which radiated uniformly in all directions. It can be shown that the directive gain depends on the aperture area A:

$$G = 4\pi A \eta_a/\lambda^2 \tag{1.15}$$

where η_a is the ***aperture efficiency***. Since $A = w^2$ for a square antenna, we see from Eqs. (1.14) and (1.15) that the gain depends reciprocally on the square of the beamwidth:

$$G = 37100/\theta_3^2 \quad (\theta \text{ in degrees})$$

$$\tag{1.16}$$

$$G = 4\pi/\theta_3^2 L_n \approx 11.3/\theta_3^2 \quad (\theta \text{ in radians})$$

(or $11/\theta_a\theta_e$ for different azimuth and elevation half-power beamwidths). (In reflector antennas the practical value for gain is reduced by the amount of spillover energy which does not reach the reflector.) These equations explain the preference for short wavelengths (high frequencies) in mobile, airborne, and other applications where w and A are limited.

By design, most of the energy radiated by the antenna is contained in the main beam, also called the ***mainlobe*** of the antenna pattern, but all antennas produce ***sidelobe*** responses, which can receive returns from unwanted targets, clutter, and jamming. The traditional measure of sidelobe response is the ***sidelobe ratio***: the ratio of the mainlobe gain to the gain of the highest sidelobes (generally the pair which are adjacent to the main lobe). For a typical horn-fed reflector antenna, the sidelobe ratio is approximately 316:1, or 25 dB (for a definition of the *decibel*, or dB, see Appendix A). For many applications, the average level of sidelobe response over sectors more distant from the main lobe than the first, or highest, sidelobes is a more significant measure of antenna performance. In modern ultralow-sidelobe antennas, the average sidelobe ratio may be 50 dB or greater. In order to reduce antenna sidelobes, the *illumination*, or excitation of the antenna aperture, must be reduced near the edges; such *tapering* of energy reduces mainlobe gain and broadens the beamwidth.

An important property of most antennas is *reciprocity*, meaning that the pattern and gain for receiving are the same as for transmitting. Thus, the transmitting feed is described as *illuminating* the aperture for projection of a beam into space; in receiving, the feed illumination refers to its acceptance of power reflected from the aperture, but the pattern in space remains the same. When different transmitting and receiving patterns are required, different feeds must be used, or coupling to the feeds must go through a switch.

Antennas can be classified by their radiation patterns, the nature of the radiating aperture, the method of feeding power to (and extracting power from) the aperture, and the method of scanning the beam. The term *phased array antenna* is generally considered to apply to an antenna having electronic scan (as opposed to mechanical scan) in at least one coordinate. A partial list of antenna classifications is shown in Table 1.2.

1.1.6 Receivers and Signal Processors

In modern radars, especially those using digital processing technology, the division between the receiver and the signal processor is indistinct. The receiver will accept low-power RF signals from the antenna, amplify these signals, and convert them to a lower frequency (intermediate frequency, or IF) more suitable for filtering and other processing, and also may perform the first stages of clutter filtering using analog circuits. Other signal processing tasks, which may be performed either digitally or by analog circuits, include:

- Filtering, to select the pulse spectrum from wideband noise and ECM;
- Pulse compression;
- Doppler filtering, including MTI or other processes;
- Estimation of target coordinates;
- Adjustment of gain, to avoid saturation in amplifiers and to control false alarms;
- Integration of successive pulses or samples in a pulse train; and
- Nonlinear processing to reject strong interfering signals.

Table 1.2 Antenna Classifications

Aperture	Parabolic reflector:
	    Paraboloid of revolution
	    Parabolic cylinder
	    Shaped paraboloid
	Lens:
	    Metallic (artificial dielectric)
	    Dielectric
	Array:
	    Dipole
	    Waveguide or cavity
	    Slotted waveguide
	    Spiral or helix

Pattern	Pencil beam Fan Beam Shaped beam (e.g., cosecant-squared) Beam cluster (e.g., monopulse) Stacked beams

Feed	Space feed:
	    Horn feed
	    Cluster or stack of horns
	    Line feed (slotted waveguide, dipoles)
	Constrained feed:
	    Series feed (e.g., waveguide with couplers)
	    Parallel feed (corporate structure)
	Lens feed

Scanning	Mechanical (one- or two-axis pedestals)
	Electromechanical (Eagle and Foster scanners, etc.)
	Electronic:
	    Frequency scan
	    Phase scan
	    Beamswitched
	Mixed (e.g., mechanical azimuth, electronic
	    elevation scan)

Important general considerations in the receiver and signal processor are the distinctions between *linear* and *nonlinear* operation. The output of a linear circuit, in the presence of two or more inputs, is simply the sum of the outputs that would have been produced by each input acting alone. In nonlinear circuits, additional outputs will be produced representing *cross-modulation* and *intermodulation* products among the input components. Circuits designed to operate linearly within their intended operating ranges include RF and IF *amplifiers* at the inputs of a receiver, *mixers* for downconversion of RF signals to IF (or IF to lower IF), *phase detectors* for conversion to *baseband*, and *A/D converters* if their smallest bit is below noise level and their largest bit is above the maximum signal level. All of these circuits become nonlinear when the input exceeds a particular design value.

Nonlinear circuits include *limiters*, *logarithmic amplifiers*, *envelope (video) detectors*, and *logical processing circuits*. These are often used in radar receivers (universally, in the case of envelope detectors), and when properly located in the receiver-processor chain they contribute to essential radar performance features. However, in the presence of strong clutter, jamming, or other interference, the nonlinear circuit can cause suppression of the desired (target) signal or generation of spurious outputs (false targets) that confuse a human operator or an automatic device using the radar data.

1.1.7 Display and Data Processing

After echoes have been filtered and processed to remove as much interference as possible, the resulting outputs are presented to the user: on a video (cathode-ray tube) display as either raw or *synthetic video* signals, or as inputs to a digital data processor. At this stage of processing, signals are remembered and carried over from scan to scan, or for periods of seconds, to establish tracks on targets, evaluation of the military situation, images of the surface, or other information needed by the user.

Direct display of raw video to an experienced operator can be useful, but most modern systems interpose a digital data processor that evaluates the output *signals* and presents or uses *data*, which is information of probable importance to the user. A typical output report will consist of coordinates of target position, velocity components, a target identification code, a data quality estimate (or target amplitude), and a time tag.

1.2 Functions and Applications

Radar functions can be divided into four broad categories:
- Search and warning (detection and acquisition);
- Tracking and measurement;
- Imaging or identification; and
- Control and communication.

Most radar applications require some combination of these four basic functions.

1.2.1 Search and Warning

The basic search and warning function requires timely detection of targets after they enter the radar's coverage volume to permit appropriate reaction by the user. Achievement of high detection probability, over the number of scans or observations available to the system, in a reasonably short time and with acceptably long intervals between false alarms, is the usual measure of performance.

1.2.2 Measurement and Tracking

Measurement and tracking can be carried out entirely in a *tracking radar*, or with the aid of external data processing, as with *track-while-scan* (TWS) systems using search radar reports over several scans. The distinction between a *tracking radar* and a TWS system is that the tracking radar antenna responds to the target, while the antenna in a TWS system does not, but scans continuously over a preassigned volume. A tracking radar can be dedicated continuously to a single target, or cycled among several targets whose position data are kept up to date in a computer, permitting the radar beam to be returned to the same target on each cycle without need for an acquisition scan. The tracking radar acquires the target initially by scanning a sector of limited size, possibly one designated by an external sensor (often a search radar). Its output data are generally precise enough for fire control of guns or of guided missiles, and in the case of instrumentation applications may serve as the primary source of target trajectory data.

When a TWS function is based on search radar data, the radar must perform target position measurement while scanning past the target, presenting the reports of target coordinates with each detection. Required single-scan detection probabilities will generally be much higher than those required to satisfy warning requirements, since the TWS processor must correlate reports over successive scans separated in time, and string them together into track files. Requirements for TWS radar scan rate and accuracy are not based on achieving timely warning (at least one detection), but on supporting the track files with adequate probability of track maintenance and minimum confusion between adjacent or crossing targets.

1.2.3 Imaging and Identification

Imaging and target identification require that the target (or target area, for ground attack) be divided (resolved) into many cells or components. The operator or computer can then recognize and select the correct target from a field of many detectable objects. Imaging refers to the formation of a two- or three-dimensional map of scatterers, often in range and azimuth coordinates.

Ground maps are generated in several different ways:

(a) With a sidelooking real-aperture radar, the azimuth beamwidth of which is narrow enough to resolve features of interest;

(b) With a *synthetic aperture radar* (SAR), in which a small, sidelooking antenna is moved over a considerable distance while coherent signal data are gathered for computer processing; and

(c) With spotlight mapping radars, in which the beam of an airborne interceptor-type radar is directed off the nose of the aircraft and held on a small region of the surface while data is obtained for SAR processing.

Target images can be obtained by coherent processing of signals from fixed radar, if the target moves in such a way that the orientation of its axis changes in a predictable way, relative to the radar line of sight. This process is called *inverse synthetic aperture radar* (ISAR), and it can be regarded as spotlight mapping which relies on the target moving instead of the radar platform. As with SAR, good range resolution is normally required, and the use of wideband waveforms is indicated.

Other means of target identification use plots of target scattering as a function of only a single coordinate. This can be range, as in the *range-profiling* approach. It can be doppler spectrum, as in identification by inspection of turbine spectral lines or audible vibration components imposed by the target on a narrow transmitted spectral line. It can be a comparison of responses at different transmitted RFs, sometimes referred to as *harmonic* identification because it may involve measurements at harmonically related RFs. Or it may be a comparison of responses at different polarizations of the transmitted and received waves. Combinations of range resolution with polarization or with measurements of monopulse angle dispersion have also been used to obtain two-dimensional characterizations of the target. Monopulse angle measurements (see Chapter 10) are normally intended to locate the center of an extended target, but by special processing they can be made to show the width of the scattering points which make up the target. When these data are taken with high range resolution an image of the target can be produced. It should be noted that the combination of range and doppler (cross-range) represents the SAR and ISAR imaging approaches.

When the image or profile of the target has been obtained, it must be matched against some library of known target characteristics in order to perform the identification or selection process. With range-azimuth or range-doppler images, this recognition can be done visually by a human operator, while with doppler spectrum the recognition may be aural or visual. For missile guidance or other automated systems, a computer algorithm may be substituted for the eyes and ears of an operator.

1.2.4 Functions Required in Typical Radar Applications

The applications of radar in modern army operations are listed in Table 1.3.

Within the air surveillance and fire control category, all of the functions discussed in the previous section must be performed, either within one radar type or distributed among two or several types. Traditionally, an air defense system would have a surveillance radar (sometimes referred to as an acquisition radar), and one or more engagement (tracking and guidance) radars to control the engagement of selected targets. Target identification would be performed by *IFF* associated with the surveillance radar, or by operational doctrine (targets approaching in other than safe corridors would be considered hostile). Outgoing interceptor missiles would be tracked by an engagement radar and command guided (using coded radar transmissions) to the target, or would home on energy reflected by the target from an illumination waveform transmitted by an engagement radar, or would use some combination of these schemes.

Table 1.3 Army Radar Applications

1. Air surveillance and fire control against: 　　　Manned aircraft, including low-observables 　　　Tactical missiles, ballistic and cruise types 　　　Helicopters
2. Battlefield surface surveillance: 　　　Intrusion detection 　　　Side-looking radar 　　　Synthetic-aperture radar
3. Counter-battery radar (e.g., Firefinder)
4. Smart munitions
5. Instrumentation radar: 　　　Velocimeters 　　　Artillery trajectory measurement 　　　Missile trajectory measurement

Other applications might involve only a single function. For example, an instrumentation radar for a surface-to-air guided missile range might have a requirement only for precision tracking, after designation to within one beamwidth of the target to be tracked.

1.3 Principles of Radar Evaluation

1.3.1 System Performance Requirements

Before a radar system design or model is *evaluated*, the requirements for its modes of operation and its performance in these modes must be known, so that the evaluation can be made with respect to these requirements. This means that a system-level specification must be available, describing the functions to be performed by the radar, the regions over which these functions are to be performed, the targets to be detected, measured, or identified, and the background environment in which this is to take place.

Evaluations are normally made at several sites, depending on the status of the radar program and the resources available to the evaluator:

- Analysis;
- Simulation;
- Tests at manufacturer's plant;
- Field test; and
- Extrapolation from field test, using both analysis and simulation.

For each type of evaluation, a different statement of radar system requirements may be necessary. In the analysis phase, for example, the target cross section is specified by its mean value in square meters, with a fluctuation model usually from the Swerling cases (see Chapter 5). Atmospheric conditions will be specified in terms of mm/h of rain, or equivalent. In simulation, the target signal is often generated by *Monte Carlo* methods from the equations representing the Swerling model, modified by the pattern of the scanning antenna. The atmospheric attenuation will be calculated from models and used to adjust the signal strength in the simulator. In field tests, a specific aircraft is used, its cross section having been measured in advance to relate it to the radar specification requirements. Weather conditions are observed by standard meteorological instruments, with rainfall estimated over the radar-target path. Extrapolation to other targets is done by scaling to their physical size, or by using measurements from a cross-section range. Extrapolation to other weather environments relies on models or direct measurement of attenuation on communication circuits or radars similar to the radar under evaluation.

The difficulty in this evaluation process is that the specification, against which the radar system may have been proposed or developed, may have stated the requirements in only one way, e.g., detect a $1.0\,\mathrm{m}^2$ aircraft target (Swerling Case 1), flying at one km altitude in 4 mm/h rain, with 0.9 single-scan probability at a range of 100 km. The procedures for analysis and simulation in this case are well defined, but the translation to a specific test target and the establishment of the rainfall rate under test conditions are more difficult. The evaluator must be able to translate the numerical specifications to the actual target and environment, and back again, if the field evaluation is to be meaningful. The relationships between models of the target and the environment, on the one hand, and the conditions of operation in the real world, on the other, will be covered in later chapters of this handbook.

1.3.2 Evaluation by Analysis

In the early stages of a radar development or procurement program, the hardware is not available for test, and evaluation must be made on paper, using analytical techniques. The necessary analyses may start from fundamental theoretical models of radar performance, or from available test data on similar radar equipment which may be adapted or improved to meet the new requirement. Some areas of radar performance are well understood, so that accurate calculations of system performance can be made from known radar parameters and models of the external environment in which the radar is intended to operate. In other areas, theoretical procedures have not been developed to the extent necessary for accurate prediction of radar performance, and simulation or field test will be required. Even in areas where adequate theory exists, there remains considerable uncertainty as to the validity of the models used to represent targets and environmental effects, and key aspects of performance must be validated by test. A thorough analytical evaluation is required, however, to identify the critical areas in which tests are most necessary to resolve uncertainties.

The need for analytical evaluation prior to testing is based on the limited test resources available, and the statistical nature of most radar performance measures. If the test resources are spread over the entire range of conditions in which the radar is expected to operate, there is little chance that the number of data points in any one region will be adequate to arrive at statistically valid conclusions. Analysis, if properly carried out, can identify the most critical regions in which the radar design is marginal or the analytical models are weak, as well as broad regions in which satisfactory performance can be expected with high confidence. A relatively few tests in these regions, repeated to give valid statistical measures of performance, can serve to validate or correct the analytical and simulation models near their boundaries of uncertainty, and give confidence to their use in extrapolation to other cases. The purpose of testing is to resolve uncertainties. Little new information is obtained when the tests merely confirm things that are already known with high confidence.

1.3.3 Evaluation by Simulation

In many areas of radar performance, the processes applied to the signals and the resulting output data are complex, involving nonlinear operations and logical paths which may not lend themselves to any reasonable mathematical analysis. While simplified models can be created to obtain approximate analytical results, simulation may be the only way to predict the actual performance of these processes with multiple inputs from targets and interference. This is especially true in areas such as target identification and multiple-target tracking. Again, however, it is important to perform as thorough an analysis as possible to guide the choice of parameters for simulation and to help interpret the results.

1.3.4 Evaluation at Manufacturer's Plant

Before a radar is delivered to a government test site, it must usually pass a series of tests, conducted jointly by the manufacturer and representatives of the government. These tests often start with evaluations of the several radar subsystems against their specifications, and may then progress to system testing against signals generated by electronic test equipment and against actual targets. In many cases, the types and operating parameters of the targets which may be provided in the vicinity of the plant are limited, compared to those covered by the specification. However, the general operating properties of the radar system may be confirmed in these tests to the extent necessary to ensure that the system is ready for full-scale field testing at a government test range.

1.3.5 Determination of Test Conditions

The techniques of analysis and simulation are seldom reliable enough to permit favorable decisions on production of radars to be made without validation by actual field test. On the other hand, designs which have fundamental flaws or limitations can often be rejected on the basis of analysis alone. Given particular areas of concern, identified by analysis or simulation, it will usually be possible to design test programs to determine whether these areas are adequately addressed by the radar design. The key to a successful test program is to reduce the number of test variables so that the test data will be definitive enough, in the statistical sense, to resolve the initial uncertainties.

In general, field test conditions are difficult to determine and describe accurately, especially those related to clutter and propagation effects. Given the limited resources usually available for testing, it is preferable to perform repeated tests under a few well defined conditions which will permit adequate statistical validation of system performance for those conditions, than to spread the test resources over all possible combinations of operating conditions and end up with too little data and questionable results in each area. It is then possible to use analytical models to extrapolate performance to conditions not too distant from those covered by test data.

The models of radar performance, target parameters, and environmental conditions, presented in the remainder of this handbook, are intended to provide guidance to the evaluator in identifying areas requiring test, and in determining the actual test conditions with sufficient accuracy to decide the critical issues of radar performance.

1.4 Textbooks on Radar

(1)   M. I. Skolnik, *Introduction to Radar Systems*, McGraw-Hill, 1982.
(2)   F. E. Nathanson, *Radar Design Principles*, McGraw-Hill, 1969.
(3)   J. L. Eaves and E. K. Reedy, *Principles of Modern Radar*, Van Nostrand, 1987.
(4)   M. I. Skolnik (ed.), *Radar Handbook*, McGraw-Hill, 1990.

1.5 References

[1.1]   E. F. Knott, J. F. Shaeffer and M. T. Tuley, *Radar Cross Section*, Artech House, 1985.
[1.2]   D. K. Barton, *Modern Radar System Analysis*, Artech House, 1988.

Chapter 2

RADAR SYSTEM DESIGN

Although this handbook is not intended to serve as a manual for designers of radar systems, the radar system evaluator should have a basic understanding of the processes by which radars are (or should be) designed, in order to plan and conduct the evaluation of systems designed by others. In this chapter, therefore, we describe the way in which radar system characteristics and design parameters are related to the requirements of the end user.

All radar requirements flow from these fundamental considerations:

1. *What functions must the radar perform?* Typical functions are search, target acquisition, tracking, imaging or target identification, and (for missile homing guidance) target illumination. What radar data are needed, and how are they to be used?

2. *Over what regime must the radar perform these functions?* What are the dimensions of the search volume and how often must this volume be scanned? What are the characteristics of the targets that the radar must be able to detect and track within this volume?

3. *How well must the radar perform these functions?* What is an acceptable probability of detecting or not detecting a given target? What is an acceptable time between false alarms? How accurately must the radar determine target position? How accurately must the radar track in range, angle, and speed? At what range must the radar be able to determine that there are two or more targets within its main beam rather than one? Over what period must the radar operate without a failure? What is an acceptable average time to repair the radar if it fails?

4. *Under what conditions must the radar operate and perform in a satisfactory manner?* Like any machine, a radar is subject to the physical environment in which it operates. This includes the local environment of the platform on which the radar is mounted. For example, an airborne radar or a radar seeker must withstand the dynamic mechanical and thermal stresses of the platform's flight environment. A fixed ground-based radar, on the other hand, may experience much less shock and vibration than an airborne radar, but may have to operate under extreme weather conditions ranging from the arctic ice and snow to the severe heat

and humidity of the tropics. Equally important are the effects of the natural environment on the radar signal itself. The radar signal, already attenuated by the moisture content of clear air, can be severely affected by rain or snow.

In addition, the atmosphere acts as a lens, causing **refraction** or bending of the radar signal over the earth's surface. Radar energy is also reflected from the surface, and this reflected energy is received along with the direct signal in a way which can enhance or cancel the signal, while causing **multipath** errors in target measurement data. Unwanted returns from terrain constitute **ground clutter** that competes with the target return, and the same effect from the sea or from weather can produce sea or   clutter. Last but not least are man-made interferences which may be intentional **jamming** by enemy forces or electromagnetic interference (**EMI**) due to operation of other radar or electrical systems in the vicinity. Natural lightning is also a potential source of EMI, especially in radars operating in the lower frequency bands.

The basic performance requirements, including the general nature of the physical, natural, and electronic environment in which the radar must operate, will be specified by the user, but before the radar designer can proceed confidently with his task he must establish a detailed dynamic model of the environment. The details of this model, and interpretation of the data resulting therefrom, often supplied by radar specialists in the laboratory or by another agency managing the radar development, control important choices to be made by the system designer, foremost among which are:
- Operating frequency,
- Waveform,
- Antenna design, and
- Signal processing approach.

2.1 Derivation of System Parameters

The user will normally have numerous requirements for data, some of which are to be met by a particular radar system. It is important to distinguish between his *primary* requirements, which justify the development and procurement of a radar system, and any *secondary* requirements which may be involved in the specification in the hope that they can be met by the same radar with minimal added cost or complexity. In this chapter, it will be assumed that the primary requirements can be classified as search, tracking, and imaging or target identification. When two or all three of these requirements are combined in a single radar, the steps outlined will be combined in arriving at radar system design requirements.

It cannot be assumed, however, that there is a realistic solution to all requirements. One of the most critical steps in radar engineering is the identification of requirements or combinations thereof which lead to impossible or impractical system designs. By eliminating, modifying, or reallocating such embellishments, a practical radar system may be obtained which meets the essential needs of the user and is affordable enough to allow procurement of the necessary number of systems.

Table 2.1 Translation of User Requirements

User Requirement	Derived Requirements	Impact on Radar Specification
Radar Function		
	Modes of operation	Operator interface: Display or through computer Availability: MTBF, MTTR
Search	Search volume	Search frame time Single-scan detection probability Unambiguous range
Tracking	Track data type Track rate Accuracy	Acquisition scan time Acquisition probability Rms error in range and angle
Imaging or identification	Resolution	Beamwidths Signal bandwidth Coherent integration time
Target Description		
Types and descriptions	RCS vs. aspect angle Fluctuation model	Average power Frequency diversity Waveform Polarization
	Target dynamics	Tracking loop
Number and spacing		Resolution

Table 2.1 (Continued)

User Requirement	Derived Requirements	Impact on Radar Specification
Electronic Environment		
Radar siting	Ground clutter model Distributed and discrete clutter sources	Waveform Signal processing
Friendly and hostile radar siting	Interfering signal environment	RF bandwidth Channel separation RF filters
Jamming threat	Jammer waveforms Power spectral density Number, type, and location	Waveform Frequency agility Sidelobe levels System logic and ECCM
Physical Environment		
Platform	Maximum weight Input power Shock and vibration Thermal environment	Power supplies Mechanical specs
Weather	Temperature extremes Wind Precipitation	Antenna, radome specs Waveform Polarization
Siting and survivability	Hardening	Mechanical design Shielding
User Constraints		
Maximum size and weight Life cycle cost Reliability goal Maintenance concept Operator skill level	Gross packaging limitations Unit cost goal Reliability model Lowest replaceable unit Command and control concept	Packaging concept Subsystem cost targets System MTBF Hardware design, packaging Displays and controls

Table 2.1 lists typical user requirements and illustrates translation of these to an intermediate set of derived requirements which are further refined by radar system engineers into features and parameters that become an integral part of the over-all radar design specification. Although representative, this chart is by no means complete, but is intended to present the major issues and processes which go into the design of a typical radar system.

An important point to keep in mind while reviewing the chart is that the process shown is by no means a "one-shot" affair, but is fundamentally iterative in nature, involving several levels of analysis, evaluation, and tradeoffs within the design agency and between the project sponsors and the radar design engineers. Every decision reached regarding the design choices for system- and subsystem-level features and parameters needs to be exhaustively reviewed before the properly balanced radar design specification can emerge. Experience plays an important role in these decisions, and ignoring past lessons learned with actual systems in the field carries with it the seeds of failure in the new enterprise.

New techniques and technologies are continuously emerging, which provide a potential opportunity for the new radar to perform some functions better, more cheaply or more reliably than in the past, or which open the way to performance not previously possible. The critical task, however, lies in demonstrating that the incorporation of any new technology into the radar design will, in the end, actually enhance the radar's capability. This can occur only if that radar system can be produced in the time frame in which it is needed, and at reasonable cost.

2.2 Effects of the Environment

The environmental issues to be discussed here are primarily those encountered by the radar waves as they propagate to the target and back. (The physical environment of the radar platform will impose other limitations which are taken into account by those who design radar equipment, and are not discussed here.) The electronic environment seen by the radar includes atmospheric attenuation, propagation effects from the environment and the underlying earth surface, backscatter from features of that surface and from airborne objects or particles, electronic countermeasures (ECM), and false tracking inputs resulting from target signals reflected from the surface (multipath). In deriving radar system parameters from the user requirements, all these environmental effects will impose constraints on the choices available to the designer. In this section we will consider briefly how these problems affect the radar parameters chosen for particular applications, and how they limit the performance that can be obtained from the resulting radar design. Details of the several environmental effects will be found in later chapters devoted specifically to those subjects.

2.2.1 Atmospheric Attenuation

Clear Atmosphere

Attenuation of radar waves in the normal, clear atmosphere is negligible at the lowest radar frequencies, becoming significant in the microwave bands, and imposing severe limits on radar operation in the millimeter-wave (mmw) bands. Although treated extensively in Chapter 8, a brief summary will indicate the basic effects of atmospheric attenuation on the derivation of system parameters.

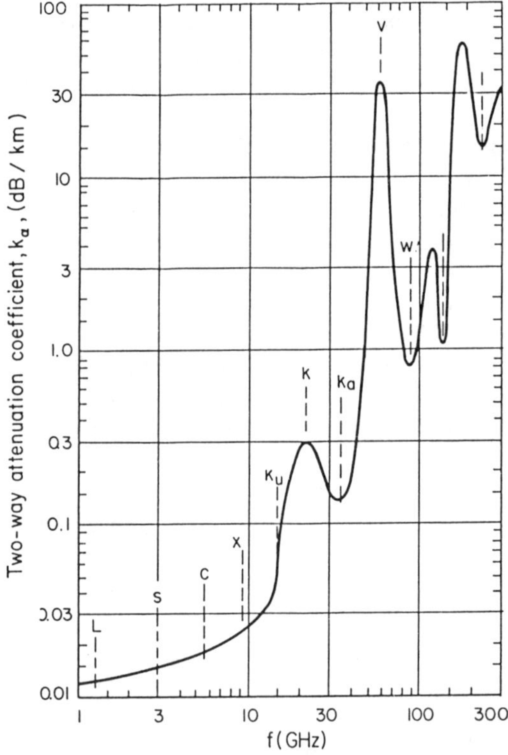

Figure 2.2.1 Clear-air attenuation coefficient vs. frequency [2.1].
©Artech House, reprinted by permission.

At sea level, the attenuation coefficient k_α (in dB/km for two-way radar propagation) is shown as a function of frequency in Figure 2.2.1. Below 1 GHz, the values are low enough that a radar beam directed toward the horizon can penetrate the entire earth's atmosphere with less than 4 dB two-way attenuation, and this loss rises only to 5 dB at 6 GHz (C-band). Only when operation at or above X-band (10 GHz) is considered will clear-air attenuation need to be considered by the designer in choosing the operating frequency of the radar.

Use of radar frequencies in the upper microwave and mmw regions is determined largely by the presence of so-called windows in the atmosphere, defined as minima in the attenuation coefficient. As shown in Figure 2.2.1, these windows are not entirely clear. Approximate values at sea level are shown in Table 2.2.1.

As a rule of thumb, if a radar encounters 6 dB or more of two-way atmospheric attenuation then that radar would be more economically designed in the next lower window or frequency band. Thus, K_a-band radars can be used economically in clear weather to ranges near 40 km, W-band to about 7 km, and higher bands to 5 km or less.

Table 2.2.1 Attenuation Coefficients for Clear Air in Radar Windows

Radar Band	Nominal Frequency (GHz)	Attenuation Coefficient k_α (dB/km)
L	1.3	0.012
S	3	0.015
C	5.5	0.017
X	10	0.024
K_u	15	0.055
K	22	0.3
K_a	35	0.14
V	60	35.0
W	95	0.8
	140	1.0
	240	15.0

Weather

Rain and wet snow produce the most serious attenuation effects on radar, while dry snow, clouds, and fog have lesser but often important effects. Figure 2.2.2 shows the attenuation coefficients for rain falling at different rates, as functions of radar operating frequency. At L-band only the most intense rain can equal the clear-atmosphere attenuation coefficient (Figure 2.2.1), and such rainfall cannot exist over any substantial path length. At S-band the clear-atmosphere value is matched by rain at 10 mm/h, doubling the total attenuation in regions having this "medium" rain condition. Higher frequencies are much more affected by rain than by the clear air.

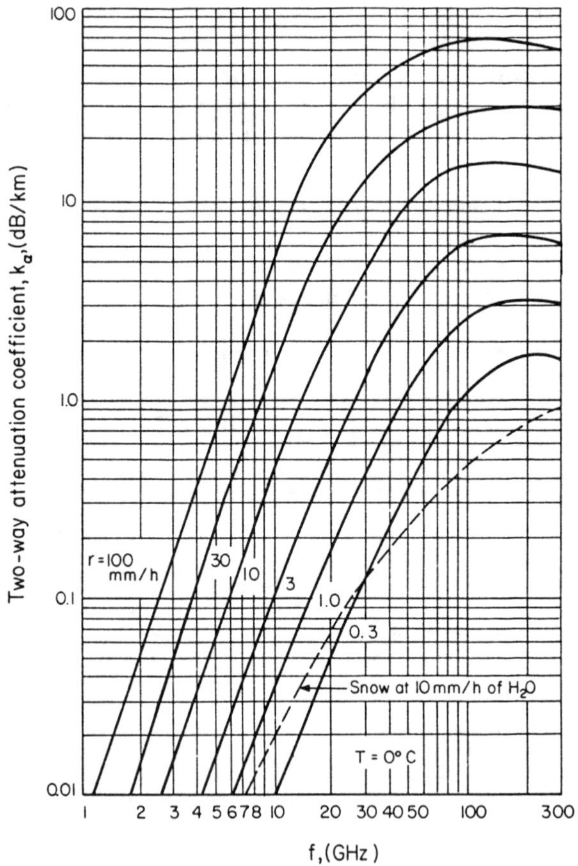

Figure 2.2.2 Two-way attenuation coefficient vs. rainfall rate [2.1].
©Artech House, reprinted by permission.

Military radars are typically designed to be capable of "significant performance" (e.g., 70% of full range) in the presence of 3 mm/h rain at altitudes below 5 km. The attenuation coefficients for the different radar bands under these conditions are shown in Table 2.2.2.

Using the criterion of 6 dB allowable attenuation, the economical ranges with 3 mm/h rain are as shown in the last column. It is, of course, possible to overcome attenuations greater than 6 dB with more power and sensitivity, but unless the application already requires the smallest possible antenna it will prove more economical to use the next lower frequency band.

Table 2.2.2 Attenuation Coefficients in 3 mm/h Rain

Radar Band	Nominal Frequency (GHz)	Two-way Attenuation Coefficient (dB/km)		Range (km) for 6 dB Attenuation
		Rain	Total	
L	1.3	0.001	0.013	
S	3	0.004	0.019	
C	5.5	0.023	0.04	150
X	10	0.11	0.13	45
K_u	15	0.25	0.3	20
K	22	0.7	1.0	6
K_a	35	1.7	1.8	3
V	60	4.0	39	0.15
W	95	6.0	6.8	0.9
	140	7.0	8.0	0.7
	240	6.5	21.5	0.28

The smaller particles suspended as clouds or fog will have less effect on the radar wave. For a density of condensed water of 0.3 g/m^2, the attenuation coefficients will be about half those for rain at 3 mm/h. Dry snow has even less attenuation, except at the highest precipitation rates. Additional data on the effects of clouds, fog, and dry snow are presented in Chapter 8, but they will not normally be the determining factors in the selection of radar parameters.

Dust and Smoke

Because of the small size of suspended particles that make up dust and smoke, this type of obscurant will have negligible effect on radar operation in any band.

Chaff

The purpose of chaff is to create a reflecting region (see Section 12.4) which will obscure real targets or appear as a false target. It is sometimes postulated that chaff (dipoles cut to resonate at radar wavelengths) may cause attenuation in addition to (or instead of) backscatter effects. The chaff densities that would be required to produce any significant attenuation, however, are extremely large. For example, a "high-density" chaff cloud is typically characterized as having a volumetric reflectivity $\eta_v = 10^{-6}$ m^2/m^3. This means that one millionth of the power entering a cubic meter cloud of chaff will be scattered, while the remainder penetrates as a coherent radar wave. If the chaff dipoles are made to be ideally absorbing, this same small fraction of power is converted to heat, or attenuated, rather than backscattered. In a 10 km path through such chaff, 1% of the incoming power will be absorbed, equivalent to a loss of only 0.04 dB, or 0.004 dB/km. It can be concluded that there is no feasible method by which chaff can be made to have attenuation effects as large as those commonly encountered in rain or fog, over any significant volume.

2.2.2 Propagation Factor

The propagation factor F is defined as the ratio of the electric field strength created at a point in space by a radio transmission, to that which would have been created by the same system operating in free space (but excluding attenuation effects of the medium). For radar, the fourth power F^4 describes the two-way power ratio. Thus, the operating range of the radar for detection, tracking, or other functions, will vary directly with the propagation factor applicable to the radar-target path. The factors which can change the free-space propagation conditions to produce greater or lesser fields include: atmospheric refraction, reflections from the underlying surface, and diffraction over this surface.

Atmospheric Lens Loss

Radar waves leaving the antenna at low elevation angles are refracted, or bent, toward the earth, resulting in curved ray paths that are often described by using an effective earth's radius that is 4/3 times the actual value. The actual curved ray can be plotted as a straight line over a spherical earth of this effective radius. The increasing refraction at lower elevations causes a dilution of power in this region, resulting from the vertical spreading of the beam as it propagates beyond the actual horizon. This *atmospheric lens loss* is a function of elevation angle, and is constant for all radar bands. Its maximum value is about 3 dB, occurring for rays directed at the horizon. This loss will have a small effect on the power required of a radar, but will otherwise not influence the system designer in choosing other radar parameters to meet user requirements.

Multipath Lobing

Transmitted radar waves which are reflected from the earth's surface may add to or subtract from the *direct wave* reaching the target. The result produces a periodic *lobing* effect in elevation, with lobes, or regions of enhanced coverage, where the phase of the reflected signal causes addition in phase, and nulls where it causes cancellation (180° out of phase). The resulting *coverage diagram* of the radar appears as in Figure 2.2.3. The peaks of the lobes correspond to a propagation factor $F = 2$, yielding 12 dB enhancement of the original radar signal. These lobes occur at elevation angles which are odd multiples of the angle

$$\theta_n = \lambda/4h_r \quad \text{radian} \tag{2.2.1}$$

where h_r is the height of the radar antenna above the reflecting surface. Nulls occur at even multiples of this angle, and the cancellation causes a large decrease in the original signal.

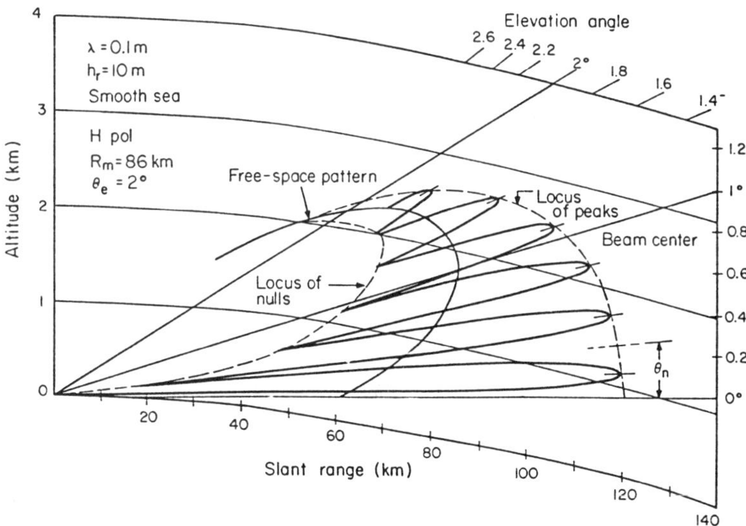

Figure 2.2.3 Radar coverage with lobing [2.1].
©Artech House, reprinted by permission.

Lobing is minimized by using an antenna with narrow elevation beamwidth, or by reducing the antenna response for elevations below the horizon. Lobing is also reduced for rough surfaces, and the effect of surface roughness is a function of radar wavelength. A given surface, viewed at a given *grazing angle*, appears rougher at short wavelengths. For example, if maximum range is desired as in the case of an early-warning radar, lobing is desirable. The antenna will then be designed to illuminate the surface and a long wavelength (for which the surface will appear smooth) will be used. Conversely, if lobing and nulling effects are to be minimized, the antenna will be designed to avoid unnecessary surface illumination and short radar wavelengths will be used to enhance the surface roughness effect.

A particular problem for the radar occurs for targets at low elevation angles, in the null that forms below the angle θ_n. Targets in this null will be affected by reflected signals having pathlength differences less than $\lambda/2$ relative to the direct path, and the effects on radar detection in this region must be calculated using diffraction theory, as discussed in the next section.

Smooth-Sphere Diffraction

Radar waves propagating near the surface of the curved earth will interact with that surface to reduce the electric field strength, even on paths that are not below the horizon. The transition from lobing to diffraction conditions takes place when the difference between the length of the reflected path and the direct path is reduced to less than $\lambda/2$, corresponding to elevation angles below the first peak of the lobing pattern. Relative elevation coverage below this angle is illustrated in Figure 2.2.4, for the same radar at different frequency bands. The geometrical or *radar horizon* is drawn for the 4/3 earth's radius, and this forms the coverage limit as radar frequency is increased. Choice of the best frequency for detection of low-altitude targets involves a trade-off between the reduction in F^4 caused by diffraction, at lower frequencies, and the increase in atmospheric attenuation, at higher frequencies.

The radar horizon for a smooth earth is at a range

$$R_h = \sqrt{2kah_r} = 4130\sqrt{h_r} \quad \text{for } h_r \text{ and } R_h \text{ in meters} \tag{2.2.2}$$

The average value for the radius of the earth is 6.5×10^6 m, and the factor $k = 4/3$ is used to describe the effects of refraction in the standard atmosphere. This factor can vary over a fairly wide range, depending on atmospheric conditions, sometimes producing ducting which permits the radar wave to propagate beyond the radar horizon. The radar horizon range for a radar with antenna height h_r, observing a target at height h_t, is found by adding the radar horizon ranges for the two heights:

$$R_h = \sqrt{2ka}(\sqrt{h_t} + \sqrt{h_r}) \tag{2.2.3}$$

The elevation angle to the radar horizon is negative:

$$\Delta E = -R_h/2ka = -\sqrt{h_r/2ka} \quad \text{radians} \tag{2.2.4}$$

Methods for calculating the propagation factor F for targets near the radar horizon are given in Chapter 8.

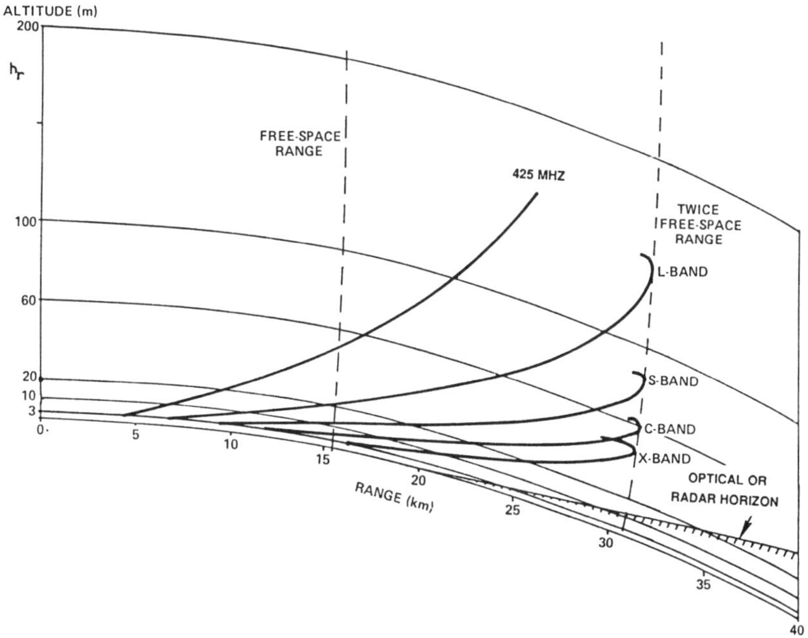

Figure 2.2.4 Radar coverage below the first lobe [2.1].
©Artech House, reprinted by permission.

Knife-Edge Diffraction

Obstacles projecting above the earth's surface introduce another type of diffraction effect: knife-edge diffraction. The object need not have a sharp contour, since at radar wavelengths almost any surface feature will produce this effect at low grazing angle. Trees, buildings, and even terrain features such as low ridges and raised ledges will establish a horizon at a *masking angle* above that of the smooth earth. The propagation factor F at this new horizon will be 0.5, giving $F^4 = -12$ dB for two-way propagation. Below the masking angle, F drops to zero more slowly than for the smooth earth (Figure 2.2.5), enabling some radars to detect targets in the shadow region. This capability is most pronounced for longer wavelengths when the obstacle projects well above the surface, but it has been exploited even at microwaves in communications systems using the obstacle gain of a mountain or ridge.

Rough Surface Effects

Many sea and land surfaces follow the spherical earth contour, but with small-scale roughness due to waves and comparable undulations on land. This roughness is often sufficient to disperse, at the higher elevation angles, the *specular reflections* that would otherwise produce lobing. As the elevation angle is reduced, propagation conditions will be intermediate between those for a smooth earth and for knife-edge diffraction. When combined with the uncertainties caused by nonstandard refraction or ducting, the effect is such as to make an accurate prediction of the characteristics of radar propagation at low elevation angles an impossible task. The smooth-sphere procedure gives a pessimistic result, while assumption of ducting or a knife edge at the horizon gives optimistic results.

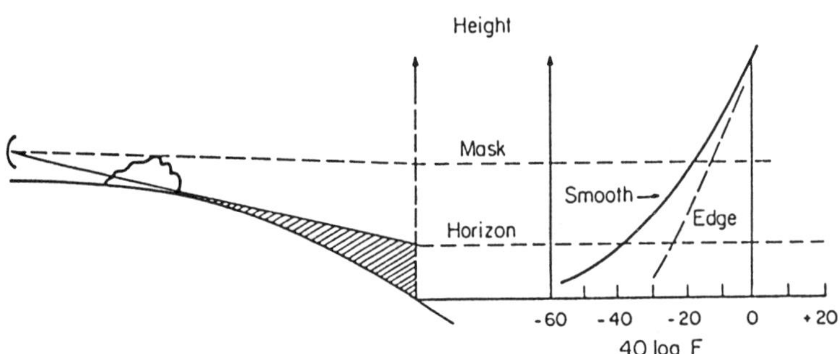

Figure 2.2.5 Knife-edge diffraction effects [2.1].
©Artech House, reprinted by permission.

2.2.3 Clutter

Radar clutter is defined as echoes (radar returns) from any objects which the radar user does not wish to detect. For the designer of air defense radar, this includes returns from *scatterers* on land and sea surfaces, rain, chaff, and birds. For the designer of a ground mapping radar, on the other hand, land is the target and aircraft in motion would be included in the list of clutter sources. Clutter may be divided into three geometrical types: surface, volume, and discrete scatterers.

The radar cross section of surface clutter is found using Eq. (2.2.7), in which the area of the radar resolution cell on the surface is multiplied by a surface reflectivity σ^0. This reflectivity, at angles well removed from both horizontal incidence and vertical incidence, can usually be represented by the so-called constant-γ model

$$\sigma^0 = \gamma \sin \psi \tag{2.2.5}$$

where ψ is the grazing angle and γ is the inherent reflectivity of the clutter surface. Figure 2.2.6 shows that the ray from an antenna at height h_r will reach a flat surface at a grazing angle such that

$$\sin \psi = h_r/R \tag{2.2.6}$$

The resulting clutter cross section, in the region well within the horizon where γ is constant, will be constant with range:

$$\sigma_c = A_c \sigma^0 = \left(\frac{R\theta_a}{L_p} \right) \left(\frac{\tau_n c}{2} \right) \sigma^0 = \left(\frac{h_r \theta_a}{L_p} \right) \left(\frac{\tau_n c}{2} \right) \gamma \tag{2.2.7}$$

where the area of the radar resolution cell, $A_c = (R\theta_a/Lp)(\tau_n c/2)$, as shown in Figure 2.2.6. Typical values of inherent reflectivity are given below, and details of this and special conditions near grazing and vertical incidence are discussed in Chapter 5.

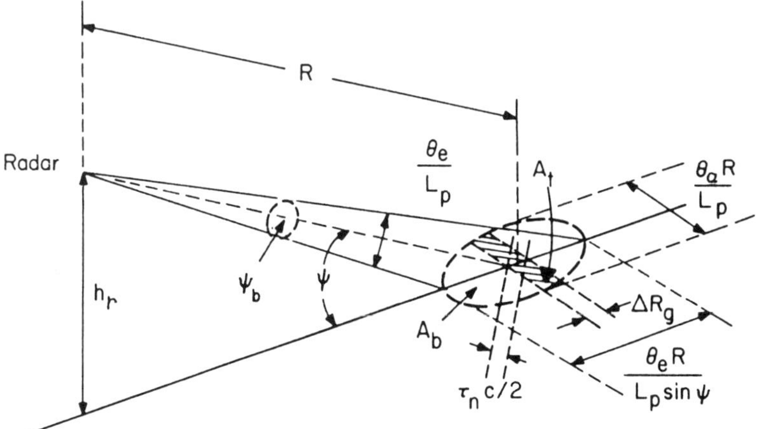

Figure 2.2.6 Geometry of surface clutter resolution cell.

The radar cross section of volume clutter is found as the product of the radar resolution volume V_c multiplied by a volume reflectivity η_v. This volume reflectivity will depend on the precipitation rate or the density of clouds or chaff, and is a strong function of radar wavelength in all cases, as will be shown below.

Discrete clutter sources include birds, man-made structures on the surface, and vehicles (other than targets) on the surface or in the air. In most cases they will occupy a portion of the radar resolution cell, and their cross sections will be independent of radar parameters other than wavelength.

Land Clutter

Typical values of inherent reflectivity for land clutter lie between 0.01 and 0.3 (-20 to -5 dB). The reflectivity $\sigma^0 F_c^4$ observed by a radar can be plotted as a function of grazing angle as shown in Figure 2.2.7. The straight-line segments in the center represent the region of constant γ and $F_c = 1$. Near vertical incidence, reflectivity rises steeply as a result of quasi-specular reflection from tilted facets of the surface, the width and height of the reflectivity lobe in this region depending on the rms slope of these surface facets. For an absolutely flat surface, this lobe would reproduce the antenna pattern of the observing radar. Near grazing incidence, the reflectivity observed by the radar or measuring instrument is affected by a propagation factor F_c^4, as described in Chapter 5, and this drops rapidly as the angle approaches zero.

Most of the wavelength sensitivity of land clutter results from the F_c term, which causes both a systematic reduction at low angles and long wavelengths, and a large variation from point to point over the terrain. This variation is described by the amplitude distribution of the clutter, and *Weibull* or *log-normal* distributions are frequently used to model the effect.

The cross section of land clutter, when viewed directly by a radar, will usually be considerably larger than that of desired targets. For example, an air surveillance radar typically has an azimuth beamwidth near $1.5° = 0.026$ radian, and a pulse width near 1 μs. Operating from a height of 30 m above a smooth land surface ($\gamma = 0.1$), the radar horizon will be at 23 km and the average clutter cross section for this radar, at ranges well within the horizon, will be 9 m^2. Peak values of the clutter, resulting from wooded hills, would be 20 to 30 dB (100 to 1000 times) greater, approaching 10^4 m^2. Since the signal-to-clutter ratio $(S/C)_o$ at the radar output must be large (e.g., +15 dB) to detect a target reliably, the radar would have to obtain an improvement of 50 to 55 dB in order to detect targets of 1 m^2 cross section in competition with these large clutter peaks. Surveillance radars which search for targets at low altitude generally realize this improvement by using effective doppler processing (e.g., moving-target indication, MTI) together with supplementary means of rejecting the clutter peaks.

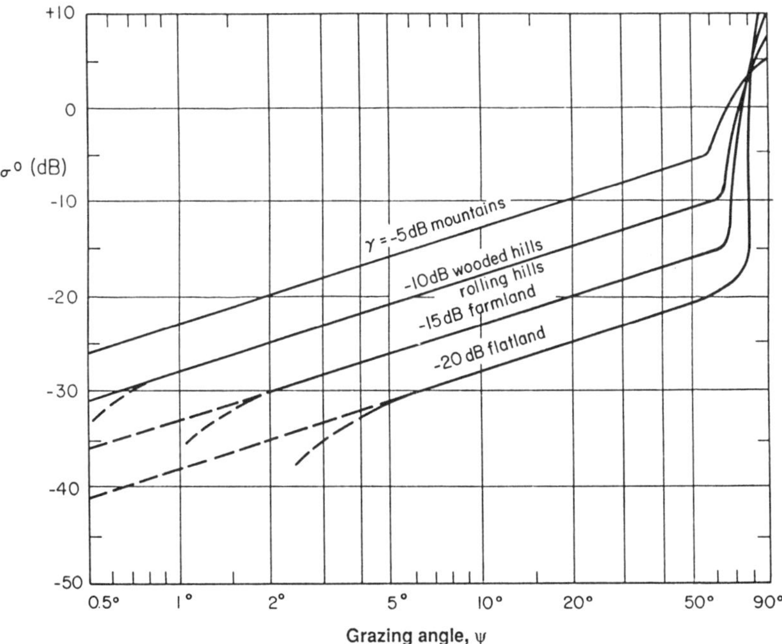

Figure 2.2.7 Reflectivity of land clutter [2.1].
©Artech House, reprinted by permission.

Sea Clutter

The inherent reflectivity of sea clutter is a function primarily of local wind speed and radar wavelength. An approximation is given by

$$10 \log \gamma = 6K_B - 10 \log \lambda - 64 \text{ dB} \tag{2.2.8}$$

where K_B is the **Beaufort wind scale number** (about one unit less than the **sea state number** for a fully developed sea). Radar polarization and direction relative to the wind are also factors, but of lesser importance. Typical values of sea clutter reflectivity σ^0 are shown in Figure 2.2.8. As with land clutter, the values rise rapidly near vertical incidence, and fall near grazing incidence as a result of the propagation factor.

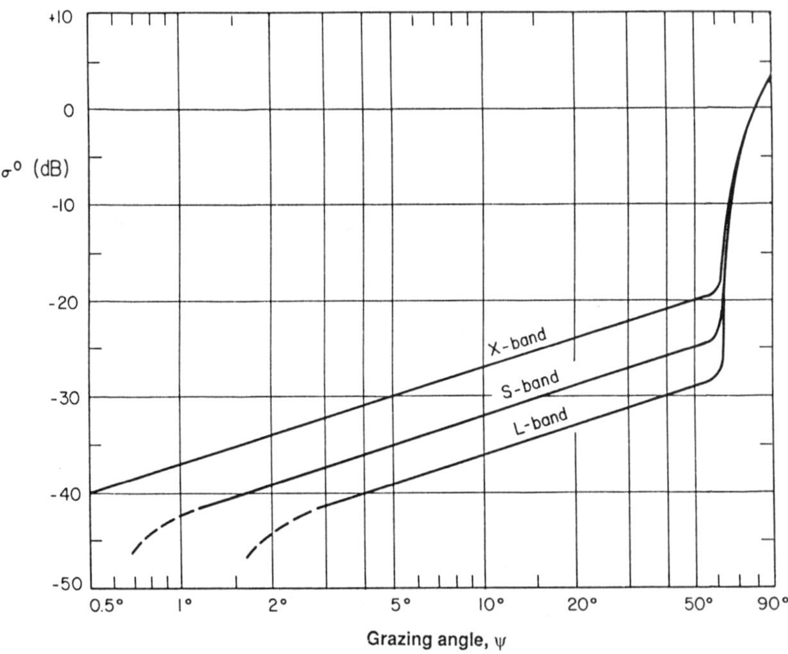

Figure 2.2.8 Reflectivity of sea clutter [2.1].
©Artech House, reprinted by permission.

Even for strong wind conditions ($K_B = 7$), sea clutter in microwave bands does not reach the reflectivity values of most land clutter. Because naval radars will often be required to detect airborne targets over land, they will generally be designed to counter the more severe land clutter. For sea surface surveillance, however, the calculations can be based on Eq. (2.2.8). For example, consider a helicopter radar designed to detect small boats ($\sigma = 10$ m^2) from an altitude of 300 m. Using 1.5° beamwidth (0.026 radian) and 0.11 μs pulsewidth, with an X-band radar, we find, for a medium wind ($K_B = 5$):

$$10 \log \gamma = 65 - 10 \log 0.03 - 64 = -19 \text{ dB} = 0.013$$

From Eq. (2.2.7), the clutter cross section is independent of range, and has a value:

$$\sigma_c = (h_r \theta_a / L_p)(\tau c/2)\gamma = (300 \times 0.026/1.33)(16.5)(0.013) = 1.26 \text{ m}^2$$

The target would be detectable with some integration to average the clutter background.

Weather Clutter

Rain is the most significant source of meteorological clutter. Values of volume reflectivity η_v for *linearly polarized* radars are shown in Figure 2.2.9 for different radar wavelengths as a function of rainfall rate. (A linearly polarized radar is one in which the electric vector is always oriented at the same angle in a plane perpendicular to the direction of wave propagation, as opposed to *circular polarization*, in which the vector rotates as the wave progresses.) At wavelengths longer than about 2 cm, the reflectivity varies inversely as λ^4 and directly as $r^{1.6}$ [1, p. 133]. Along with its attenuation, the reflectivity of rain imposes strict limits on the performance of radars in the higher microwave and millimeter-wave (mmw) bands.

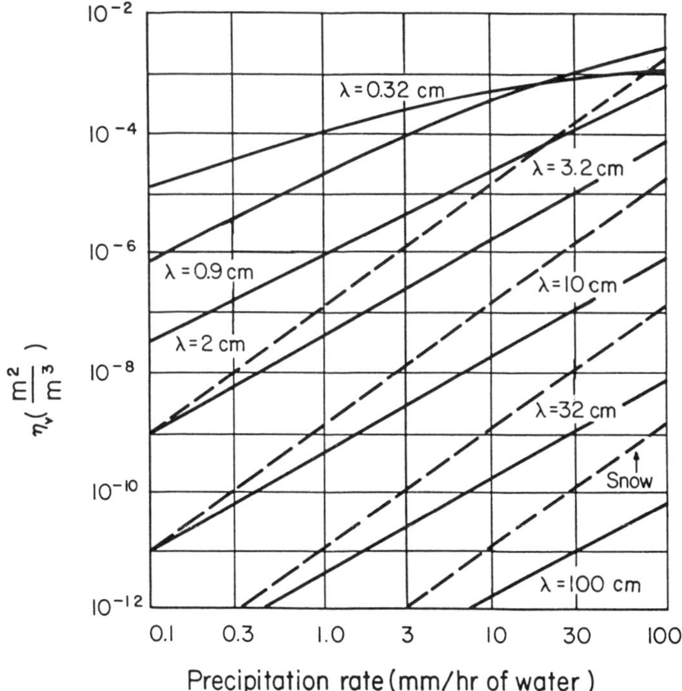

Figure 2.2.9 Reflectivity of rain and snow [2.1].
©Artech House, reprinted by permission.

For example, consider an X-band tracking radar with a 1° pencil beam and a 0.1 μs pulse, tracking a 1 m^2 target at 50 km in 3 mm/h rain. The radar resolution volume will be, from Eq. (2.4.9):

$$V_c = (R\,\theta_3/L_p)\,(\tau c/2) = (50 \times 18/1.33)^2 \times (15) = 6.8 \times 10^6 \text{ m}^2$$

The clutter reflectivity, shown in Figure 2.2.9, is $\eta_v = 3 \times 10^{-7}$ m^2/m^3, leading to a clutter cross section $\sigma_c = 2$ m^2 and $S/C = 0.5 = -3$ dB. Approximately 18 dB improvement is required to make the target reliably detectable. This could be obtained by using circular polarization in the radar, since the rain reflectivity drops by about 20 dB while the reflected target signal loses only about 3 dB.

If the radar frequency is increased to K_u-band ($\lambda = 0.02$ cm) to decrease the beamwidth, maintaining the same antenna size, the rain reflectivity increases by $(3/2)^4 = 5$, while the resolution volume decreases by $(3/2)^2 = 2.2$, resulting in a degradation in S/C by the factor 2.2, or 3.5 dB. Note from Figure 2.2.8, however, that the rate of increase in reflectivity slows near $\lambda = 0.32$ cm. This is because the rain droplets are already in the *resonant region* (in which the cross section reaches its peak, see Section 5.2), and their cross sections cannot increase further. The result is that for the short-range systems at which this band can be used there will be no further increase in reflectivity and the clutter cross section seen by an antenna of given size will decrease as the beamwidth decreases.

When the precipitation is in the form of dry snow, its reflectivity increases somewhat, compared to the same amount of water falling as rain. This is shown by the dashed lines on Figure 2.2.9. If the snow passes through a melting layer, the reflectivity increases markedly. This is called the *bright-band effect*.

Reflectivity of clouds and fog is much less than that of rain, because of the smaller size of the suspended particles of condensed water. For clouds which are not forming raindrops, the maximum reflectivity is 10 dB less than the values shown in Figure 2.2.9 for 0.1 mm/h rain. Additional data are presented in Chapter 8.

An additional problem posed by weather clutter is that it is not fixed, but moves with the wind at whatever altitude regions it occupies. Hence it cannot easily be canceled by MTI and similar systems, unless they are designed to reject all echoes (including potential targets) over a band of velocities up to 20 or 30 m/s.

Chaff

Chaff consists of aluminum or metalized glass fibers which are cut to resonate (or act like small antennas) at the radar wavelength, and yield a large reflectivity per unit weight. The cross section produced by a given weight W of chaff can be estimated as [1, p. 134]:

$$\sigma_c = 2 \times 10^4 \lambda W \text{ m}^2 \qquad\qquad (2.2.9)$$

where W is in kg and λ in m is the wavelength at the center of a broad (decade) band covered by the chaff. This equation is valid for wavelengths as long as 0.3 m, beyond which it becomes difficult to maintain the dipole in a straight configuration without extra stiffness and greater weight. One kg of chaff is sufficient to produce, over wavelengths from 0.3 to 0.03 m (frequencies 1 GHz to 10 GHz), a reflectivity $\eta_v = 10^{-7}$ m^2/m^3 (or 100 m^2/km^3) in 32 km^3 of space. Such a density has a more serious effect in most of these bands than heavy rain. In addition, because each chaff dipole responds only to the linear polarization aligned with its random orientation in space, the clutter cannot be reduced with circular polarization.

The inability to rely on the use of circular polarization to reject chaff, the high velocities at which it moves in the upper troposphere, and its relatively low cost present major problems to the designer of military radar, and therefore makes chaff a preferred ECM approach in many applications.

Birds and Insects

Individual birds have cross sections in the order of 0.001 m^2 (or -30 dBm^2). Larger birds such as ducks, geese, and seagulls may reach -20 dBm^2, and flocks of birds will have cross sections proportional to the number of birds in the flock. Under most conditions, there will be an average of one large bird ($\sigma > 0.01$ m^2) in flight per km^2 of area, with higher densities during seasons of migration. The altitudes of the birds are usually low, from the surface to about 1 km, but higher altitudes are used in migratory flights. These birds may be regarded as discrete targets in cases where the radar resolution cell area is less than 1 km^2. The birds move with velocities ± 15 m/s relative to the surface wind speed. Birds have, on occasion, saturated the display scopes of air traffic control en route radars.

Insects have much smaller cross sections, typically near 10^{-7} m^2 in microwave bands. Few radars need consider insects as a source of interference.

Land Vehicles

Automobiles, trucks, and tanks can represent targets in some radar applications, or in others (air surveillance) they may be considered as moving clutter. Cross sections range from 1 to 20 m^2, but the propagation factor F^4 is usually low enough for ground-based radars that detection is minimized. Vehicles present a significant problem for airborne radars, and for ground-based radars when a road passes over a hill, within line-of-sight range. The relatively high velocities of these vehicles present the radar designer with a problem, as with rain, chaff, and birds, because conventional MTI may be inadequate to reject the moving clutter without also causing rejection of valid targets.

2.2.4 Electronic Countermeasures (ECM)

Noise Jamming

The most common type of ECM is noise jamming, which is intended to prevent detection of the target echo signal by raising the receiver noise level. The normal receiver noise is described (Section 3.1) in terms of the equivalent temperature T_s of the input termination (the antenna and its associated transmission line). The effect on the radar is to increase the effective input temperature from T_s, determined by the receiver and the natural environment, by an amount

$$T_j = \frac{P_j G_j A_r F_j^2}{4\pi k B_j R_j^2 L_{\alpha j}} \quad \text{kelvins, K} \tag{2.2.10}$$

where $P_j G_j$ is the effective radiated power (ERP) of the jammer, F_j is the pattern-propagation factor of the radar receiving antenna in the jammer direction, B_j is the bandwidth of the jammer spectrum, R_j is the jammer range, and $L_{\alpha j}$ is the one-way propagation loss from the jammer. When more than one jammer is present, a temperature is calculated for each, and these temperatures are all added to T_s to find an input temperature $T_i = T_s + T_{j1} + T_{j2} + \dots$.

The radar detection range on echo (quiet) targets, in the presence of noise jamming from remote platforms (*stand-off jamming, SOJ*), is called the ***burn-through range***, and is best evaluated by calculating the increase in input temperature from Eq. (2.2.10). When the SOJ is in the sidelobe region (small values of F_j^2), the radar may achieve good range performance, particularly when low or ultra-low sidelobe levels are provided by refined antenna design.

When the jammer range decreases along with the target (either for *escort jamming, ESJ*, or *self-screening jamming, SSJ*), a burn-through range may be calculated using equations presented in Chapter 5. These calculations will not be discussed here, because the resulting burn-through range is so small as to be of little consequence in the design of radar systems.

In *barrage noise jamming*, the bandwidth B_j extends over the entire tuning band of the radar, typically 10% of the radar frequency. The *spot noise jammer* concentrates its power on a band little wider than the instantaneous signal band-width of the radar, achieving higher noise density at the risk of missing the frequency at which the radar is actually receiving. Spot jammers, along with other specialized jammers, require intercept receivers to control their frequency, and these receivers must operate during quiet (look-through) periods in which no jamming is emitted. When several tunable radars are within line-of-sight of the jamming platform, in the same band, barrage jamming will normally be used.

The noise jammer can mask target echoes, but in the process it permits the radar to obtain an accurate reading of jammer angle (azimuth, for 2D search radars, azimuth and elevation, for 3D and tracking radars). The angle location procedure in search radars results in a *jammer strobe*, while in trackers it provides *track-on-jam* data. The jammer objective is met, however, if the jammer is masking other targets (SOJ or ESJ), or if the intended victim system is unable to operate without range and doppler information denied by the jamming. Since most missile guidance systems are designed to work in the track-on-jam or *home-on-jam* mode, noise jamming is not effective as an SSJ technique.

Deception Jamming

The deceptive jammer uses specialized waveforms to produce multiple false targets or a target at an angle other than the jammer angle. The jammer waveform in many cases is an amplified, or repeated, version of the incident radar signal. Against search radars, the multiple false targets are intended to saturate the data system at the radar output. Against tracking and guidance radars, false targets delay acquisition of the actual target, and prevent accurate fire control or guidance on the jamming platform. There are no general equations which are applicable to deception jamming, other than to calculate the attainable J/S ratio, and hence it will not be discussed further here. Information on its effect on trackers is covered in Chapter 12.

Decoys

Decoys are a specialized form of deception jamming in which a signal source is established outside the jammer or decoy carrying platform. They may be passive, behaving simply as additional targets, or active, as repeaters or noise jammers. The basic purpose of the decoy is to dilute the effectiveness of a defensive system, to delay acquisition of data on real targets, or to protect those targets by diverting missiles or gunfire to the decoy. These techniques are discussed in Chapter 12.

2.2.5 Multipath Error

Reflections from the surface beneath the radar-target path produce not only clutter and changes in the propagation factor, but errors in measurement of target position. While these errors occur in all measured coordinates, they are especially severe in elevation angle, where a second source of target echo is established below the real target. The traditional view of multipath effects has been based on a well-defined *specular reflection*, representing an image of the target below the surface. However, for rough surfaces this model must be modified to account for reduction of this reflection mode and its replacement by *diffuse reflection* from broad regions of the surface.

Specular Reflection

A surface which is flat enough to produce specular reflection is one which meets the *Rayleigh criterion* for a smooth surface:

$$\sigma_h \leq \lambda/(8 \sin \psi) \tag{2.2.11}$$

where σ_h is the rms departure of the surface from a plane, within the *first Fresnel zone* surrounding the point of specular reflection, and ψ is the grazing angle at that point. The first Fresnel zone is the region on the surface within which the reflected path is not more than $\lambda/2$ longer than the path from the specular point. It is usually an extended ellipse surrounding that point, as shown in Figure 2.2.10. When the condition for a smooth surface is met, most of the echo signal reflected from the surface will appear as an image target at the specular point, at an elevation angle equal to the negative of the actual target elevation. When the image lies within $1.5\theta_e$ of the target (i.e., when the target elevation is below $0.75\theta_e$), large errors will be produced in the elevation data. The magnitude of these errors is predicted by equations given in Chapter 11. In extreme cases, when the reflection coefficient of the surface is near unity, tracking may be lost as a result of these elevation errors. Various techniques are available to reduce the specular multipath error, preserving the radar's ability to track at low angle, but the accuracy of data is always impaired at elevation angles near and below the elevation beamwidth θ_e.

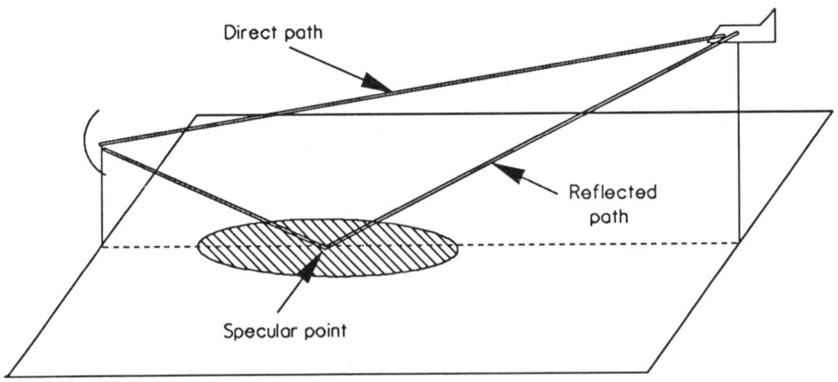

Figure 2.2.10 Fresnel zone surrounding point of specular reflection.

Diffuse Reflection

Diffuse reflection is produced by any deviation of the surface from a plane or smooth sphere, but it becomes intense when the surface roughness σ_h is greater than given by Eq. (2.2.11). Diffuse reflections surround and ultimately replace the specular image, and the region on the surface from which the stronger reflections originate is called the glistening surface. This surface extends over an elevation sector $\pm 2\beta_0$ from the image point, where β_0 is 1.4 times the rms slope of the surface. Usually the surface terminates at the horizon below the target, and extends to a depression angle far enough below the target that it fills the lower half of the radar beam. The surface also extends in azimuth by an amount which depends on the target elevation. The result is to produce both elevation and azimuth error components which are independent of the signal-to-noise ratio (snr) and other radar-target parameters.

Tracking errors produced by diffuse reflections, over the sea or barren terrain, will usually be several hundredths of the radar beamwidth, in both elevation and azimuth, when the target elevation is below 1.5 beamwidths. Detailed analysis can be made, as described in Appendix F, but in the absence of detailed data the following errors can be assumed as minimum values for low-elevation targets:

$$\sigma_\theta = 0.05\,\theta_e, \quad \sigma_a = 0.02\,\theta_a \tag{2.2.12}$$

Multipath over Vegetation

Land surfaces covered by vegetation will show greatly reduced reflections, primarily of a diffuse nature. Most of the incident power is absorbed by the vegetation, and the scattered power is quite widely distributed over surface areas beyond the normal glistening surface. Tracking errors due to this source are on the order of 0.01-0.02 radar beamwidths (θ_3) in both elevation and azimuth.

2.3 Target Detection

Regardless of the function specified by the user, a basic radar requirement is detection of the target. The *detection probability* P_d is the probability that the radar signal processor (electronic or human) will report the presence of a target when it appears during observation of a given resolution cell of the radar. In most cases, this observation will result from the scan of an antenna beam across the target position, and during the *observation time* t_o each range cell within the beam will be inspected for evidence of a target. In attempting to achieve the greatest sensitivity to targets, the threshold for detection will be set as low as possible, subject to the requirement that the false-alarm rate not be too great. The measure of this rate is its reciprocal, the *false-alarm time* t_{fa}, defined as the average time between reports which are not targets but the result of noise and other interference.

The level at which the detection threshold is set, and hence the sensitivity of the radar to targets, depends on the false-alarm probability P_{fa}, which is the probability that a given decision, in the absence of a target, will produce a report at the processor output. This probability is related to the false-alarm time and the rate at which decisions are made by the processor:

$$P_{fa} = t_o / n_d t_{fa} \tag{2.3.1}$$

where n_d is the number of cells in which decisions are made during the observation time t_o. From P_d and P_{fa}, the requirement for target signal-to-noise ratio (snr) may be established. This required snr is denoted by the detectability factor D, and will typically be in the range 10 to 30 (+10 to +15 dB), when evaluated at the output of a processor working over t_o. Methods for determining D are discussed in Chapter 4, where it is shown that the actual requirements depend on the type of target fluctuation, the use of diversity, and the efficiency of the processor. The excess requirement over that for an ideal system on a steady target may be considered as a loss factor to be overcome by higher power and aperture area in the radar.

2.4 Search Function

2.4.1 Detection Specification

The primary objective of a search radar is to scan a given volume of space and detect all targets within that space in time for the using system to respond to them. The volume is defined in terms of contours in range, azimuth, and elevation (or height), within which the targets must be detected. The detection performance may be defined on the basis of the *single-scan detection probability* P_d, often referred to as the *blip/scan ratio*. Alternatively the *cumulative probability of detection* P_c may be specified at a contour within the search volume. This is the probability that at least one detection will have been achieved by the time the target reaches the contour. For k scans conducted rapidly enough that snr, and hence P_d, remains constant,

$$P_c = 1 - (1 - P_d)^k \tag{2.4.1}$$

This equation expresses the fact that the probability $(1-P_c)$ of missing the target on all k scans is equal to the probability $(1-P_d)$ of missing it on one scan, raised to the power k.

If the target changes range significantly over k scans,

$$P_c = 1 - \prod_{i=1}^{k} (1 - P_{d_i}) \tag{2.4.2}$$

where P_{di} is the detection probability on scan i, and Π is the product function.

When the search function is primarily to provide early warning of an approaching target, specification of a contour of P_c is appropriate. The radar designer can then choose a *search frame time* t_s and a range R_m at which some particular detection probability P_d is to be obtained, using the procedures described in Chapter 9. In cases of low target velocity and long detection range, low values of P_d and long frame times t_s are permitted.

Often, however, the search function must provide data to establish and maintain track files in a *track-while-scan* (TWS) operation, where each report is associated with a string of reports from the same target. In order to perform the association, the TWS computer must estimate the target velocity and establish a *correlation gate* in space for each target, based on the velocity and the frame time t_s. A correlation gate is an area in range and azimuth in which a detection on the next scan would be expected, for a particular target. In such systems, especially when many targets are present, P_d must be higher than for the cumulative detection criterion, typically 0.7 to 0.9, and frame times must be shorter, if confusion among tracks is to be avoided.

2.4.2 The Search Radar Equation

The radar equation in Chapter 3 establishes a relationship describing the ability of a radar to detect a target in a benign environment, in which the background interference consists of thermal noise from the receiver and similar sources. Also derived is a *search radar equation* which relates the power-aperture product of a search radar to the coverage volume.

2.4.3 Resolution

The actual differences between the systems listed in the previous paragraph will lie in the ability to resolve the target from other targets and clutter, to measure position of the target, and to perform the integration of signals efficiently. After having established the power-aperture product of the search radar, the next step will normally be to conduct trade-offs of the radar resolution properties in angle. These will be based on Eq. (1.13), written in the form

$$\theta_a = K_\theta \lambda/w, \quad \theta_e = K_\theta \lambda/h \quad \text{radians}$$

where $K_\theta \cong 1.2$ is the constant relating beamwidth to aperture dimensions in wavelengths [1, pp. 151-155] and $wh = A$. In addition to possible requirements in resolving multiple targets, which may set an upper limit on θ_a, the governing considerations will be jamming and clutter. Since a single jammer can effectively mask targets lying within about one radar beamwidth on each side of the jammer angle, beamwidths greater than about 2° (in azimuth) are seldom used in military radars.

The clutter power received by the radar, in a given frequency band, will also be proportional to beamwidths. From Section 2.2.7, we find the cross section of surface clutter competing with a target as:

$$\sigma_c = A_c = (R\theta_a/L_p)(\tau_n c/2)\sigma^0 \tag{2.4.3}$$

where A_c is the area within the spatial resolution cell, τ_n is the width of the processed pulse, and σ^0 is the clutter reflectivity (depending on wavelength, grazing angle, and the nature of the surface). Volume clutter (precipitation or chaff) will have a cross section

$$\sigma_c = V_c \eta_v = (R\theta_a/L_p)(R\theta_e/L_p)(\tau_n c/2)\eta_v \tag{2.4.4}$$

where η_v is the volume clutter reflectivity (depending on wavelength and rainfall rate or chaff density). Thus, for a given clutter environment, the wavelength will affect the clutter power both through beamwidth and clutter reflectivity parameters.

2.4.4 Wavelength and Waveform

Radar wavelength affects the resolution properties of the system in all four coordinates, and resolution improves as wavelength decreases. In range, resolution depends directly on radar signal bandwidth, since the width of the processed output pulse τ_n will be approximately the reciprocal of the bandwidth:

$$\tau_n = 1/B \quad \text{(sec)} \tag{2.4.5}$$

Range resolution is not limited by the transmitted pulsewidth, τ, but may be improved through the use of pulse compression techniques, in which the product of signal bandwidth and transmitted pulsewidth is made large: $B\tau \gg 1$. The available bandwidth, in most cases, is limited to about 10% of the radar frequency, so the highest resolutions are available only in the upper microwave and mmw bands.

In doppler, the frequency resolution depends on the duration of the signal which can be processed coherently. The term coherent refers to the ability of the system to maintain a phase reference matching that of the signal. The reference may be maintained either physically, as an oscillator signal, or mathematically, in the signal processor. If a single pulse of width τ is transmitted, the frequency resolution is:

$$B = 1/\tau \quad \text{(Hz)} \tag{2.4.6}$$

The response to a pulse in both time and frequency is described by the ambiguity function, as described in Chapter 7.

If a train of pulses (or a continuous signal) is transmitted with controlled phase over an interval t_f, the frequency resolution is:

$$B_f = 1/t_f \quad \text{(Hz)} \tag{2.4.7}$$

The corresponding resolutions in radial velocity are:

$$\Delta_v = B\lambda/2$$

$$\tag{2.4.8}$$

$$\Delta_{vf} = B_f\lambda/2$$

2.4.5 Ambiguity

The *response* of the radar receiver and processor (as measured by the output voltage as a function of target position in any of the four coordinates) may consist of more than one lobe, leading to ambiguity when a target is detected. The secondary lobes may consist of sidelobes of low amplitude, adjacent to the mainlobe, as in the typical antenna response, or they may be almost as large as the intended response, and separated widely from it, as in the antenna pattern of an interferometer. In the time-frequency plane, there is a particularly serious problem for pulsed radar, resulting from the sampling effect of the pulse waveform at a given pulse repetition frequency, f_r. Echoes with delay $t_d + it_r$, where $t_r = 1/f_r$ is the pulse repetition interval, will be superimposed on those with delay t_d, and those with doppler shift $f_d + if_r$ will be superimposed on those with shift f_d. As a result, all echoes within the ambiguity function of the single pulse in the waveform will appear to occupy an area $f_r t_r = 1$ in the radar receiver response (Figure 2.4.1).

Translation from delay and frequency into range and radial velocity is done using

$$R = t_d c/2, \quad R_u = t_r c/2 = c/2f_r$$

$$v = f_d \lambda/2, \quad v_b = f_r \lambda/2$$

$$(2.4.9)$$

Targets at ranges $R + iR_u = R + ic/2f_r$ (where $i = 0, 1, \ldots$) will produce echoes at apparent range R, leading to ambiguity in range indication. Similarly, targets with radial velocities $v + v_b = v + if_r\lambda/2$ will appear with the doppler frequency $f_d = 2v/\lambda$. Increasing f_r to reduce the range ambiguity problem will increase the problem in doppler, and vice versa. Only an increase in wavelength can reduce the total ambiguity problem, since the unambiguous area in range-velocity space is:

$$R_u v_b = (c/2f_r)(f_r\lambda/2) = c\lambda/2 \qquad (2.4.10)$$

The minimum wavelength for a given ambiguity area is:

$$\lambda_{min} = 4c/R_u v_b \qquad (2.4.11)$$

Thus, a radar operating at $\lambda = 1.3$ m can be unambiguous in range to 150 km and simultaneously unambiguous in velocity to 600 m/s, qualifying as both low and high prf according to the criteria defined in Appendix A. For these same regions in range and velocity, an S-band radar ($\lambda = 0.1$ m) must have 13 ambiguous responses in range, velocity, or both.

In addition to considerations of resolution and ambiguity, the radar frequency to be used will be subject to constraints imposed by atmospheric attenuation and

    RADAR SYSTEM DESIGN

propagation over the earth's surface, as described in Section 2.2. Attenuation places an upper limit on the frequencies which can be used, while propagation may place a lower limit on radars intended to detect targets near the surface.

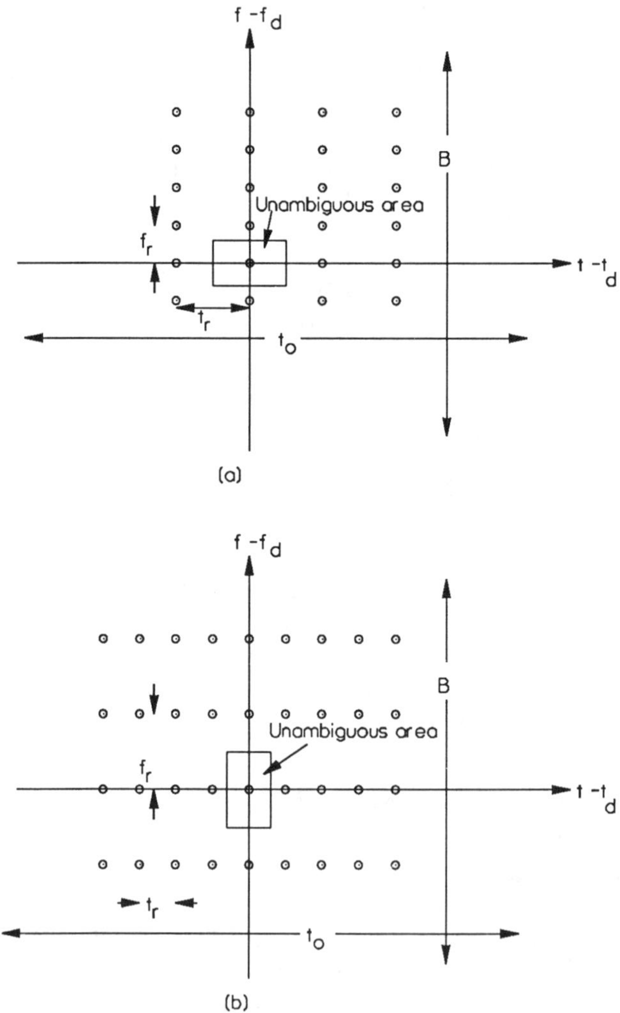

Figure 2.4.1 Delay and doppler ambiguities of pulse trains. (a) Low-PRF pulse train. (b) Higher-PRF pulse train.

2.4.6 Wavelength of a Search Radar

Considerations of coverage, resolution, ambiguity, target cross section, clutter reflectivity, and propagation combine with physical limitations to constrain severely the choice of wavelength for a search radar. There are no general solutions, but an example based on a requirement for $P_{av}A_r = 1.4 \times 10^4$ W · m^2 will illustrate the issues involved and the procedures used. Assume that the radar is to be mobile on a tracked vehicle, and to operate in an environment of wooded hills, rain, and stand-off barrage jamming (see Section 2.2.3). If the available prime power can be taken as 50 kW, this limits the RF output power to about 2500 W (over-all efficiency is seldom better than 5%, considering RF amplifier efficiency and other demands on the power source). An antenna 3.5 m × 3.5 m, giving a 2° × 2° beam at S-band, is consistent with the physical limitations of the radar platform. Total attenuation in rain would be 0.02 dB/km, or 3 dB if rain exists over the entire 150 km path.

Assuming an RF signal bandwidth of 1 MHz, the clutter area, for land clutter at $R = 20$ km, is given by Eq. (2.2.7) as

$$A_c = (20 \times 35/1.33)(150) = 8 \times 10^4 \, \text{m}^2$$

and the clutter volume for rain or chaff at $R = 150$ km is

$$V_c = (150 \times 35/1.33)2(150) = 2.3 \times 10^9 \, \text{m}^3$$

Corresponding mean clutter cross sections for land ($\gamma = 0.1$) and rain ($\eta_v = 2 \times 10^{-9}$ m^2/m^3) are 12 m^2 = +11 dBm^2 and 5 m^2 = +7 dBm^2, respectively. Land clutter peaks would typically be 20 dB or more above the mean, or +31 dBm^2. To achieve output $S/C = D = +13$ dB, the improvement required with respect to input S/C must be +44 dB for land, +20 dB for rain.

To obtain detection at $R_m = 150$ km, the unambiguous range of the waveform should be $R_u = 180$ km, leading to $f_r = 840$ Hz, $v_b = 42$ m/s. The ability of MTI or other processing to reject clutter is dependent on the ratio of this blind speed to the clutter mean velocity v_c and its spread σ_v. As will be shown in Chapter 7, a doppler processor for this PRF and wavelength cannot reliably reject moving rain, so circular polarization would be used to obtain the 20 dB improvement. An increase in signal bandwidth to 10 MHz could also be used to obtain 10 dB improvement, but further increases would not increase S/C ratio for targets whose scatterers are distributed over the target length $L_r = 15$ m. For land clutter, the number of pulses available for processing would be

$$n = f_r t_o = 840 \times 10 \times 2/360 = 47$$

which is adequate to support three groups of 16-pulses, for coherent processing using the *fast Fourier transform* (FFT). This would provide for PRF diversity to fill blind speeds, and RF diversity to reduce target fluctuation loss. Increase in signal bandwidth would not substantially improve detection in land clutter, because the clutter peaks would remain at about the same level, although there might be regions of low clutter between these peaks.

With regard to jamming, assume three stand-off barrage jammers at a range $R_j = 200$ km, each with $P_j G_j = 50$ kW, distributed over $B_j = 200$ MHz. The effective temperature produced by each jammer, when viewed by the victim radar's mainlobe would be (Eq. 2.2.10):

$$T_j = \frac{5 \times 10^4 \times 6 \times 1}{4\pi \times 1.4 \times 10^{-23} \times 2 \times 10^8 \times (2 \times 10^5)^2 \times 1.4} = 1.5 \times 10^8 \text{ K}$$

This level is +52 dB relative to thermal noise at $T_s = 1000$ K. An effective antenna design would provide average sidelobes of -60 dB with respect to the main lobe, reducing the jamming below noise level when the antenna points more than 2° or 3° away from any jammer. The combined temperature of all three jammers would then be -3 dB relative to thermal noise. If sidelobes cannot be made this low, *adaptive cancellation* is needed, or radar performance degradation in the jamming environment must be accepted.

Note that the environmental model assumed does not include chaff. If chaff were included, and if that chaff were spread over extended regions in altitude and range, the velocity spectrum of the clutter would be so wide that the S-band low-PRF waveform could not achieve chaff rejection with an unambiguous range of 150 or 180 km. It would be necessary to use a longer wavelength or range-ambiguous PRFs or CW radar to solve the problem.

2.4.7 Three-Dimensional Radar

The 3D radar is a search radar which obtains elevation angle data, in addition to range and azimuth, on each target as it scans. This can be done by scanning a pencil beam rapidly in elevation, or by generating a stack of contiguous elevation beams which cover the entire elevation search sector. In either case, if the azimuth search sector is large, azimuth scanning is best performed by mechanical rotation of the antenna. Elevation data are obtained by comparing the amplitudes of signals received in adjacent beams, and interpolating to find the elevation of the target within a fraction of the beamwidth. There are several reasons for using a 3D search approach:

(a) It permits the use of a larger receiving antenna aperture than would be possible, at a given frequency, if the entire sector were covered by one beam;

(b) It discriminates against clutter and jamming arriving in the lowest beam, permitting targets at higher elevations to be detected more reliably;

(c) It reduces the volume of rain or chaff clutter which competes with a target in any given beam;

(d) It provides elevation data on targets, permitting the ground range to be calculated from slant range and making it possible to merge tracking data from different radars in a network; and

(e) It provides altitude data on targets, permitting them to be designated to tracking radars, intercepted by manned aircraft, or controlled to avoid collisions.

If 3D operation is required, the choice between elevation scanning and stacked-beam approaches is made primarily on the basis of the observation time t_o required for adequate processing and doppler resolution of targets in clutter. In electronically scanned, pencil-beam radars, pseudo-random scanning and raster scanning in azimuth and elevation are possible, with the same resolution benefits and limitations that apply to rotating search radars.

2.5 Tracking Function

A tracking radar must initially acquire its target, using target designation data from an external source or from a preliminary search mode of operation. The requirements imposed by this acquisition mode are, in many ways, similar to those for a search radar, and the considerations described in the previous section will apply. Generally, however, the sector to be searched will be small, relative to a search radar, and a single, narrow tracking beam can perform the required scan in a few seconds. It should be noted, from Eq. (2.4.7), that the acquisition range will be inversely proportional to the fourth root of the solid angle to be scanned.

Consider a tracking radar which is to acquire a 1.0-m^2 target at 100 km, on the basis of designation that places the target within a $10° \times 10°$ sector. Using the same assumptions as with the search example of Section 2.4.3, but with $\psi_s = 0.03$ steradians and $t_s = 5$ s, we find:

$$P_{av} A_r = 220 \text{ W} \cdot \text{m}^2$$

An antenna diameter near $D_a = 1$ m with a transmitter of about 500 W average power will be required.

2.5.1 Acquisition Procedure

The target must be acquired in four-dimensional radar space, but in most cases the scanning will be done only in angle, and all range and doppler cells will be covered in parallel (i.e., simultaneously using multiple receiver channels) during each beam dwell. The angle scan should be adjusted to have a high probability of including the actual target position, and yet coverage of positions in which the target is almost certainly not located must be avoided, to conserve energy and reduce the time required to search. The correct procedure will balance the probabilities of failing to acquire due to two problems: target position beyond the scan sector, and target signal insufficient to generate a detection during the time-on-target within the scan sector.

As an example, assume that the probability of acquisition P_a of a designated target, within 5 s, is to be 0.9. The target is designated in azimuth and elevation, with an accuracy $\sigma_\theta = 2°$. In order to achieve $P_a = 0.9$, the designer should allow 0.05 probability that the target lies outside the scan sector, and 0.05 probability that the signal within the scan sector is inadequate for detection ($P_d = 0.95$). The scan sector should be a circle of diameter $\Delta_\theta = 5\sigma_\theta = 10°$, and the power-aperture product of the radar should be sufficient to achieve $P_d = 0.95$ when scanning this sector in 5 s. If the target is fluctuating, it will prove desirable to scan the volume two or more times in this interval, with $P_c = 0.95$ over the total number of scans.

In the tracker example, if the radar operates at X-band ($\lambda = 0.03$ m) with a resulting pencil beamwidth of $2°$, the designer might choose a rectangular *raster scan* covering $10° \times 10°$, using five lines of five beam positions each. He could also use a spiral scan with a radius increasing by $2°$ per rotation. In either case, the region would best be scanned about four times in 5 s. If a cumulative probability of detection $P_c = 0.95$ is required over four scans of the sector, the detection probability P_d per scan can be found from:

$$P_c = 1 - (1-P_d)^4 = 0.95$$
$$1-P_d = 0.05^{1/4}$$
$$P_d = 1 - 0.47 = 0.53$$

Each scan would take 1.25 s, and would cover 25 beam positions. The resulting observation time would be $t_o = 1.25/25 = 0.05$ s.

The radar designer often cannot predict the accuracy of the designation data for a particular target. In that case, he must use the largest expected error and overdesign the tracking radar to achieve adequate P_a with this error. When the designation is more accurate, acquisition will take place more rapidly if the scan starts at the designated point and proceeds outward. If several scans are to be used, the first scan should be of small radius, expanding on subsequent scans until the full sector has been covered. The same procedure applies in cases where no external designation is received. The local search should start, and concentrate, on regions where the target is most likely, extending later and with reduced energy density to regions less likely to contain the target.

2.5.2 Number of Targets and Data Rate

The typical mechanically steered tracker can provide data on only a single target at a time, and the data are continuous (sampled at the radar pulse repetition rate, for monopulse radar, or the beam scanning rate, for conical scan radar). The output data rate will be determined by the bandwidth β_n of the antenna servo system, typically a few hertz. The bandwidth requirements in such a case are determined by the errors generated by target accelerations, as described in Chapter 11.

For example, assume that a target acceleration $a_t = 3g = 30 \, \text{m/s}^2$ is to be tracked with a lag error ε_a less than 5 m. The required *acceleration error constant* K_a and servo bandwidth can be found from:

$$K_a = a_t/\varepsilon_a = 6s^{-2}, \quad \beta_n = \sqrt{K_a/2.5} = 1.5 \, \text{Hz}$$

Data bandwidths of this sort impose no significant constraints on choice of PRF or tracking method. If a *conical scan tracker* were to be used, any scan rate greater than about $f_s = 10\beta_n = 15$ Hz would suffice, and the PRF would be at least another factor of ten higher: $f_r \geq 150$ Hz.

When the radar must provide data on more than one target, its beam and power must be shared among the targets, either uniformly or in a way that will optimize the quality of the data. The electronically steered *phased array* system is the usual response to such a requirement. The beams are formed at a rate f_b equal to the product of data rate and number of targets (or to the sum of the data rates of individual targets, where they differ). If beam steering is essentially instantaneous, the dwell time per beam will be $t_o = 1/f_b$. Both the energy per target and the doppler resolution will be reduced as the number of targets and data rate are increased. A practical limit is reached when the beam rate is equal to the reciprocal of the delay time to the target, for which case only one pulse may be exchanged with each target and doppler processing is impossible.

For example, assume that a phased array system tracks 100 targets, each with 10 pulses per second, with one pulse per beam dwell. The beam forming rate is f_b = 1000 Hz, and the dwell time $t_o = 1/f_b = 1$ ms, providing coverage to 150 km maximum range. The 10 Hz rate on each target can support a maximum tracking loop bandwidth of 5 Hz, which is more than adequate to provide the acceleration error constant $K_a = 6$ as derived above. Although it is conceivable that the beam can transmit a pulse to one target, switch to another position for reception of a second target, and get back in time to receive the echo from the first, this mode of operation is not practical in a great majority of cases, and leads to unrealizable systems when it is attempted.

2.5.3 Resolution and Accuracy

The resolution requirements imposed on a tracking radar system are more stringent than for search, because the tracking error in a benign environment (without intentional jamming) will be some fraction of the resolution cell width, and this error will increase rapidly if a source of interference or a second target enters this cell. The basic expression for error in measurement of an arbitrary coordinate z is

$$\sigma_z = \frac{z_3}{k_z \sqrt{2(S/I_\Delta)n_e}} \tag{2.5.1}$$

where z_3 is the width of the resolution cell, k_z is a measurement slope constant (near unity), S/I_Δ is the ratio of sum channel signal to difference channel interference, and n_e is the number of independent samples of the interference integrated by the tracking filter. In sequential measurement radars (e.g., *conical scan)*, sum and difference channels are formed in the signal processor, while in *monopulse radars* they are formed (for angle channels) in microwave networks. Monopulse radars ($k_z = k_m = 1.5$) have achieved errors as small as $\theta_3/1000$ in suitably benign environments with strong signals. In most cases, however, it can be assumed that the error will be hundredths of the resolution cell width, and this sets a maximum cell width for a required accuracy figure.

For example, consider the multiple-target phased-array radar described earlier, tracking 100 targets at 10 Hz per target. Most sources of interference will be uncorrelated over time intervals of 0.1 s, so with a tracking loop bandwidth of 5 Hz there will be $n_e = 2$ independent interference samples averaged by the tracker. If the tracking error is to be $\theta_3/100$, Eq. (2.5.1) can be solved for the required signal-to-interference ratio:

$$S/I_\Delta = (\theta_3/k_m\sigma_\theta)^2/2n_e = 1100 = +30.5 \text{ dB}$$

In more difficult environments, especially in the presence of multiple targets, ECM, and clutter, the resolution cell size will be set by the requirement for exclusion of significant interference sources from the target cell.

2.5.4 Electronic Counter-Countermeasures

The ECCM considerations for tracking radar are somewhat different from those of a search radar, since mere detection and warning of target presence is not sufficient. Target position data are required, and jammers can introduce large errors in these data or deny them entirely in some cases. For example, the on-board (self-screening) jammer can almost always deny range and doppler data to the tracking radar, even while providing a strong beacon which assists the angle tracker. Most military systems using tracking radars must rely on the angle data to carry out their function without accurate range or doppler. This is done using so-called track-on-jam and home-on-jam techniques, which are especially effective in homing missile systems for which range is of secondary importance as long as the target is within the reach of the missile.

When denial of range and doppler data fails to thwart the radar system, the jammer must use an angle deception technique. This can be a simple matter for a jammer located other than on the target, since all that is required is to mask the target echo, transferring the track-on-jam point from the target to the jamming source. It is a more difficult matter for the self-screening jammer, which must introduce signals into the tracker which disturb both normal track and the track-on-jam operation.

2.6 Imaging and Target Identification

2.6.1 Terrain Mapping

The statement of requirements for the terrain mapping function will normally include the following:

Range limits, R_0 and R_m;

Azimuth sector width A_m or cross-range width $R_m \sin A_m$;

Resolution Δ_r in range, Δ_x in cross-range;

Time t_s to form the map; and

Altitude h_a and speed v_a of the platform.

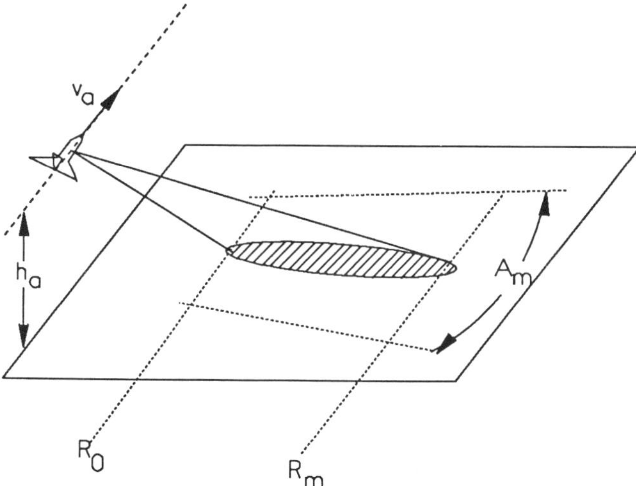

Figure 2.6.1 Typical terrain mapping geometry.

With these data and the minimum terrain reflectivity value $\gamma = \sigma^0/\sin\psi$ to be detected, the power-aperture product required for the mapping radar can be found from Eq. (3.2.3) by substituting the following:

$$\psi_s = A_m \left(\frac{h_a}{R_0} - \frac{h_a}{R_m} \right)$$

$$\sigma = \Delta_r \Delta_x \sigma^0 = \Delta_r \Delta_x \gamma h_a / R_m$$

The result is

$$P_{av}A_r = \frac{4\pi A_m[(R_m/R_0) - 1]R_m^4 kT_s DL_s}{\Delta_r \Delta_x \gamma t_s} \tag{2.6.1}$$

The assumptions here are that the antenna beamwidths are no wider than the elevation sector (h_a/R_0)-(h_a/R_m) and azimuth sector A_m, so that transmitted energy is not lost outside the mapped area, and that the radar platform moves far enough in t_s to achieve the desired resolution. If wider beams are used, the required *PA* product will be increased in proportion to the beam solid angle.

Terrain mapping can be carried out in different ways, depending on the resolution required:

(a) Real-aperture sidelooking radar (SLR), in which a wide-aperture antenna is placed along the side of the vehicle, producing a narrow beam $(\theta_3 \approx \lambda/t_f v_a)$ and either scanned in azimuth or allowed to scan across the terrain as the vehicle moves.

(b) *Synthetic aperture* sidelooking radar (SAR), in which a small antenna is placed on the side of the vehicle, and coherent processing over a long period t_f is used $(\theta_3 \approx \lambda/t_f v_a)$. The beam produced by the physical aperture may be fixed or scanning slowly in azimuth.

(c) Synthetic aperture *spotlight mapping*, in which an antenna of medium size is kept pointed at a designated target region while coherent processing is carried out as in (b).

The range resolution will be determined by signal bandwidth, in accordance with Eq. (2.4.10). Cross-range resolution for synthetic aperture processing (with focusing) will be

$$\Delta_x = R_m \lambda/2m v_a t_f \tag{2.6.2}$$

where m is a synthetic aperture beamwidth factor whose value lies between 0.5 and 1, and $t_f \leq t_s$ is the coherent processing time. From this, assuming $m = 0.5$ as a practical value and using the entire time t_s for processing, the maximum wavelength of the radar can be determined:

$$\lambda_m = \Delta_x t_s v_a/R_m \tag{2.6.3}$$

At this maximum wavelength, the minimum antenna dimensions are given by

$$w_0 = \lambda_m/A_m = \Delta_x v_a t_s/R_m A_m$$

$$\tag{2.6.4}$$

$$h_0 = \lambda_m/(E_m - E_0) = \Delta_x v_a t_s/h_a((R_m/R_0) - 1)$$

Shorter wavelengths can be used, and the minimum antenna dimensions (used in the nonscanning system) will scale to the actual wavelength. Larger antenna apertures can be used at the shorter wavelengths, with multiple or scanning beams and reduced t_f.

As an example, consider a mapping requirement for a 10° sector extending 25 to 50 km from a platform at 5 km altitude. A resolution cell 3 m × 3 m is to be obtained, in 10 s, with 10 dB snr on terrain for which the reflectivity $\gamma = 0.03$. The system temperature will be assumed $T_s = 1000$ K, with $L_s = 30 = 15$ dB. Application of Eq. (2.4.7) gives the minimum power-aperture requirement:

$$P_{av}A_r = 22 \text{ W} \cdot \text{m}^2$$

The maximum wavelength and antenna dimensions will be proportional to the platform velocity. For $v_a = 200$ m/s,

$$\lambda_m = 0.12 \text{ m}, \quad w_0 = 0.67 \text{ m}, \quad h_0 = 1.2 \text{ m}$$

For this antenna, with 50% efficiency, the average power will be

$$P_{av} = 22/0.5A = 54 \text{ W}$$

Options at other wavelengths would include scaling the height h to a shorter wavelength (e.g., $\lambda = 0.03$ m, $h = 0.3$ m), leaving the width fixed, and scanning the beam to get $t_f = 2.5$ s per beam. Four contiguous maps would be obtained to cover the area. It is also possible to extend the mapping time to the point where a sidelooking strip map would be obtained, with maximum resolution given by

$$\Delta_x = w/2m \approx w \tag{2.6.5}$$

As the coherent processing times increase to meet this resolution requirement, the radar system stability and signal processing requirements become more difficult to meet.

2.6.2 Inverse Synthetic Aperture Radar (ISAR)

The processing techniques of SAR may be applied when the radar is fixed and the target is moving so as to produce a rotation of the target about the line of sight from the radar. The target motion can be the linear motion (but not radial) of a target which is stable in inertial space, or the target can have a real rotational component. For real or apparent target rotation rate ω, with coherent processing over time t_f, the resolution for scattering sources on the target is:

$$\Delta_x = \lambda/\omega t_f \geq \lambda/2 \tag{2.6.6}$$

Thus, for targets which rotate 180° or more, features with dimensions $\lambda/2$ can be resolved, while for small rotation angles $\omega t_f < 1$ radian, the resolution cell will be greater than λ. This resolution is available at any range for which adequate snr can be obtained by the radar, and the only requirement is that the rotation angle be accurately determined at the radar, either through measurement of the trajectory of a stabilized target or through observation of regular rotational motion. Resolution in the radial dimension is obtained with wide signal bandwidths, producing a two-dimensional image of the target scatterers.

Because the target rotation angle is not under the control of the radar, observations may be required over long time periods, and signal coherence must be maintained. This generally requires that the target be observed continuously with a tracking radar, and the use of short wavelengths provides better resolution with shorter tracking times.

2.6.3 Range Profile Recognition

Target classification or identification can sometimes be carried out using high resolution only in the range coordinate. These data are available almost instantaneously (a single pulse of high snr will be adequate), and can be obtained either by a tracking radar or a search radar. Additional data on the attitude of the observed target, obtained from a track file giving target velocity, may be needed to interpret the range profile in terms of linear dimensions along the axis of the target.

2.6.4 Doppler Spectrum

Many targets produce recognizable spectra, when observed over long enough periods with coherent radar. Examples of spectral features which can be recognized include:

(a) Body rotation or attitude instability;
(b) Propeller, rotor, or turbine components;
(c) Vibrations of the target surface; and
(d) Explosive expansion of ordnance, or destructive effect of a weapon.

2.6.5 Inherent Resonances of Target

When the radar can operate at wavelengths comparable to the dimensions of the target, resonance effects can be observed which may permit target identification or classification. Several frequencies, preferably covering the fundamental resonance and its harmonics, must be used by the radar, and this requirement limits the applicability of the technique.

2.6.6 Polarization Properties

The scattering of radar waves from many targets is sensitive to the polarization of the waves, the amplitudes and phases of the echoes being different for horizontal, vertical, circular, or the more general elliptical polarizations. If the radar can transmit and receive different polarizations, and can compare the amplitude and phases for different combinations of transmit and receive polarizations, these observed properties can be compared with a computer memory-stored library of measured data on targets of interest, permitting target classification or identification. There are no general rules for determining the effectiveness of this technique, and particular targets must be extensively analyzed to determine the potential applicability of the technique to each target type.

2.7 References

[2.1] D. K. Barton, *Modern Radar System Analysis*, Artech House, 1988.

Chapter 3

RADAR RANGE EQUATION

This chapter will discuss the basic approach to radar detection of target objects, and will derive the radar range equations that establish the average volume within which target detection can be obtained. Subsequent chapters will consider the characteristics of noise and other interference components entering the receiver, the required ratio of target signals to this interference, and the means by which this ratio can be obtained in the actual environments where radars operate.

3.1 Derivation of the Range Equation

The radar equation is a basic relationship which permits the calculation of the target echo signal strength from the known parameters of the radar, the propagation path, and the target. The radar equation may be extended to express the ratio of signal power to noise power (snr) at the output of a radar receiver, and this ratio may then be used to calculate the expected performance of the radar as a detection or measurement device. If the exact values of the parameters and propagation factors used in the radar equation are known, the equation will yield an exact value of received signal and snr. In practice, however, knowledge of the radar and target parameters is imperfect, and a series of loss factors is included to make the results of the equation agree with the actual performance. The calculation or estimation of these loss terms is an important step in predicting and evaluating the performance of a practical radar.

3.1.1 Received Signal Power

Assume first that the power P_t of the radar transmitter is radiated *isotropically* from the antenna (see Figure 3.1.1). Since the surface area of a sphere at radius R from the radar is $4\pi R^2$, this uniform distribution of transmitted power over the surface or the sphere will produce a power density given by

$$I_s = \frac{P_t}{4\pi R^2} \quad \text{W/m}^2$$

(3.1.1)

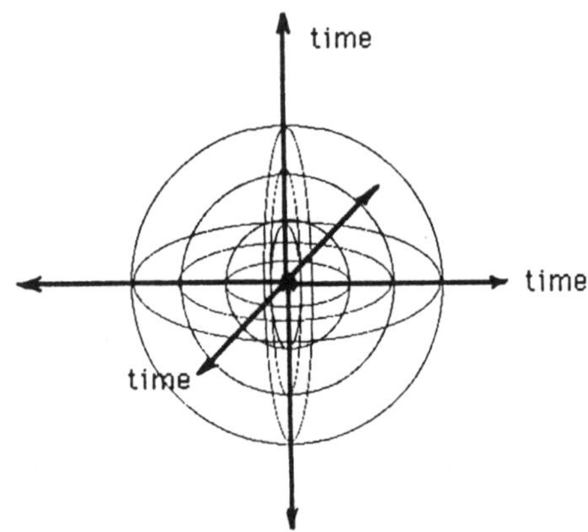

Figure 3.1.1 Isotropic radiation.

The concept of power density is illustrated in Figure 3.1.2. If a directive transmitting antenna with *power gain* G_t is now used in place of the omni-directional or isotropic antenna, the radiation becomes focused and the power density along the axis of its beam will be increased to

$$I_s = \frac{P_t G_t}{4\pi R^2} \quad \text{W/m}^2$$

(3.1.2)

for P_t in W and R in m.

Imagine now a spherical target of radius a at range R on this beam axis. The target has a projected area of a circle, $\sigma = \pi a^2$, which will intercept a fraction of the incident radar power given by

$$P_i = I_s \sigma = \frac{P_t G_t \sigma}{4\pi R^2} \quad \text{W}$$

(3.1.3)

This power P_i will be reradiated isotropically by the sphere, producing at the radar a power density equivalent to that of an isotropic transmitter of power P_i located at the target:

$$I_r = \frac{P_i}{4\pi R^2} = \frac{P_t G_t \sigma}{(4\pi R^2)^2} \quad \text{W/m}^2 \tag{3.1.4}$$

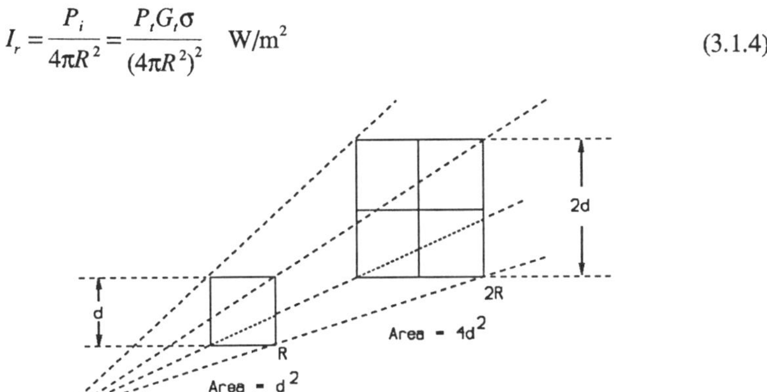

Power density for $P_t = 1$ watt, $d = 1$ m
at R: $P_t/d^2 = 1$ W/m²
at $2R$: $P_t/4d^2 = 1/4$ W/m²

Figure 3.1.2 Power density vs. range.

The radar antenna may be characterized by its ***effective aperture area*** A_r, which will capture from the reradiated wave power:

$$S = I_r A_r = \frac{P_t G_t A_r \sigma}{(4\pi)^2 R^4} \quad \text{W} \tag{3.1.5}$$

This is the radar echo, and when certain losses are included, is the power available to the radar receiver. If the receiving antenna is described by its gain G_r, we may use the following relationship:

$$A_r = G_r \lambda^2/4\pi \ \text{m}^2 \tag{3.1.6}$$

to arrive at the received power:

$$S = \frac{P_t G_t G_r \lambda^2 \sigma}{(4\pi)^3 R^4} \quad \text{W} \tag{3.1.7}$$

This is the equation for received signal power at the radar antenna, for a spherical target of cross section σ. An actual target, viewed at a particular ***aspect angle***, may be replaced by an equivalent spherical target that produces the same signal power. The target is then characterized by a radar cross section which is not necessarily related to its physical area, but which describes the ability to scatter power toward the radar.

The preceding equation may be restated in the form of the product of three factors:

$$S = (P_t G_t)\left(\frac{\sigma}{4\pi R^2}\right)\left(\frac{A_r}{4\pi R^2}\right)$$

(3.1.8)

where the first factor is the effective radiated power (ERP) of the radar transmission in the direction of the target, the second is the fraction of this ERP intercepted by the effective spherical target cross section, and the third is the fraction of the resulting scattered power captured by the receiving aperture. Figure 3.1.3 summarizes the received signal power discussion.

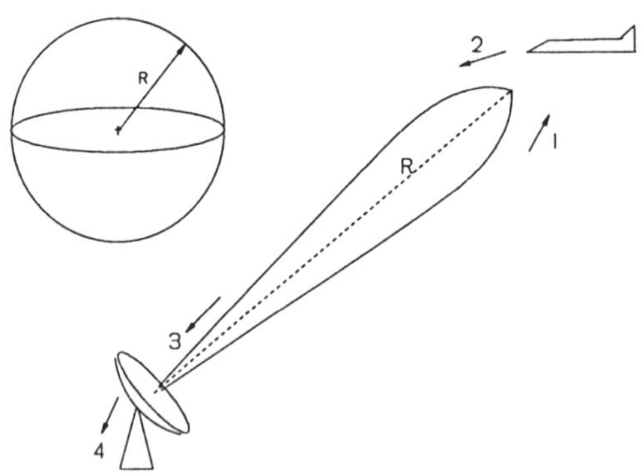

Figure 3.1.3 Derivation of recovered signal power.

As an example, consider a radar transmitting $P_t = 100$ kW with an antenna gain $G_t = 10^4$. The target cross section is $\sigma = 1$ m^2 at a range $R = 90$ km, and from (3.1.6) the receiving aperture ($G_r = 10^4$ at $\lambda = 0.1$m) is $A_r = 8$ m^2. The three terms of (3.1.8) and resulting received power are:

$$S = (10^9)(10^{-11})(8 \times 10^{-11}) = 8 \times 10^{-13} \text{ W} \tag{3.1.9}$$

Notice that this very small signal is less (by a factor of 1000) than 1 quintillionth of the transmitter ERP.

3.1.2 Receiver Noise

The ability of the radar to detect the target echo depends not only on the signal power S, but also on the competing noise present in the radar receiver. The sources of this noise lie both within the radar itself and in the outside environment, as shown in Figure 3.1.4.

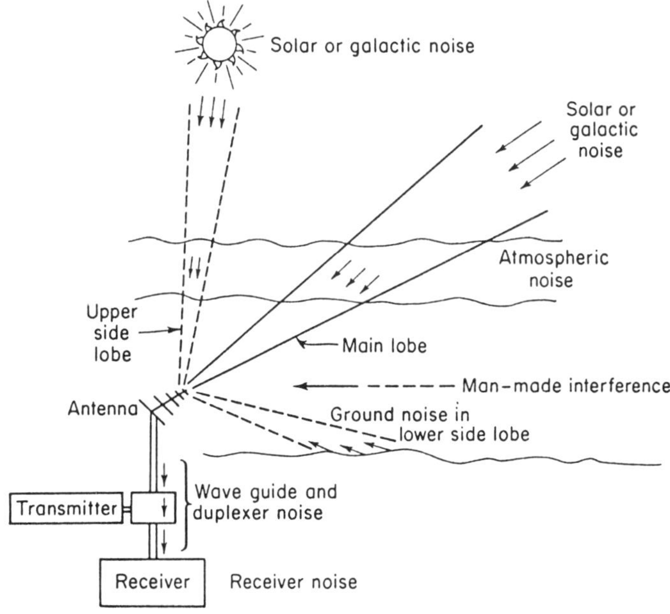

Figure 3.1.4 Sources of radar noise [3.1].
© Artech House, reprinted by permission.

It is convenient to summarize the effect of all of these noise components in terms of a single source at the antenna output terminal, represented by an input termination resistor with a temperature T_s, which produces in the receiver a noise with spectral density:

$$N_0 = kT_s \quad \text{W/Hz} \tag{3.1.10}$$

where k is **Boltzmann's constant**, 1.38 x 10 $^{-23}$ W/(Hz-K), and T_s is in **kelvins**, giving N_0 in W/Hz. This system input noise temperature, using the method of Blake [3.2], may be divided into three major components.

$$T_s = T_a + T_r + L_r T_e \tag{3.1.11}$$

where

$$T_a = \frac{0.88T_a' - 254}{L_a} + 290 \tag{3.1.12}$$

$$T_r = T_{tr}(L_r - 1) \tag{3.1.13}$$

$$T_e = T_0(F_n - 1) \tag{3.1.14}$$

Here, the temperature T_a is the contribution from the antenna, T_r is that of the RF components connecting the antenna to the receiver, and T_e that of the receiver itself. T_a' is the apparent temperature of the sky and viewed at the radar frequency, plotted in Figure 3.1.5, L_a is the dissipative loss within the antenna, T_{tr} and L_r are the physical temperature and loss of the input RF components, T_0 is the reference temperature of 290 K, and F_n is the **noise factor** of the receiver. For antennas pointing at absorbing ground surface, $T_a = 290$ K, and $T_s = 290\ L_r F_n$. However, for antennas which look into cool space, $T_s \to 290\ (L_r\ F_n{-}1)$. The difference in system performance can become appreciable when $L_r\ F_n \to 1$, and is one dB for $L_r\ F_n = 5.0$ (or 7dB). Note that all terms in (3.1.10) to (3.1.13) are expressed either in degrees or as power ratios, not in dB. Thus a loss $L_r = 1.0$ dB must be expressed as a ratio 1.26 when used in any of these equations.

In effect, (3.1.10) to (3.1.14) represent the system noise in terms of an input termination resistor heated to a temperature T_s, followed by an ideal (noise-free) receiving system having the gain and bandpass characteristics of the actual system. This permits us to analyze the system without becoming involved in details of the receiver noise sources and gain structure, which are instead summarized in F_n and T_e. The output signal-to-noise ratio can be calculated by assuming that the input signal S competes with an effective input noise power N, where:

$$N = N_0 B_n = kT_s B_n \tag{3.1.15}$$

Both signal and noise are passed through the receiver having a noise bandwidth B_n.

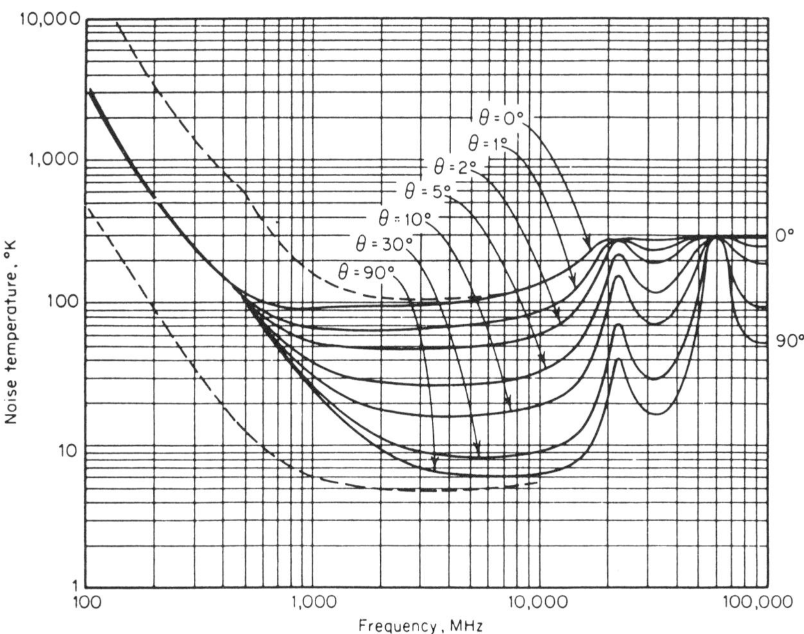

Figure 3.1.5 Noise temperature of an idealized antenna (lossless, no earth-directed sidelobes) at the earth's surface as a function of frequency for a number of beam elevation angles [3.2].
© Artech House, reprinted by permission.

3.1.3 Noise Density and Power Example

To illustrate the receiver noise calculation, consider an S-band receiver consisting of a low-noise RF amplifier, $F_n = 5$ dB, connected through $L_r = 1$ dB to an antenna pointed at $\theta = 1°$, with a low internal antenna loss $L_a = 0.2$ dB and normal sidelobe characteristics. The receiver is matched to receive 1 µs pulses ($1/\tau = B_n = 1$MHz). The calculation of noise temperature is as follows:

$$T_a' \quad \text{(from Figure 3.1.5)} \qquad\qquad = \ 65 \text{ K}$$

$$T_a = (0.88 \times 65 - 254)/1.047 + 290 \qquad = \ 102 \text{ K}$$

$$T_r = 290(1.26 - 1) \qquad\qquad\qquad = \ 75 \text{ K}$$

$$L_r T_e = 1.26 \times 290(3.16 - 1) \qquad\qquad = \ \underline{790 \text{ K}}$$

$$T_s \qquad\qquad\qquad\qquad\qquad\qquad\qquad = \ 967 \text{ K}$$

$$N_0 = kTs = 1.38 \times 10^{-23} \times 967 \qquad = \ 1.33 \times 10^{-20} \text{ W/Hz}$$

$$B_n = 1 \text{ MHz} \qquad\qquad\qquad\qquad\qquad = \ 10^6 \text{ Hz}$$

$$N = kT_s B_n = 1.33 \times 10^{-14} \quad \text{W} \qquad = \ -138.8 \text{ dBW}$$

If used to receive the signal calculated in the previous radar example, with the pulsewidth assumed $\tau = 1$ µs, the received pulse power, energy, and signal-to-noise energy ratio are:

$$S = 8 \times 10^{-13} \quad \text{W}$$

$$E_1 = S\tau = 8 \times 10^{-19} \quad \text{W-s}$$

$$E_1/N_0 = \frac{8 \times 10^{-19}}{1.33 \times 10^{-20}} = 60.02$$

$$10 \log_{10} 60.02 = +17.8 \quad \text{dB}$$

3.1.4 Signal-to-Noise Ratio

The optimum filter for detection of a signal in white noise is the *matched filter* [3.3]. The theory of the matched filter states that the maximum output snr is equal to the ratio of total received energy to noise spectral density:

$$(S/N)_{mf} = E/N_0 \qquad\qquad (3.1.16)$$

For a single received pulse, when the receiving filter is matched to the pulse spectrum, this becomes:

$$(S/N)_{mf} = E_1/N_0 = S\tau/N_0 = \frac{P_t \tau G_r \lambda^2 \sigma}{(4\pi)^3 R^4 k T_s} \qquad\qquad (3.1.17)$$

Filters other than matched filters will produce lower snr values:

$$S/N = E_1/N_0 L_m \qquad\qquad (3.1.18)$$

where $L_m > 1$ is the *matching loss* of the filter to the pulse spectrum. For *continuous wave* (CW) or *coherent pulse radars*, which integrate over an observation time t_o in a *predetection filter*, the received energy ratio and matched filter output snr is equal to the energy ratio evaluated over t_o:

$$(S/N)_{mf} = E/N_0 = \frac{P_{av}t_o G_t G_r \lambda^2 \sigma}{(4\pi)^3 R^4 kT_s} \tag{3.1.19}$$

The matching loss for a practical pulsed system of this sort can be expressed as the product of two terms $L_m L_{mf}$, where the first represents the mismatch of the IF filter to the spectrum of each individual pulse, and the second denotes the mismatch loss of the *integrating filter* to the envelope of the pulse train, extending over t_o.

The advantage of approaching radar range calculation through the matched-filter snr is that the effects of modern waveforms and processing, involving pulse compression and integration of multiple-pulse bursts, are readily expressed by the approximately matched filter with small matching loss factors. The energy transmitted in these waveforms, $P_t\tau$ per pulse and $P_{av}t_o$ per target observation, is easily established, as is the noise spectral density $N_0 = kT_s$. All that must be known about the actual receiving filters and processing steps is how closely they approximate the matched filter. Since the matching losses seldom exceed 2 or 3 dB in a practical system, there is little opportunity for serious error or misinterpretation. Alternative procedures, which involve expressions that use an input bandwidth B_n and sequent processing gain factors, are subject to misinterpretation and large errors.

3.1.5 Loss Factors

So far, the radar equation has been derived on the basis of *free-space propagation* conditions and ideal radar operation. In practice, a number of other factors must be included in arriving at the energy ratio available to the receiver. (See also Appendix C, Ideal Assumptions and Loss Factors.)

(a) *Signal attenuation prior to the receiver.* This comprises: transmission line loss, L_t, between the transmitter tube (at which P_t or P_{av} is measured) and the antenna terminal (at which G_t is measured); antenna losses (usually included in G_t, G_r, or T_s); receiving line and circuit losses (included in T_s through the term L_r); atmospheric attenuation, L_α; and atmospheric noise (included in T_a' for clear air, but requiring an increase for larger L_α in precipitation).

(b) *Surface reflection and diffraction effects.* In Chapter 8 the pattern-propagation factor F is defined, describing the ratio of one-way field amplitude at range R to that which would have been obtained under free space conditions in the center of the beam. Hence, the radar equation for a target at a given point in space will contain the factor F^4 in the numerator, accounting for the two-way power ratio.

(c) *Antenna pattern and scanning.* At any given instant, the factor F^4 will give the combined effects of two-way propagation and antenna pattern variation from the peak of the beam. In many cases, however, it is convenient to average the pattern effects over the scan of the antenna, and to include the average loss in received energy as a loss factor L_p in the denominator of the radar equation, either as a separate factor or as an increase in the required beam-center snr to achieve detection. If L_p is included in this way, for a scan in either coordinate, the pattern-propagation factor F will be calculated for a target on the center of the beam in that coordinate. Details in evaluation of L_p and F will be discussed later.

Applying the appropriate loss factors as described above, the equations for available snr at the input of the receiver can be written as follows. For the single pulse case,

$$\frac{E_1}{N_0} = \frac{P_t \tau G_t G_r \lambda^2 \sigma F^4}{(4\pi)^3 R^4 k T_s L_t L_\alpha} \tag{3.1.20}$$

For the complete observation over t_o seconds,

$$\frac{E}{N_0} = \frac{P_{av} t_o G_t G_r \lambda^2 \sigma F^4}{(4\pi)^3 R^4 k T_s L_t L_\alpha} \tag{3.1.21}$$

For a pulsed radar, we can see that (3.1.21) gives an energy ratio equal to nE_1/N_0, where $n = f_r t_o$ is the total number of pulses received during the observation. This follows because the product $P_t \tau f_r$ is equal to the average power P_{av}.

3.1.6 Solution for Maximum Range

In Chapter 4 we will derive expressions for the snr required to achieve target detection with given probabilities of detection and false alarm, and for some additional loss factors which enter into target detection. This will permit us to establish the ***detectability factor*** D_x , which is the input energy ratio required to achieve detection. For the present, we will assume that this factor is known, and solve for maximum range by setting the energy ratio in (3.1.20) or (3.1.21) equal to the corresponding value $D_x(n)$ or $D_x(1)$. The notation $D_x(n)$ indicates n samples (or pulses) of signal, each with this energy ratio, will be integrated after ***envelope***

detection before being applied to the target detection threshold. Thus, in (3.1.20), where only one coherently integrated sample is available at the envelope detector, the value $D_x(1)$ will be large relative to the value $D_x(n)$, which applies to each one of many received pulses.

In terms of the single-pulse values, the radar range R_m is expressed as:

$$R_m^4 = \frac{P_t \tau G_t G_r \lambda^2 \sigma F^4}{(4\pi)^3 k T_s D_x(n) L_t L_\alpha} \tag{3.1.22}$$

while for coherent integration (over the total observation, time t_o),

$$R_m^4 = \frac{P_{av} t_o G_t G_r \lambda^2 \sigma F^4}{(4\pi)^3 k T_s D_x(1) L_t L_\alpha} \tag{3.1.23}$$

Equation (3.1.22) is essentially the form used by Hall [3.4] in his classic IRE paper, while (3.1.23) is a more general form. The latter equation may, of course, be applied to radars using ***post-detection integration***, if the appropriate integration loss factors are included in calculation of $D_x(1)$. This will be discussed further in Chapter 4.

In most CW and pulsed doppler radars, the ***coherent integration time*** $t_f = 1/B_f$ of the pre-detection filters will be less than the total observation time t_o of the signal. Hence, during t_o, there will be $n' = t_o/t_f = t_o B_f$ independent outputs from the filter, and these may be integrated after envelope detection, i.e., post-detection integration. The resulting performance is only slightly less than that of the same system using pre-detection filters matched to t_o. If we use, in (3.1.22), the energy $P_{av} t_f$ transmitted in each sample and the detectability factor $D_x(n')$, we can write:

$$R_m^4 = \frac{P_{av} t_f G_t G_r \lambda^2 \sigma F^4}{(4\pi)^3 k T_s D_x(n') L_t L_\alpha} \tag{3.1.24}$$

It is this expression for mixed integration (partially pre-detection and partially post-detection) which will be applicable to most practical doppler radar problems.

3.1.7 Range Calculation Using the Blake Chart

In the absence of an IEEE or other universal standard for the radar range calculation procedure, the nearest to an accepted procedure is represented by the range calculation work sheet or Blake chart, Figure 3.1.6. The version shown here is a modification of that presented by Blake [3.2], which permits either (3.1.22) or (3.1.24) to be used, and hence is more readily applicable to CW and pulsed doppler radars. The modified chart also directly uses the terms from these equations, in basic units, rather than requiring conversion of P to kW and λ to c/f with corresponding conversion constants. In the chart, the value $D_x(n)$ is represented by the product $D(n)ML_pL_x$, corresponding to Blake's $V_0C_BL_pL_x$. Calculation and definition of the detectability and matching factors $D(n)$ and M will be discussed in Chapter 4.

1. Record detection and false-alarm probabilities, target case, hits per scan, radar height and target elevation angle.

2. Compute the system input noise temperature T_s, following the outline in Section A.

3. Enter range factors known in other than decibel form in Section B.

4. Enter logarithmic and decibel values in Section C, positive values in the plus column and negative values in the minus column. For detection on display, substitute V for D and C_B for M.

Detection probability, $P_d =$ ______; False-alarm probability, $P_{fa} =$ ______; Case ____ target; $n =$ ______ hits

Radar antenna height, $h_r =$ __________ m;   Target elevation angle, $\theta_t =$ ______ °

A. Computation of T_s:	B. Range Factors		C. Decibel Values	Plus (+)	Minus (-)
$T_s = T_a + T_r + L_rT_e$	P_t or P_{av} (W)		10 log P (dBW)	.	.
(a) Compute T_a	τ or t_f (s)		10 log τ (dBs)	.	.
For $T_{tg} = T_{ta} = 290$, $T_g = 36$,	G_t		$G_{t(dB)}$	.	.
$T_a = (0.876T_a'-254)/L_a + 290$	G_r		$G_{r(dB)}$	.	.
$L_{a(dB)}:$ ______, $L_a:$ ______	σ (m²)		10 log σ (dBm²)	.	.
$T_a' =$ ______K	λ (m)		20 log λ (dBm²)	.	.
$T_a =$ ______K	T_s (K)		-10 log T_s (dBK)	.	.
(b) Compute $T_r = T_{tr}(L_r-1)$	D		$-D_{(dB)}$	.	.
$L_{r(dB)}:$ ______, $L_r:$ ______	M		$-M_{(dB)}$		.
$T_r =$ ______K	L_p		$-L_{p(dB)}$		.
(c) Compute $T_e = T_0(F_n-1)$	L_x		$-L_{x(dB)}$		.
$F_{n(dB)}:$ ______, $F_n:$ ______	L_t		$-L_{t(dB)}$		.
$T_e:$ ______K, $L_r:$ ______	Range-equation constant (dBW•s/K)			+75.62	
$L_rT_e =$ ______K	5. Obtain column totals			.	.
(d) Add. $T_s =$ ______K	6. Enter the smaller total below the larger			.	.
7. Subtract to obtain net decibels $X = 40$ log R_{km} (dBm⁴)				.	.

8. Calculate $R_{0(km)} =$ antilog $(X/40)$ $R_{0(km)} =$ ☐

9. Calculate the pattern-propagation factor $F = \sqrt{F_tF_r}$ $F =$ ☐

10. Multiply R_0 by the pattern-propagation factor to obtain $R' = R_0{\times}F$ $R' =$ ☐

11. From the appropriate curve, determine the atmospheric absorption loss $L_{\alpha(dB)}$, corresponding to R'. This is $L_{\alpha(dB)(1)}$. $L_{\alpha(dB)(1)} =$ ☐

12. Find the range factor $\delta_1 =$ antilog $(-L_{\alpha(dB)(1)}/40)$. $\delta_1 =$ ☐

13. Multiply R' by δ_1. This is a first approximation of range, R_1. $R_1 =$ ☐

14. If R_1 differs appreciably from R', find a new value of $L_{\alpha(dB)}$ corresponding to R_1. This is $L_{\alpha(dB)(2)}$. $L_{\alpha(dB)(2)} =$ ☐

15. Find the range increase factor δ_2 corresponding to the difference between $L_{\alpha(dB)(1)}$ and $L_{\alpha(dB)(2)}$. $\delta_2 =$ ☐

16. Multiply R_1 by δ_2 to obtain the maximum radar detection range R_m in km $R_{m(km)} =$ ☐

Figure 3.1.6 Range calculation worksheet

The "range-equation constant" of 75.6 dB shown in the plus (+) column of the worksheet simply accounts for the constant terms in the range equation, in dB:

$$-30 \log 4\pi - 10 \log k - 120 = 75.6 \text{ dB}$$

and the 120 dB gives the range R_0 in km rather than in m. For other units of range, the range-equation constant becomes:

195.6 dB for R_0 in m

64.9 dB for R_0 in nmi

When the chart is used to solve (3.1.23) for pulsed doppler or CW radar, the peak power P_t is replaced by average power P_{av}, pulsewidth τ by the averaging time $t_f = 1/B_f$ of the predetection filter, and the number of pulses $n = f_r t_o$ by the number of filter output samples $n' = t_o / t_f$, for which (3.1.24) applies.

The major benefit of the standardized chart is that it makes clear what loss and signal integration assumptions have been used in performing the range calculation, in addition to knowing the values of the major radar parameters. A shortcoming is that the loss term L_x includes a large number of individual contributing losses, not all of them well specified in most radar programs.

3.1.8 Example of Blake Chart Calculation

The Blake chart of Figure 3.1.7 has been filled in using the parameters from the previous radar example. We will take the required detection probability $P_d = 0.9$, and false-alarm probability $P_{fa} = 10^{-6}$, and assume a *Case 1 target* with $n = 24$ *hits per scan*. Chapter 4 will show that the detectability factor $D_1(n) = 11.0$ dB for this case (with simple video integration). Allowances for practical receiver matching to the pulse spectrum, $M = 0.8$ dB, for transmission line loss, beamshape loss, $L_p = 1.3$ dB, and miscellaneous processing loss, $L_x = 3$ dB, will also be assumed. This results in a requirement $D_x = +16.1$ dB. The transmission loss is 1.0 dB and the sum of all gains and losses allows an initial range estimate to be solved using either Table 2, or by the following calculation:

$$40 \log R_0 \qquad = + 78.6 \text{ dB}$$
$$\log R_0 \qquad = +1.965 \text{ dB}$$
$$R_0 \qquad = 10^{1.965} = 92.3 \text{ km}$$

Detection probability, $P_d = $ **0.9** ; False-alarm probability, $P_{fa} = $ **10^{-6}** ; Case **1** target; $n = $ **24** hits.					
Radar antenna height, $h_r = $ **10** m; Target elevation angle, $\theta_t = $ **1** °					
A. Computation of T_s:	B. Range Factors		C. Decibel Values	Plus (+)	Minus (-)
$T_s = T_a + T_r + L_r T_e$	P_t or P_{av} (W)	10^5	10 log P (dBW)	50.0	.
(a) Compute T_a	τ or t_f (s)	10^{-6}	10 log τ (dBs)	.	60.0
For $T_{tg} = T_{ta} = 290$, $T_g = 36$,	G_t	10^4	$G_{t(dB)}$	40.0	.
$T_a = (0.876 T_a' - 254)/L_a + 290$	G_r	10^4	$G_{r(dB)}$	40.0	.
$L_{a(dB)}$: **0.2** , L_a: **1.047**	σ (m²)	1.0	10 log σ (dBm²)	0.0	.
$T_a' = $ **65** K	λ (m)	0.1	20 log λ (dBm²)	.	20.0
$T_a = $ **102** K	T_s (K)	967	-10 log T_s (dBK)	.	29.9
(b) Compute $T_r = T_{tr}(L_r - 1)$	D	12.6	$-D_{(dB)}$	.	11.0
$L_{r(dB)}$: **1.0** , L_r: **1.26**	M	1.2	$-M_{(dB)}$		0.8
$T_r = $ **75** K	L_p	1.35	$-L_{p(dB)}$		1.3
(c) Compute $T_e = T_0(F_n - 1)$	L_x	2.0	$-L_{x(dB)}$		3.0
$F_{n(dB)}$: **5.0** , F_n: **3.16**	L_t	1.26	$-L_{t(dB)}$		1.0
T_e: **625** K, L_r: **1.26**	Range-equation constant (dBW•s/K)			+75.62	
$L_r T_e = $ **790** K	5. Obtain column totals			205.6	127.0
(d) Add. $T_s = $ **967** K	6. Enter the smaller total below the larger			127.0	.
7. Subtract to obtain net decibels $X = 40$ log R_{km} (dBm⁴)				+78.6	.

8. Calculate $R_{0(km)} = $ antilog ($X/40$) $R_{0(km)} = $ **92.3**

9. Calculate the pattern-propagation factor $F = \sqrt{F_t F_r}$ (Secs. 6.2, 6.3) $F = $ **1.0**

10. Multiply R_0 by the pattern-propagation factor to obtain $R' = R_0 \times F$ $R' = $ **92.3**

11. From the appropriate curve, determine the atmospheric absorption loss $L_{a(dB)}$, corresponding to R'. This is $L_{a(dB)(1)}$. $L_{a(dB)(1)} = $ **1.3**

12. Find the range factor $\delta_1 = $ antilog ($-L_{a(dB)(1)}/40$). $\delta_1 = $ **0.9279**

13. Multiply R' by δ_1. This is a first approximation of range, R_1. $R_1 = $ **85.6**

14. If R_1 differs appreciably from R', find a new value of $L_{a(dB)}$ corresponding to R_1. This is $L_{a(dB)(2)}$. $L_{a(dB)(2)} = $ **1.2**

15. Find the range increase factor δ_2 corresponding to the difference between $L_{a(dB)(1)}$ and $L_{a(dB)(2)}$. $\delta_2 = $ **1.0058**

16. Multiply R_1 by δ_2 to obtain the maximum radar detection range R_m in km $R_{m(km)} = $ **86.1**

Figure 3.1.7 Example of range calculation using worksheet.

Assuming for the present, that a reflection-free surface is present under the radar signal path, with the target at the center of the elevation beam $\theta_t = \theta_b = 1°$, we have $F = 1$ and $R' = R_0 = 92.3$ km. Atmospheric attenuation data may be found in Chapter 8, indicating a loss at this range of $L_a = 1.3$ dB. This decreases the radar detection range by a factor of 0.9279, to 85.6 km. The chart allows for one iteration to account for the slight reduction in L_a at this range, leading to a final range $R_m = 86.1$ km.

This calculation will be repeated for a coherent integration process in Chapter 4. Further discussion of propagation effects is given in Chapter 8.

3.1.9 Computer Program

The PC software accompanying this handbook includes a range calculation program based on the method of the Blake chart.

3.2 The Search Radar Equation

3.2.1 Derivation for Uniform Search

The potential performance of a search radar, distributing its energy uniformly over an assigned solid angle, can be determined from its average power, receiving aperture area, and system temperature, *without regard to frequency and waveform*. The steps in deriving the equation for optimum search performance, starting from (3.1.23) are:

(a) Assume uniform search (see Figure 3.2.1) without overlap of an assigned solid angle ψ_s in a frame t_s, using a rectangular beam that has a solid angle:

$$\psi_b = \theta_a \theta_e = \frac{4\pi}{G_t L_n} \ll \psi_s = A_m(\sin\theta_m - \sin\theta_0) \tag{3.2.1}$$

and also assume that all signal energy reaching A_r during t_o is integrated for one detection decision.

(b) Express the observation time for a target as

$$t_o = \frac{t_s \psi_b}{\psi_s} = \frac{4\pi t_s}{G_t \psi_s L_n} \tag{3.2.2}$$

Note that the definition of two-coordinate beam-shape loss, included below in search loss L_s, is consistent with (3.2.2).

(c) Substitute (3.1.6), (3.2.1), and (3.2.2) into (3.1.23), and assume $F = 1$ (free-space propagation), to obtain the search radar equation

$$R_m^4 = \frac{P_{av} A_r t_s \sigma}{4\pi \psi_s kT_s D_0(1) L_s} \tag{3.2.3}$$

where the total search loss L_s includes L_n and all the loss factors from the radar equation and the equation relating the detectability factor $D_x(1)$ to the basic detectability factor $D_0(1)$ for the *steady target*.

Note that neither wavelength nor waveform appears directly in (3.2.3); although the combination of wavelength and aperture size must permit (3.2.1) to be satisfied for concentration of search energy within ψ_s, and the loss terms will vary with wavelength, waveform, and scan procedure.

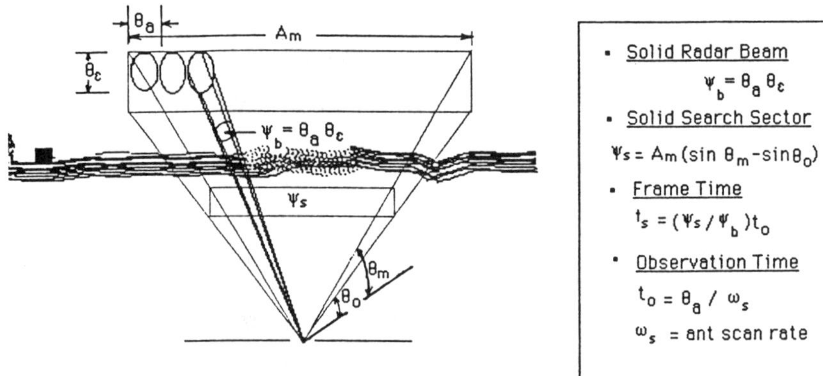

Figure 3.2.1 Area search definitions.

3.2.2 Effective Angle for Cosecant-Squared Coverage

In air surveillance applications, the required radar coverage often follows the type of contour shown in Figure 3.2.2, with a maximum range R_m, a maximum altitude h_m, and a maximum elevation angle θ_2. Full-range coverage is required only from the horizon to $\theta_1 = \sin^{-1}(h_m/R_m)$ while at shorter ranges the principal requirement is to maintain radar coverage up to altitude h_m. This coverage can be produced by an antenna pattern having a *cosecant-squared gain function:*

$$G(\theta) = G(\theta_1)(csc^2\theta/csc^2\theta_1), \quad \theta_1 < \theta < \theta_2 \tag{3.2.4}$$

If such a pattern is used for transmitting and receiving, the effective elevation sector angle is calculated from

$$\theta_m = \theta_1(2 - \theta_1 \cot \theta_2) \tag{3.2.5}$$

and the effective receiving aperture is reduced by the ratio θ_m/θ_1. The effective sector angle approaches $2\theta_1$, for $\theta_2/\theta_1 \gg 1$.

There are two other approaches to $\csc^2$ search coverage. In a *stacked-beam radar,* elevation beams for receiving, using the full aperture A_r (and hence with constant G_r), can be used from the horizon to θ_2, and the broad transmitter pattern can be shaped as $\csc^4\theta$. The effective elevation search sector is then calculated from

$$\sin\theta_m = \frac{4}{3}\sin\theta_1\left[1 - \frac{1}{4}\left(\frac{\sin\theta_1}{\sin\theta_2}\right)^3\right].\qquad(3.2.6)$$

This effective elevation sector angle approaches $(4/3)\theta_1$ for $\theta_2/\theta_1 \gg 1$. The same equation applies when a scanning *pencil beam* is used with constant G_tG_r and with $P_t\tau$ adjusted according to $\csc^4\theta$.

In calculating the extra time required to perform this search, for a scanning beam with reduced power, and with interpulse period matched to the coverage contour, the integration leads to a different value:

$$\sin\theta_m' = \sin\theta_1\left[1 + \ln\left(\frac{\sin\theta_2}{\sin\theta_1}\right)\right]\qquad(3.2.7)$$

where $\ln$ = natural logarithm.

The solid angle ψ_s for calculation of power-aperture product or detection range using (3.2.3), will be evaluated from (3.2.1) with $\sin\theta_m$ given by (3.2.5).

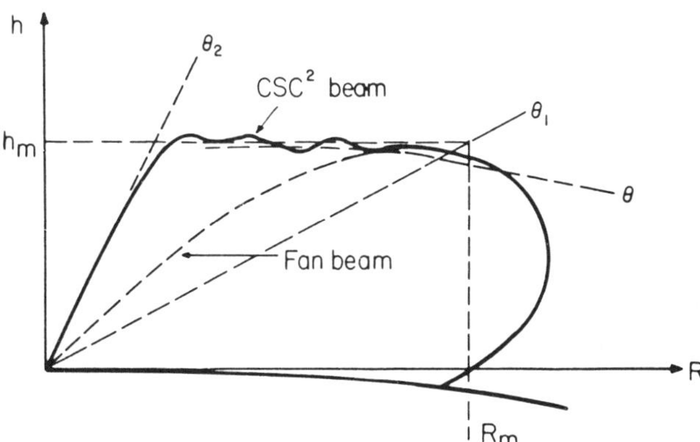

Figure 3.2.2 Cosecant-squared coverage pattern [3.1].
© Artech House, reprinted by permission.

3.2.3 Significance of the Search Radar Equation

The search radar equation states, in essence, that the performance of the radar (as measured by the snr which can be developed on each target during the frame time) is dependent on the product of average power and receiving aperture, divided by the search rate in steradians per second.

The radar's performance is also dependent on such factors as target cross section, system temperature, and quality of the detection performance (as measured by the probabilities of detection and false alarm, which determine detectability factor and some loss components). Apart from the "brute force" factors of power and aperture, the radar characteristics enter the equation primarily through the several components of the total search loss factor L_s.

The number of transmitting beam positions into which the search sector is divided does not enter the equation, because the higher gain resulting from a smaller transmitting beam is cancelled by the resultant reduced time-on-target. In order to use a given receiving aperture at shorter wavelengths, the resulting smaller beamwidth may make it necessary to cover the transmitting beam angle by using several smaller receiving beams. It makes no difference, however, whether these receiving beam positions are all illuminated simultaneously or sequentially by the transmitting beam, as long as there is efficient integration of the received energy from each beam position. In principle, the transmitting beam can illuminate the entire search sector, with multiple receiving beams staring at each point within this sector and integrating the received echoes over the entire frame time t_s. In practice, integration over time periods in excess of a few tens of milliseconds is quite difficult and likely to be accompanied by large losses.

Since the search radar equation *leaves open the choice of wavelength and waveform* for the radar, in achieving the required snr under idealized conditions, we must look to other requirements to establish the constraints that dominate the system design process. These other requirements are imposed primarily by the clutter and multiple-target environments in which most radars must operate. Once the general scale of the radar has been determined from the search radar equation, the details of design will be determined by these environmental factors, not by the need for a specific value of snr.

3.2.4 Search Loss Budgets

Before proceeding to the environments of clutter and jamming, it is appropriate to define the components of the search loss factor L_s, as well as the loss components that normally appear as contributors to other terms in the search radar equation, some of which have already been discussed (see also Appendix C). These losses are shown in Figure 3.2.3 and are defined as follows. *The first five components represent actual reductions in the signal power or energy ratio available to the receiver.*

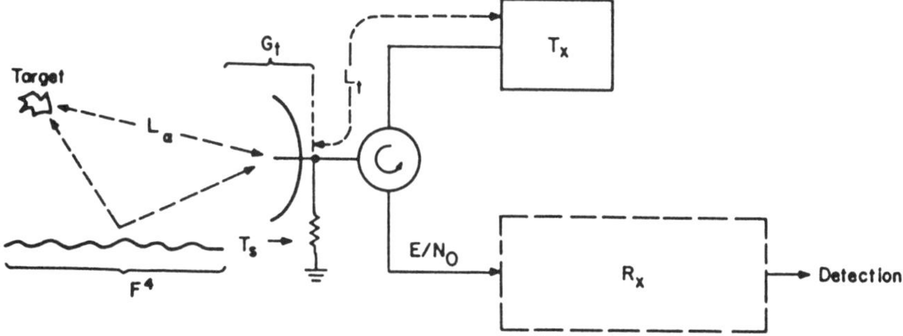

Figure 3.2.3 RF and other loss factors in radar system [3.1].
© Artech House, reprinted by permission.

    (a) **Transmission line loss,** L_t. This was included in (3.1.20) as a component which reduces the actual signal power received from the target.

    (b) **Antenna losses,** L_a **and** L_n. The first was considered in connection with (3.1.12). In the receiving path, L_a is normally included as a component of T_s. In the transmitting path, L_a may be included as a component of L_n in (3.2.2), reducing the gain without expanding the solid angle covered by the transmitting beam, or L_a may be included in L_t. The minimum value of L_n for a constrained-feed array, with no spillover and with ohmic losses included in L_t, is 0.48 dB. This corresponds to an expression for antenna gain:

$$G_t = 11.3/\theta_a \theta_e \quad \text{for } \theta \text{ in radians}$$

$$G_t = 37,100/\theta_a \theta_e \quad \text{for } \theta \text{ in degrees} \tag{3.2.8}$$

For horn-fed reflectors, $G_t = 7.6/\theta_a \theta_e$, or for θ in degrees, $G_t = 25,000/\theta_a \theta_e$, corresponding to $L_n = 1.6$, or 2 dB.

It is important to note the difference between the antenna losses used in the search radar equation and those in the Blake chart or (3.1.22). In the Blake chart or (3.1.22) antenna gains and observation time are used to calculate received signal energy, and loss appears in the gain and noise temperature factors. In the search radar equation, the transmitting gain and observation time are assumed to be related by (3.2.2), and the factor L_n must be included in the total search loss factor L_s.

(c) **Atmospheric attenuation**, L_α. This is the two-way loss in passage of the radar wave through the atmosphere to the target, discussed in Chapter 8. Atmospheric loss is included as a separate factor in the Blake chart and in (3.1.22), but as a component of total search loss L_s in the search radar equation.

(d) **Receiving line loss**, L_r. This loss is included as a contributor to the system temperature, T_s, according to (3.1.11) to (3.1.13), rather than as a component of L_s.

(e) **Beamshape loss**, L_p. This loss is a component of L_s , representing the reduction in signal energy received from the average target during the search scan, as compared with what would have been achieved if the beam were rectangular in shape or were to point directly at the target for a time interval t_o. In the search radar equation, beamshape loss is normally assigned a value of 2.5 dB to represent the fact that the target may be off the beam axis in both angle coordinates. In the Blake chart, beamshape loss is stated separately from other losses, while in (3.1.22) beamshape loss is included in the detectability factor $D_x(1)$.

The remaining loss components represent inefficiencies in processing the received energy, and may depend strongly on the objective of the radar (e.g., the required detection and false-alarm probabilities). Hence, these losses are sometimes characterized as statistical losses to distinguish them from the pure attenuation factors of the first group.

(f) **Filter matching factor**, M. This factor represents the amount by which the input signal energy must be increased to compensate for the fact that the ideal (matched) filter is not used in the receiver. When the actual filter bandwidth is narrower than that of the matched filter, the factor M takes into account the resulting reduction in output false-alarm rate, and hence is somewhat less than the loss L_m, which describes only the snr at the filter output. When the filter is wider than optimum, the factor M takes into account the usual post-detection bandwidth reduction, which increases the number of signal-plus-noise samples available for integration during each input pulse, again resulting in a lower loss than would be found by considering only the snr at the filter output. The matching factor is stated as a separate loss in the Blake chart, and included as a component of $D_x(1)$ in (3.1.22). In the search radar equation, it appears as a component of total search loss L_s.

(g) **Integration loss,** L_i. This is the loss resulting from failure to integrate all signal pulses coherently in the receiving filter. The integration loss actually is caused by the passage of the signal through the envelope detector at a lower level than would have been attained through coherent (pre-detection) integration, and the *cross-modulation* of noise and signal in the envelope detector prevents the subsequent integrator from regaining the snr corresponding to input energy ratio of the pulse group. In the Blake chart and (3.1.22), the integration loss appears as a component of the detectability factor $D(n)$, while in (3.1.23) the loss is assumed to equal unity for coherent integration. In the search radar equation, all received energy is assumed to be integrated for one detection decision with detectability factor $D_0(1)$, and the integration loss must be included as a component of total search loss L_s.

(h) **Collapsing loss,** L_c. This loss results from use of any *video integration* process in which extra noise samples are combined with the desired signal-plus-noise samples. Examples are integration in range gates or display elements wider than the signal pulse, collapsing of data from separate receivers or filters into a single video integration channel, or extension of the integration period beyond the duration of the signal pulse train. In the Blake chart or (3.1.22), collapsing loss is included as a component of L_x or in $D_x(1)$, while in the search radar equation this loss is a component of L_s.

(i) **Fluctuation loss,** L_f. All practical radar targets fluctuate in amplitude as a function of time and radar frequency. Fluctuation loss is the radar engineering term for the fade margin necessary to achieve adequate detection performance when the target drops below its average amplitude. In the Blake chart or (3.1.22), the fluctuation loss is included as a component of the detectability factor $D(n)$ or $D_x(1)$, while in the search radar equation it becomes a component of the total search loss L_s.

(j) **Processing losses,** L_x. Many other steps in the practical signal processing chain involve compromises in approaching the optimum process. These compromises result in requirements for greater input energy ratio in order to achieve the desired detection performance level, and the total increase is described as processing loss. It may be divided into components resulting from straddling of the signal by range gates, doppler filters, or integration windows, from action of the constant-false-alarm rate (CFAR) circuits, from MTI or doppler filter stopbands, from coarse quantization in digital systems, and from sundry other departures from the ideal processing steps. Processing losses are stated separately in the Blake chart and as a component of $D_x(1)$ in (3.1.22), but in the search radar equation these losses become components of total search loss L_s.

(k) **Scan distribution loss, L_d**. This loss is unique to the use of the search radar equation as a performance reference. Scan distribution loss describes the relative inefficiency of scanning the search volume more than once, when the detections from successive scans are combined in the simple cumulative detection process, rather than by integration before the detection threshold.

(l) **Other losses**. In particular systems there may be several other sources of loss, relative to the idealized performance represented by the search radar equation. These will be discussed in subsequent chapters.

3.2.5 Example of Search Radar Calculation

As an example of a search radar range calculation, consider the S-band radar for which the Blake chart calculation was performed. The *reflector antenna* gain of 40dB corresponds to a beam solid angle:

$$\psi_b = \frac{4\pi}{G_t L_n} = \frac{4\pi}{10^4 \times 1.6} = 0.0007 \text{ sr}$$

for example, $\theta_a = 1.3°$, $\theta_e = 2.0°$. The effective receiving aperture is

$$A_r = \frac{G_r \lambda^2}{4\pi} = 10^4 \times 10^{-2}/4\pi = 8 \text{ m}^2$$

The search solid angle, for a single elevation beam, is from (3.2.1)

$$\psi_s = 2\pi \sin 2° = 0.22 \text{ sr}$$

If the system is to achieve the $n = 24$ hits per scan assumed in calculation of the detectability factor for the Blake chart, at a rotation rate $\omega_a = 2\pi/t_s = 1.05$ r/s = 60°/s, for $t_s = 6$ s we have

$$t_o = t_s \theta_a/360° = 6 \times 1.3/360 = 0.0217 \text{ s}$$

$$f_r = n/t_o = 24/0.0217 = 1108 \text{ Hz}$$

The radar duty factor $D_u = f_r \tau = 0.0011$ and the average power, $P_{av} = P_t D_u = 110$ W.

The search loss budget, for targets at the peak of the elevation beam, is as follows:

$$L_t = 1.0 \text{ dB} \qquad\qquad M = 0.8 \text{ dB}$$

$$L_n = 2.0 \text{ dB} \qquad\qquad L_i = 3.2 \text{ dB}$$
$$L_\alpha = 1.2 \text{ dB} \qquad\qquad L_f = 8.4 \text{ dB}$$
$$L_p = 1.3 \text{ dB} \qquad\qquad L_x = 3.0 \text{ dB}$$

Total search loss, $L_s = 20.9$ dB or a factor of 123. Solving (3.2.3) for range assuming a 1 m^2 target then gives:

$$R_m^4 = \frac{110 \times 8 \times 6 \times 1}{4\pi \times 0.22 \times 1.38 \times 10^{-23} \times 967 \times 20.9 \times 123} = 5.56 \times 10^{19} \text{ m}^4$$

$$R_m = 86.4 \quad \text{km}$$

For targets distributed uniformly between the horizon and two-degree elevation, the beamshape loss for the elevation coordinate adds 1.3 dB to the loss budget, reducing the average range to $R_m = 80.2$ km.

In a more realistic search radar example, the elevation coverage would be extended to about 10°, requiring the use of five vertically stacked beams. The transmitter power could be split equally among these beams, with corresponding reduction in G_t by a factor of five. If the 80.2 km average detection range were to be maintained, average power would be increased by this same factor of five (maintaining the ratio P_{av}/ψ_s in (3.2.3), or the $P_tG_t\tau$ product in the Blake chart and (3.1.21). If this average power were obtained by increasing the peak power or pulsewidth, this would also maintain the number of hits per scan at the azimuth scan rate of 60°/s. Numerous other options are available, such as using a pulse segmented into five sub-pulses at different frequencies, to generate the five beams with a *frequency-scanning antenna*. In this case the peak power of the transmitter would remain at 100 kW, and the pulsewidth would be 5 μs. A wideband receiver front end would then feed five 1.0 MHz receiver channels tuned to the frequencies of the sub-pulses, effectively creating five radars operating in parallel, each transmitting 100 W average power and covering a two-degree elevation sector 80.2 km in range.

3.3 Evaluation of Detection Range

The evaluation of radar detection range, predicted by (3.1.22)-(3.1.24), requires that two groups of factors be tested and confirmed.

(a) The factors that establish the actual input snr, as given in (3.1.18) and (3.1.19), including RF loss factors discussed in Section 3.1.5; and

(b) The factors that relate input snr to detection probability, through the detectability factor introduced in (3.1.22)-(3.1.24) and discussed in detail in Chapter 2.

In establishing the evaluation program, the testing of (a) can be relatively straightforward, since the signal, noise, and loss values are measurable using test equipment, at the subsystem or system level. Flight tests may be run with targets of known RCS, and the received signal measured as a function of range. The use of a calibration sphere, carried by a meteorological balloon, can provide a known target for some classes of radar.

The evaluations of (b) are more complex, and these will be discussed briefly after the detectability factor and the losses related to it are explored in Chapter 2

3.4 References

[3.1]   D. K. Barton, *Modern Radar System Analysis*, Artech House, 1988.

[3.2]   L. V. Blake, *Radar Range-Performance Analysis,* Artech House, 1986.

[3.3]   D. O. North, "An analysis of the factors which determine signal/noise discrimination in pulsed carrier systems," *RCA Laboratories Tech. Rpt. PTR-6C,* June 25, 1943; reprinted in *Proc IEEE* **51,** No. 7, July 1963, pp. 1016-1027.

[3.4]   W. M. Hall, "Prediction of pulse radar performance, "*Proc. IRE* **44**, No. 2, February 1956, pp. 224-231; reprinted in D. K. Barton (ed.), *Radars,* Vol. 2, *The Radar Equation*, Artech House, 1984.

[3.5]   W. M. Hall, "General radar equation," *Space/Aeronautics R and D Handbook*, 1962-63; reprinted in D. K. Barton (ed.), *Radars*, Vol. 2, *The Radar Equation*, Artech House, 1974.

[3.6]   D. K. Barton and H. R. Ward, *Handbook of Radar Measurement*, Prentice-Hall, 1969; Artech House, 1984.

Chapter 4

TARGET DETECTION

A fundamental issue concerning target detection is that the user have confidence that if a target which the radar is designed to detect is present within the radar coverage volume, the radar will reliably detect it. Conversely, if no such target is present, the radar should not report one. The problem of target detection involves successfully achieving a balance between these two requirements; the radar receiver must have the sensitivity to detect very small signals, but the high receiver sensitivity required to do so means that interference sources such as noise, clutter, and other natural phenomena (e.g., birds) are detected as well. The result is the presence of unwanted targets and *false alarms*. Unwanted targets may be dealt with, with varying degrees of success, by post-detection processing techniques such as doppler filtering (in a coherent radar) or some form of automated logic. False alarms, on the other hand, are the consequence of noise-like interference exceeding the detection threshold established for the radar, and can never be eliminated, although their occurrence can be made very infrequent.

Detection is a statistical process. Detection of a target that is actually present in the radar's coverage within a radar frame time either occurs or does not, but because of the statistical nature of the interference, and that of the signal plus interference, we can speak only of the *probability* of detection and the *probability* of false alarm. This statistical nature of the detection process dictates that we be quite specific if we are to correctly characterize the detection range of a radar. A radar may have an ***instrumented range*** of x kilometers, but this is *not* its detection range. A radar may have, on one or more occasions, detected a target of unknown cross section at range y kilometers, but this is *not* its detection range. As Schleher [4.1] clearly points out, to describe the radar's performance in terms of its detection range requires that we specify the probability of detection, probability of false alarm, target radar cross section, and the statistical model that describes the target fluctuations. Further, it is important to be clear about how detection is defined in each case, e.g., is it *single-scan*, or the *cumulative probability* of detection associated with multiple scans?

The probabilities of detection and false alarm are determined by the amplitude distribution of the noise, and that of the signal plus these unwanted components. In this chapter we will define the ***probability density function*** (pdf), the statistical means by which these distributions are characterized, in particular the pdf values which result from thermal (***Gaussian***) noise, added to the signal. This case will then be generalized for application to interference having other distributions. Procedures will then be developed by which the detection probability, P_d, can be calculated when the available snr and accetable   false-alarm probability P_{fa} are known. In addition, procedures will be given to find the required signal-to-noise ratio, or detectability factor, D, when the required values of P_d and P_{fa} are given. Both snr and D will refer to the signal-to-noise power ratio at the input to the envelope detector, the difference between them being that snr is the result of applying the radar range equation, e.g., (3.1.20) at a given range, while D is a *required value* of signal-to-noise calculated from the values chosen for P_d and P_{fa}. The formal definition of detectability factor is, from [4.2]:

> In a pulsed radar, the ratio of single-pulse signal energy to noise power per unit bandwidth that provides stated probabilities of detection and false alarm, measured the intermediate-frequency amplifier and using an intermediate-frequency filter matched to the single-pulse, followed by optimum video integration.

Thus, the detectability factor D_x used in (3.1.22) - (3.1.24) is the energy ratio actually required per pulse, or for the central pulse of a group modulated by the antenna pattern, increased by the amount necessary to account for mismatch of the filter, beamshape loss, and other nonoptimum conditions in the processing.

4.1 Noise Statistics

Thermal noise is the result of a natural random process which is responsible for its two basic characteristics:

(1) Its power spectral density (psd, in W/Hz) is uniform over the receiver passband, a condition referred to as ***white*** noise.

(2) Its amplitude is varying continuously, which means that we can describe the noise only in statistical terms such as its mean, variance, and standard deviation.

The mathematical function that allows us to describe the behavior of the noise and that of the signal plus noise and thus determine required probabilities of detection and false alarm is the probability density function (pdf). Because of its importance to understanding the discussion on probability of detection and false alarm that follows, we start with an abbreviated derivation of the probability density function, using the random process of thermal noise as the example.

Following the concept described by Raven [4.3], consider a group of noise generators with outputs $x(t), x'(t), x''(t)...$, as illustrated in Figure 4.1.1. The actual output of any specific noise generator may, with equal probability, be any one of the functions of the entire group. If the group is expanded to infinity, we have an informal definition of a random process.

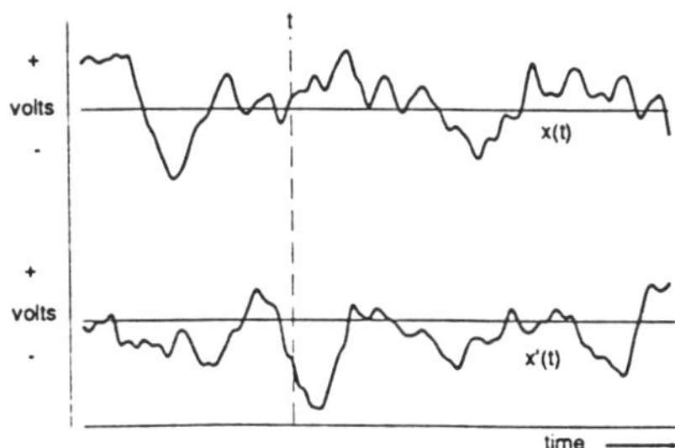

Figure 4.1.1 Group of noise generated outputs vs. time.

If we were to sample the output of a number of these noise generators at some time t, and plot the resulting data in terms of a *histogram*, we would get a graph similar to that shown in Figure 4.1.2, which shows output noise voltage versus the number, or frequency of occurrence of each voltage level. (The voltage levels represented here have been grouped into large class intervals δv wide for convenience; thus each bar of the histogram shows the number of occurrences of a noise voltage between a value v and $v + \delta v$.)

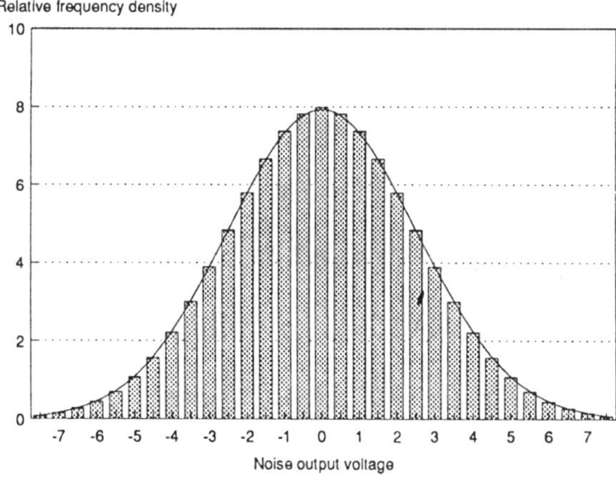

Figure 4.1.2 Histogram of random noise voltage.

If the total number of measurements is increased without limit, corresponding to the output of an infinite number of noise generators at time t, and the class interval is correspondingly decreased, the histogram becomes a *frequency distribution curve*, and the y-axis would represent the *frequency density*. The curve $y = f(x)$ that fits the data, shown overlayed on the histogram, is characteristic of many natural phenomena which have continuously varying values. This is the *Gaussian distribution*, which describes the distribution of a random variable and takes the form[1]

$$y = C \exp(-h^2 X^2) \tag{4.1.1}$$

here X is the deviation from the mean ($X = x\text{-}\mu$); h is the precision constant, equal to $1/(\sqrt{2\pi}\,\sigma)$; σ is the standard deviation, and C is a term which describes the maximum height of the curve. The total number of measurements can be obtained from the histogram by adding the areas (nX class interval) of each bar. Similarly, the area under the curve of Figure 4.1.2 must also equal the total number of occurrences. Thus,

$$\int_{-\infty}^{\infty} y\,dX = n \tag{4.1.2}$$

From (4.1.1):

$$C \exp(-X^2/2\sigma^2)dX = n \tag{4.1.3}$$

It can be shown that $\exp(-X^2/2\sigma^2)dX = 2\pi\sigma$, and therefore that $C = n/2\pi\sigma$, where, as above, σ is the standard deviation of the total population.

With these steps the equation for the Gaussian frequency distribution curve can be expressed as

$$y = \frac{n}{\sigma\sqrt{2\pi}} \exp\left[\frac{-(x-\mu)^2}{2\sigma^2}\right] \tag{4.1.4}$$

If we normalize the expressions in (4.1.2) through (4.1.4), we note two important results:

(1) The area under the curve y equals 1, and now represents the *normal probability distribution* rather than the frequency distribution.

(2) y, given by (4.1.4), now represents the *probability density* for the deviation $(x\text{-}\mu)$, or the rate of change of probability with change in the value of x.

We recall that μ is the mean and is constant for any given distribution. A change in μ moves the position of the curve along the x-axis, but does not alter its shape. The shape of the curve (see Figure 4.1.3) depends only on the value of σ, the standard deviation.

---

[1] The derivation of the Gaussian probability density function given here is that of [4.4] which is also an excellent introductory statistics text.

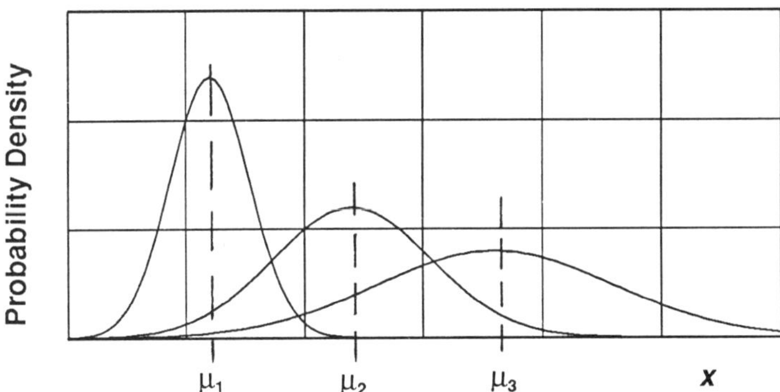

Figure 4.1.3 Normal distributions with different means and standard deviation.

In Section 3.1, thermal noise was described by an equivalent input spectral density N_0, resulting from a receiving system input termination at a temperature T_s. For simplicity in system analysis, the receiving system (see Figure 4.1.4) was modeled as a cascade of ideal components, converting the input signal of energy E, along with noise density, N_0, to intermediate frequency (IF) in a filter of noise bandwidth B_n with unity gain and without adding any further noise. As a result, the noise at the output of the IF system has a power $N = N_0 B_n$. This noise is described as band-limited Gaussian noise. In an actual receiver, of course, there would be a large gain for both signal and noise components, but it is convenient to set this gain to unity so that we may use the symbols S and N when expressing the components applied to the phase and envelope detectors.

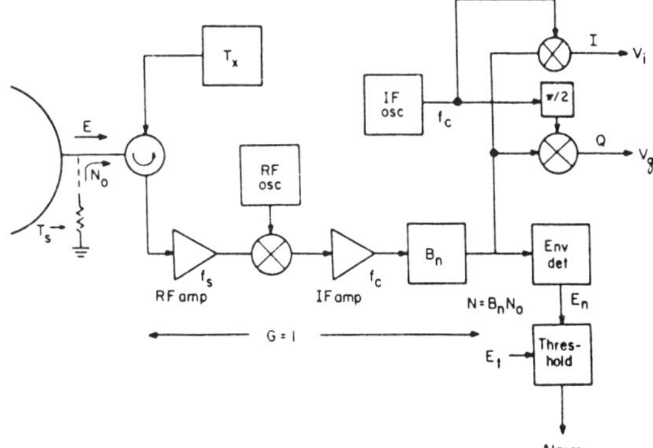

Figure 4.1.4 Idealized receiver block diagram [4.11].
© Artech House, reprinted by permission.

The noise statistics can be measured by connecting phase-sensitive detectors to the output, as shown in Figure 4.1.4. The in-phase (I) and quadrature (Q) detectors will each produce a bipolar output v characterized by a Gaussian power density function:

$$dP_v = \frac{1}{\sqrt{2\pi N}}\exp(-v^2/N)dv \qquad (4.1.5)$$

where dP_v is the probability that the instantaneous value of the noise lies between v and $v + dv$, and N is the mean noise power. Note that this expression is the exact equivalent of (4.1.4), with the term y, the position of a point on the curve, replaced in (4.1.5) with the differential form consistent with the units P/volt, i.e.,

$$y = dP_v/dv \qquad (4.1.6)$$

and the term σ^2 in (4.1.4) has been replaced with N, since the mean square value of a Gaussian process (σ is the noise voltage) about the mean is equivalent to the mean noise power. Note also that the average value of the noise voltage is zero, so that the term $(x-\mu)^2$ in (4.1.4) becomes v^2 in (4.1.6). The pdf distribution for this representation of the IF noise is shown in Figure 4.1.5(a).

The *linear envelope detector* in Figure 4.1.4 will produce a noise output with the *Rayleigh distribution* shown in Figure 4.1.5(b), and expressed as

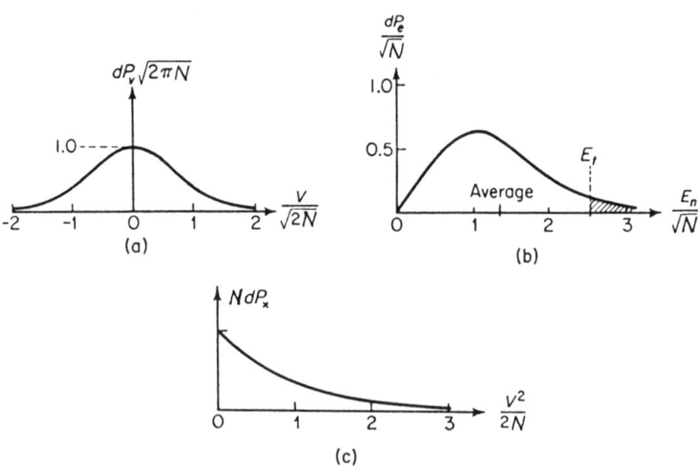

Figure 4.1.5 Probability density functions of noise. (a) Gaussian distribution of IF noise components. (b) Distribution of IF noise voltage envelope. (c) Distribution of IF noise power [4.11]. © Artech House, reprinted by permission.

$$dP_e = \frac{E_n}{N} \exp(-E_n^2/2N)dE_n, \quad E_n > 0 \tag{4.1.7}$$

If a *square-law detector* is used as the envelope detector in Figure 4.1.4, the output will be a voltage equal to the square of the envelope of the IF noise, and therefore proportional to the IF power $x = E_n^2$. This process has the *exponential* power density function shown in Figure 4.1.5(c), where

$$dP_x = (1/N)\exp(-x/N)dx, \quad x > 0 \tag{4.1.7}$$

With this background, we are in a position to describe the concepts of probability of false-alarm and of detection.

4.2 False-Alarm Probability

In the simple detection process of Figure 4.1.4, an amplitude threshold is set at a voltage level E_t so that noise out of the detector which is below this level will not register an alarm. The probability of false alarm can be represented by the hatched area under the noise pdf curve of Figure 4.1.5(b), which is found from

$$P_{fa} = \int_{E_t}^{\infty} (E_n/N) \exp(-E_n^2/2N)dE_n = \exp(-E_t^2/2N) \tag{4.2.1}$$

From this expression, the required threshold voltage can be found as a function of the desired probability of false alarm:

$$E_t = \sqrt{2N \ln(1/P_{fa})} \tag{4.2.2}$$

If the receiver output is applied continuously to the threshold, independent samples of noise at a rate B_n will yield an average *false alarm rate* $B_n P_{fa}$, with the corresponding false-alarm time given by

$$t_{fa} = 1/(B_n P_{fa}) \tag{4.2.3}$$

It is this false-alarm time that establishes the requirements on the detection threshold, and is of primary interest to the use of the radar.

4.3 Detection of One Signal Sample

Let us now introduce at the receiver input, a sinusoidal signal $E_s \cos(2\pi f_s t)$, the duration which is such that its peak amplitude at the IF output (see Figure 4.1.4) reaches the value E_s. The output envelope of *signal plus noise* at the time of peak signal output, will have a pdf described by Figure 4.3.1, termed **Rician** after S. O. Rice, whose early work [4.5] provided the basis for modern signal detection theory. The probability of detection P_d, for a sample taken at the time of peak signal output, is the area under this curve and above the threshold E_t. As shown, this area differs depending upon the signal-to-noise power ratio $S/N = E_s^2/2N$.

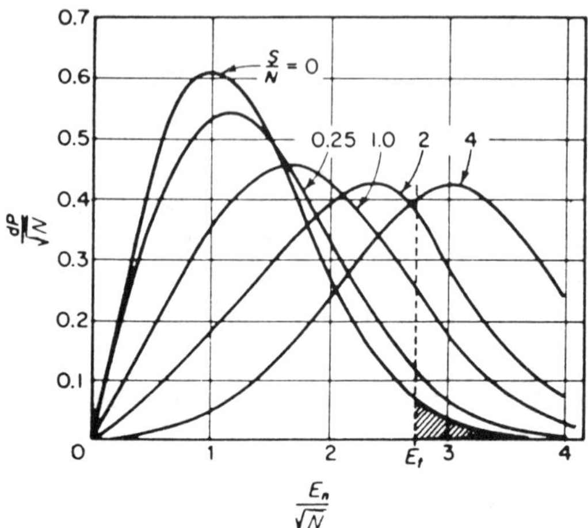

Figure 4.3.1 Probability density functions of envelope of signal plus noise [4.11]. © Artech House, reprinted by permission.

Using (4.2.1) to express E_t as a function of P_{fa}, we can use Rice's results to plot P_d as a function of S/N for different values of P_{fa}. The resulting family of curves (Figure 4.3.2) *serves as the basis for the calculation of detection probabilities and detectability factors for all types of radar signals and detection procedures.* This plot is used directly for the single-pulse detection process.

For example, assume that $P_d = 0.9$ is required, with $P_{fa} = 10^{-6}$, a typical radar requirement. For a single pulse of steady, sinusoidal signal, the required snr can be described as the detectability factor $D_0(1)$, where the subscript "0" denotes a Case 0 (steady, nonfluctuating) target and the argument (1) denotes the single-pulse process. From Figure 4.3.2 we find that $D_0(1) = +13.2$ dB.

4.3.1 Losses in a Non-Ideal System

Detector Loss

The envelope detector is used in radar when the phase of the received signal is unknown. Ideally, if the signal phase were known *a priori*, the IF oscillator of Figure 4.1.4 would be operated in phase with the signal, and the I-channel phase detector would produce a positive output reaching a peak equal to $\sqrt{2E_s}$. The in-phase noise component v_i would add or subtract from this level according to its Gaussian pdf distributions. Calculations of P_{fa} and P_d for this *ideal* system can be made by using tabulated values for the integral of the normal distribution (or of the error function), and its inverse.

When the ideal detectability factor $D_c(1)$ is compared to $D_0(1)$ for envelope detection, it is found that the lack of *a priori* phase data has caused a detector loss $C_x(1)$, which indicates the loss in information for target detection purposes. This loss, which describes the slightly higher input snr required for the envelope detector case, can be defined and expressed as follows [4.6]:

$$C_x(1) \equiv D_0(1)/D_c(1) \approx \frac{(S/N)+2.3}{(S/N)} \qquad (4.3.1)$$

$$C_x(1) \equiv D_0(1)/D_c(1) \approx \frac{D_0(1)+2.3}{D_0(1)} \qquad (4.3.2)$$

For this single-pulse case, $C_x(1)$ is quite small, amounting only to 0.8 dB for $P_d = 0.5$, $P_{fa} = 10^{-4}$, and to 0.4 dB for $P_d = 0.9$, $P_{fa} = 10^{-6}$. The effect is sometimes described as *small-signal suppression*, and it becomes increasingly important when we consider post-detection integration of pulses with small snr. The detector loss is plotted in Figure 4.3.3 as a function of snr at the envelope detector.

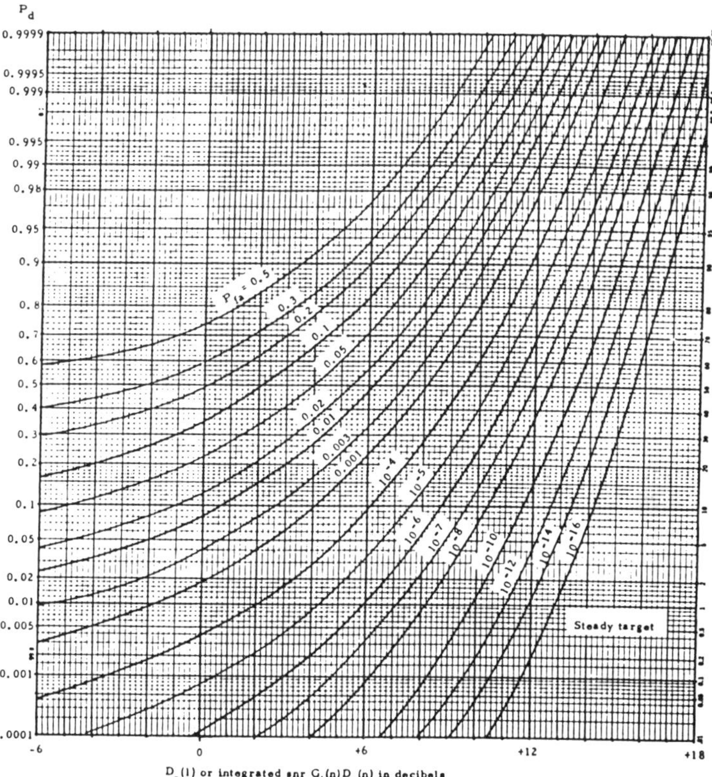

Figure 4.3.2 Detectability factor for a steady target [4.11].
© Artech House, reprinted by permission.

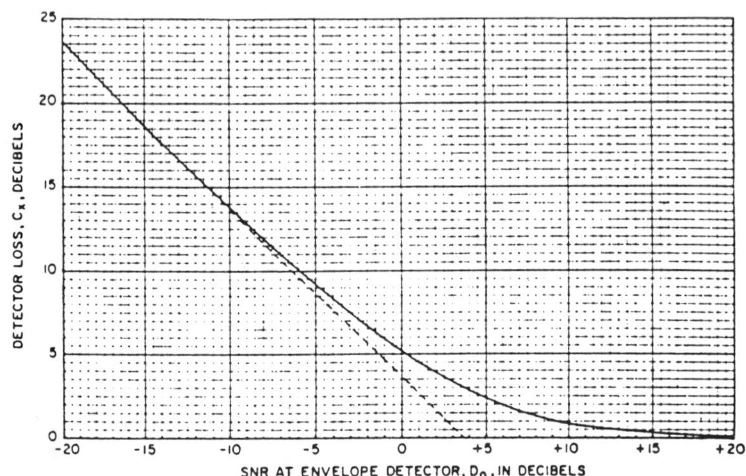

Figure 4.3.3 Envelope detector loss vs. snr [4.11].
© Artech House, reprinted by permission.

Matching Loss

The equations for detection probability involve the voltage ratio $E_s/\sqrt{2N}$, or the equivalent power ratio snr at the input to the envelope detector. This ratio can achieve its maximum value when the IF filter is matched to the signal. For other filters, there will be a *matching loss* L_m with respect to the pulse spectrum, and also possibly a loss L_{mf} with respect to the spectrum of the envelope of the pulse train. The IF output snr will then be:

$$\text{snr} = E_1/N_0 L_m \qquad \text{for a single pulse}$$
$$\text{snr} = E/N_0 L_m L_{mf} \qquad \text{for the entire observed signal}$$

The effect of these mismatch losses is that the energy ratio, D_x at the receiver input must be increased in order to meet the snr requirements at the envelope detector.

4.4 Integration of Pulse Trains

When a train of radar pulses is received from the target over the observation time t_o, there will be $n = f_r t_o$ pulses available for processing. Similarly, if a continuous signal, e.g., CW, is received for a time t_o in a receiver whose bandwidth $B_n > 1/t_o$, there will be $n = B_n t_o$ samples of signal and independent noise available. There are four distinct ways in which the information from these n pulses or samples may be processed to improve detection performance:

(a) *Coherent integration*, in which the pulses are added prior to envelope detection;

(b) *Noncoherent (or video) integration*, in which each pulse is envelope detected, and the resulting video pulses are added together prior to application of thresholding;

(c) *Binary integration*, in which each pulse is applied to a threshold, and the number m of threshold crossings is used as the criterion for an output alarm; and

(d) *Cumulative detection*, in which $m = 1$ is the alarm criterion.

The methods are listed in order of declining efficiency of integration, and also declining complexity of implementation.

4.4.1 Coherent Integration

If the IF filter is matched to the entire signal observed over t_o coherent integration is obtained in that filter. This requires that the signal have a predictable phase relationship (coherence) over this period, and that the phase response of the filter be such as to bring all signal components into the same phase while they are added. If the signal is a pulse train, then the filter must be matched to the pulse-to-pulse phase change which is the doppler frequency of the target. For an ideal coherent system the coherent integration gain, which is the ratio of the effective snr for detection purposes to that of the single pulse, is exactly n. For a system with a filter spectral envelope which fails to match the pulse spectrum there will be a mismatch loss L_m, and for one having passbands that fail to match the received spectrum's fine line shape, there will be a further loss L_{mf}. Then, ignoring the effects of possible post-detection low-pass filtering to recover a portion of the matching loss, the required total energy ratio will be

$$E/N_0 = D_c(1)L_m L_{mf} \tag{4.4.1}$$

The energy ratio for each of n equal pulses will be

$$E_1/N_0 = D_c(n)L_m L_{mf} \tag{4.4.2}$$

4.4.2 Noncoherent (Video) Integration

If n signal samples are to be added *without* controlled phase relationships, the samples must first be passed through the envelope detector to remove the random phase of each sample. The effective (integrated) snr is increased in the video integrator by the factor n, but not before the pulses experience an increased detector loss, given by (4.3.1) with the reduced value $S/N = D_0(n)$:

$$C_x(n) = \frac{D_0(n) + 2.3}{D_0(n)} \tag{4.4.3}$$

where $D_0(n)$ denotes the detectability factor *(required snr) for each pulse of the n-pulse train.* As the signal input to the envelope detector decreases, the small-signal suppression loss increases, and subsequent video integration cannot restore the snr represented by the total signal energy ratio. As a result, there is a requirement for more energy per pulse. This increased energy per pulse (in dB) is termed the **integration loss.** It is important to recognize, however, that *the loss is not the result of the integration,* but of the location of the integrator after the envelope detector, and the resulting increased small-signal suppression in the nonlinear process of envelope detection. This integration loss may be expressed as

$$L_i(n) \equiv nD_0(n)/D_0(1) = C_x(n)/C_x(1) \tag{4.4.4}$$

This loss is thus the ratio of total signal energy required of the n-pulse train to that which would have been required if only a single pulse had been transmitted and processed, or if a matched filter for the entire n-pulse train had been used.

The **integration gain** for the noncoherent process of

$$G_i(n) \equiv D_0(1)/D_0(n) = n/L_i(n) \tag{4.4.5a}$$

and *the effective snr of the pulse train, to be used in entering Figure 4.3.2 to find detection performance,* is $D_0(1)_{\text{eff}} = D_0(n)G_i(n)$. The n-pulse detectability factor is thus

$$D_0(n) = \frac{D_0(1)L_i(n)}{n} \tag{4.4.5b}$$

To calculate $C_x(n)$ and the integration loss or gain, $D_0(n)$ must be known. The family of curves in Figure 4.4.1 represents the solution to this problem when the required P_d and P_{fa}, and hence $D_0(1)$, are given along with n. For example, assume that $P_d = 0.9$, $P_{fa} = 10^{-6}$, and $n = 24$ are given. The total signal energy ratio and the ratio per pulse for a system using a filter matched to the spectrum of each pulse, are found from the graphs as follows:

$P_d = 0.9,\ P_{fa} = 10^{-6}$	Fig. 4.3.2 $\rightarrow D_0(1) =$	+13.2 dB
$n = 24$	Fig. 4.4.2 $\rightarrow +L_i(n) =$	<u>+3.2 dB</u>
Total energy ratio,	$nD_0(n) =$	+16.4 dB
	$-10 \log n =$	<u>-13.8 dB</u>
Single-pulse energy ratio,	$D_0(n) =$	+ 2.6 dB

The calculation may also be made easily without graphs, using procedures described in [4.6].

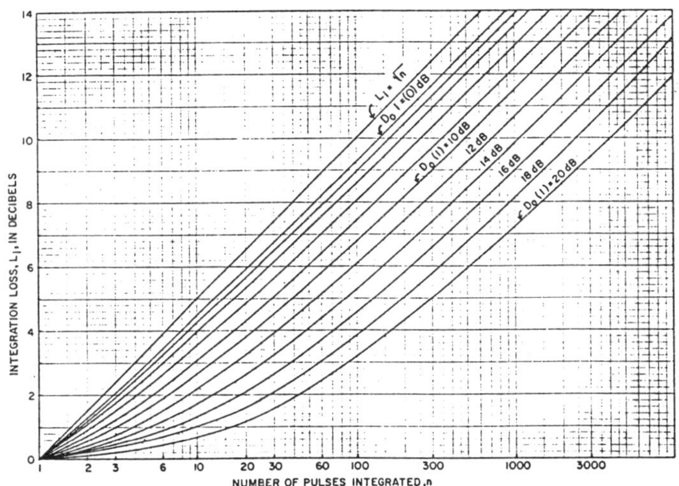

Figure 4.4.1 Integration loss vs. number of pulses integrated after envelope detection [4.11]. © Artech House, reprinted by permission.

The problem must sometimes be worked in the other direction, where snr is known and detection probability is to be obtained. Marcum's technique [4.8] has been programmed for calculators and personal computers to solve this [4.9, 4.10].

4.4.3 Binary Integration

The video integrator is often implemented in digital form, by placing an *analog-to-digital (A/D) converter* after the envelope detector. With fine enough granularity in the converter, the results of the previous section will be obtained. A simplification in equipment results when the number of converter bits is reduced to one, corresponding to a threshold detector operating directly on the output of the envelope detector, followed by an *accumulator* which counts up to m threshold crossings before generating an output alarm. The resulting performance is within about 1.5 dB of ideal video integration, comparing favorably with most practical video integration circuits. The range of optimum m is quite broad, lying near the value determined by Schwartz [4.12] as

$$m_{opt} = 1.5\sqrt{n} \tag{4.4.6}$$

A comparison of required snr for optimum binary integration $D_b(n)$, and $D_0(n)$ is shown in Figure 4.4.2, along with a comparison of integration losses associated with each technique. A description of the technique to find P_d with a binary integrator is contained in [4.11, p. 74].

4.4.4 Cumulative Detection

The least efficient way of combining information for n signal samples is to make independent detection decisions on each, and to rely on the cumulative probability of detection to reach the required output value. This is equivalent to operating a binary integrator with $m = 1$. The single-pulse false-alarm probability in this case must be set at P_{fd}/n, while single-pulse p_d is given by

$$p_d = 1 - (1 - P_d)^{1/n} \tag{4.4.7}$$

where P_d is the required probability of detection. When n pulses are distributed over separate antenna scans, the extra loss in the cumulative process, compared to video integration, can be described as *scan distribution loss*. It is apparent that the gain from cumulative detection of n signals on a steady target is small, but when fluctuating targets are considered, there may be significant gain.

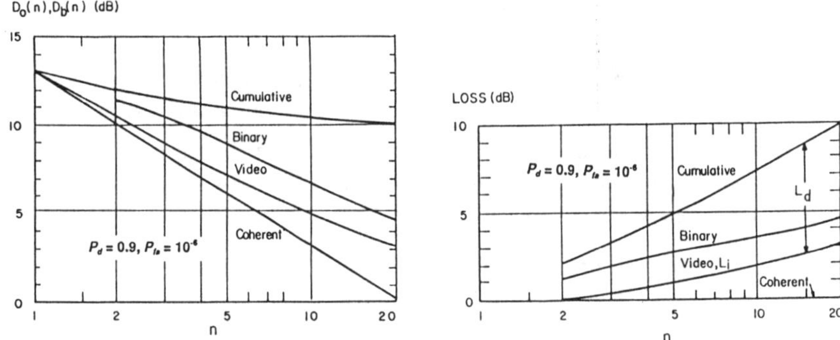

Figure 4.4.2 Comparison of integration methods. (a) Detectability factors. (b) Integration loss [4.11]. © Artech House, reprinted by permission.

4.4.5 False-Alarm Time

When integration is used, the number of independent detection decisions is reduced. In a coherent system, such as a pulse doppler radar, in each interval t_o, a decision is made in each range-doppler resolution cell. If there are n_t range gates or cells, and n_f doppler filters, the decisions are made at a rate:

$$v = n_t n_f / t_o \tag{4.4.8}$$

The average false-alarm rate is $v P_{fa}$, and the average time between false alarms is

$$t_{fa} = 1/v P_{fa} = t_o / n_t n_f P_{fa} \tag{4.4.9}$$

For a coherent system, using all doppler cells, we have $n_f = n$, and

$$t_{fa} = \tau / P_{fa} \approx 1/B_n P_{fa} \tag{4.4.10}$$

The numbers of cells will be reduced, and false-alarm time increased, if only certain ranges and dopplers are used for detection.

For a ***noncoherent*** system in which $n_f = 1$ and all range cells are used for detection, we have $n_t = t_r/\tau = 1/f_r\tau$, $t_o = nt_r = n/f_r$, and

$$t_{f_a} = n\tau/P_{f_a} \approx n/B_n P_{f_a} \text{ (matched system)} \tag{4.4.11}$$

4.4.6 Losses Due to Pulse Integration

Integrator Weighting Loss

When video integration is used, the optimum weighting of the n pulses is matched to their individual snrs. The calculations of integration gain and loss were based on pulses of equal snr, with uniform weighting over the train. Practical integrators are more likely to use uniform weighting of the most recent k pulses, where $k < n$, or other, nonlinear processes. Such weighting will introduce an additional loss, which will be small if the time constant is properly selected [4.13]. The often used value [4.13, 4.14] of beamshape loss $L_p = 1.6$ dB is actually the combined beamshape and weighting loss for uniform weighing of $0.84n$.

Collapsing Loss

Practical radars seldom preserve their full RF signal resolution through the integration and thresholding processes, for which reduced video bandwidth and broadened range gates may prove economical. The term ***collapsing loss*** is applied to describe *the increase in required input snr when noise from adjacent resolution cells or channels is combined with the signal and noise of the channel actually containing the target.* The n video samples of signal plus noise will then be integrated along with m extra samples of noise alone, giving a ***collapsing ratio*** ρ defined by

$$\rho \equiv (n + m)/n \tag{4.4.12}$$

The effect, when using a square-law detector, is the same as if the signal energy had been redistributed over $\rho n = n + m$ pulses, leading to a larger value of integration loss $L_i(\rho n)$. The collapsing loss is then defined as

$$L_c(\rho, n) = L_i(\rho n)/L_i(n) \tag{4.4.13}$$

Equations for collapsing ratio are given in Table 4.1 for different processes. The collapsing loss is used to calculate the actual increase in required energy ratio, denoted by the matching factor M, in cases where *the IF bandwidth is greater than its matched value,* if the video bandwidth is matched to the pulse.

Table 4.1 Equations for Collapsing Ratio

Cases for which P_{fa}/ρ *remains constant:*	
(a) Restricted CRT sweep speed s, where d is spot diameter and τ is pulsewidth.	$\rho = \dfrac{d + s\tau}{s\tau}$
(b) Restricted video bandwidth B_v, where $B = 1/\tau$ is IF signal bandwidth.	$\rho = \dfrac{2B_v + 1/\tau}{2B_v} = \dfrac{2B_v + B}{2B_v}$
(c) Collapsing of coordinates onto the display, where $2\Delta_r/c$ is the time-delay interval displayed per display cell, $\omega_e t_v$ *and* $\omega_a t_v$ are elevation and azimuth scans during the integration time t_v, and θ_e and θ_a are the beamwidths.	$\rho = \dfrac{2\Delta_r}{c\tau}$ or $\rho = \dfrac{\omega_e t_v}{\theta_e}$ or $\rho = \dfrac{\omega_a t_v}{\theta_a}$

Cases for which P_{fa} *remains constant:*	
(d) Excessive IF bandwidth $B > 1/\tau$ followed by matched video.	$\rho = 1 + B\tau$ (use L_c in place of L_m)
(e) Receiver outputs mixed at video, where m is the number of receivers.	$\rho = m$
(f) IF filter followed by gate of width τ_g and by video integration.	$\rho = \dfrac{1}{B_\tau} + \dfrac{\tau_g}{\tau}$

4.5 Detection of Fluctuating Targets

4.5.1 Single-Sample Detection

Steady targets almost never exist in the real world except as calibrating devices. Therefore, having established procedures for calculating detection performance on a steady target signal, we must now extend the result to the fluctuating target. Essentially, all radar targets produce echo signals having amplitudes that are Rayleigh distributed (exponential distribution of power, or cross section). In his classic work on fluctuating targets [4.15], Swerling designated such targets as Case 1 (when the fluctuation was slow) or Case 2 (when it was fast). Swerling also considered a Case 3 and Case 4 but the only practical application of these models lies in the use of Case 3 to represent a Case 1 target observed by dual-diversity radar.

The pdf and detection equations for the Case 1 target are shown in Table 4.2, for the single-sample case. The signal-plus-noise pdf is the same as that for noise alone, except that the mean power is now $S+N$ rather than N. The expressions for P_d and $D_1(1)$ are exact, as well as being simple.

Table 4.2 Equations for Fluctuating Target (Case 1)

pdf of IF signal-plus-noise envelope (Rayleigh)

$$dP_s = \left[\frac{E_n}{(S+N)}\right]\exp\left[\frac{-E_n^2}{2(S+N)}\right]dE_n \tag{4.5.1}$$

Detection probability

$$P_d = \int_{E_t}^{\infty} dP_s = \exp\left[\frac{-E_t^2}{2(S+N)}\right] = \exp\left[\frac{-\ln P_{fa}}{(1+S/N)}\right] \tag{4.5.2}$$

Detectability factor

$$D_1(1) = \frac{\ln P_{fa}}{\ln P_d} - 1 \tag{4.5.3}$$

Fluctuation Loss

Curves for the Case 1, single-pulse detectability factor are shown in Figure 4.5.1. When these are compared to Figure 4.3.2, we will see that a larger average signal is required for the fluctuating target than for the steady target, if high P_d is to be obtained. The extra signal requirement is similar to the fade margin required in communications channels, but in radar this requirement is referred to as a *fluctuation loss*. It is defined [4.2] as

> The apparent loss in radar detectability or measurement accuracy for a target of given average echo power return due to target fluctuation. It may be measured as the increase in required average echo return power of a fluctuating target as compared to a target of constant echo return, to achieve the same detectability or measurement accuracy.

Thus, in our notation, we have

$$L_f(1) = D_1(1)/D_0(1) \tag{4.5.4}$$

This loss, plotted in Figure 4.5.2 for the single-pulse case, is primarily a function of P_d, but also depends weakly on P_{fa}.

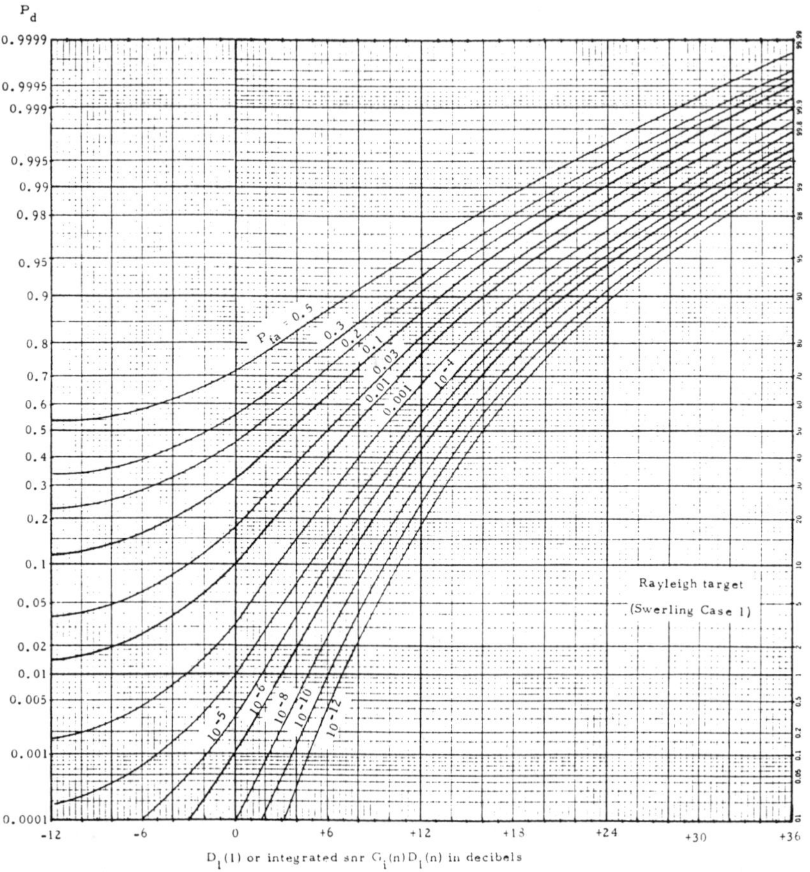

Figure 4.5.1 Detectability factor for single-pulse Case 1 target [4.11].
© Artech House, reprinted by permission.

4.5.2 Integration of Case 1 Signals

The result of integrating n pulses or samples of fluctuating signal plus noise will depend on whether there is only one independent sample of the signal, or more than one. If the target amplitude remains constant for the n-pulse train, either video or coherent integration can be used as with the steady target. The integration gain will be the same as for the steady-target case, namely $n/L_i(n)$, where $L_i = 1$ for coherent integration, or as given in (4.4.4) for video integration. It is important to note that the integration loss depends not on the average snr, $D_1(n)$, but rather on the effective steady-target value $D_0(n) = D_1(n)/L_f(1)$. This is because the detector loss increases as the target fade depth becomes larger. To achieve a high P_d, the signal reduced by fading must be adequate, not just its average value. Thus, we can write the simplified equation for the n-pulse detectability factor of the Case 1 target as

$$D_1(n) = D_0(n)L_f(1) = \frac{D_0(1)L_i(n)L_f(1)}{n} \qquad (4.5.5)$$

For example, in calculating the Case 1 detectability factor for $P_d = 0.9$, $P_{fa} = 10^{-6}$, $n = 24$, we follow the procedure of Section 4.4 but with the addition of a fluctuation loss:

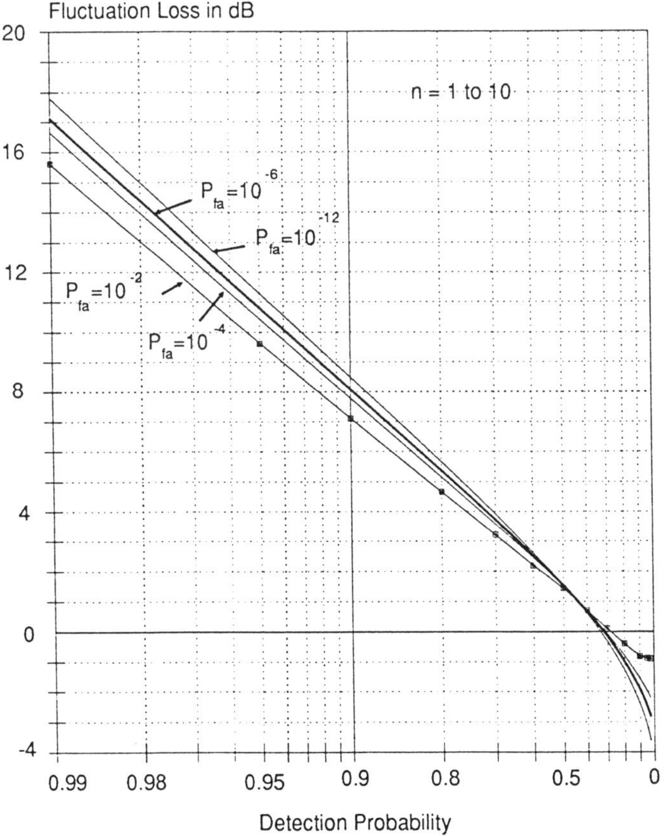

Figure 4.5.2 Fluctuation loss for Case 1 target [4.11].
© Artech House, reprinted by permission.

$$P_d = 0.9, P_{fa} = 10^{-6} \qquad \text{Fig. 4.3.2} \rightarrow D_0(1) = \quad +13.2 \text{ dB}$$
$$n = 24 \qquad \text{Fig. 4.4.2} \rightarrow +L_i(n) = \quad \underline{+3.2 \text{ dB}}$$
$$\text{Total energy ratio,} \qquad nD_0(n) = \quad +16.4 \text{ dB}$$
$$-10 \log n = \quad \underline{-13.8 \text{ dB}}$$
$$\text{Single-pulse energy ratio,} \qquad D_0(n) = \quad +2.6 \text{ dB}$$
$$\text{Fluctuation loss,} \qquad +L_f(1) = \quad \underline{+8.4 \text{ dB}}$$
$$\text{Average single-pulse,} \qquad D_1(n) = \quad +11.0 \text{ dB}$$

Note that the total average energy ratio $nD_0(n)L_f(1)$ is increased to 24.8 dB for the Case 1 target.

If the average input snr is given, and calculation of P_d is required, the problem becomes more difficult. Both fluctuation and integration losses depend on the final value of P_d, which is unknown. In [4.15] an iterative procedure was used. In hand computations, a convenient approach is to calculate $D_1(n)$ for $P_d = 0.5$, compare this with the available snr, and make one or two further calculations of $D_1(n)$ for P_d either higher or lower than 0.5. When these are plotted on probability paper, interpolation will quickly yield the correct P_d for the given snr. Alternatively, the method of [4.7, pp. 406-207] or the many curves of [4.16] may be used.

Integration with Independent Signal Samples

In many cases, the target amplitude does not remain constant over the observation time t_o, but varies between the beginning and the end of the interval. Although Swerling solved this problem for the case of very rapid fluctuation, such that n independent signal samples were received (Case 2), there was no solution for the general case where $n_e < n$ independent signal samples were integrated along with n noise samples. An empirical expression was developed [4.6] to make possible calculation for such cases. Thus,

$$L_f(n_e) = [L_f(1)]^{1/n_e} \tag{4.5.6}$$

This agrees closely with the Case 2 results, when $n_e = n$, and also permits the intermediate cases to be evaluated. The general expression for detectability factor with n pulses, within which there are n_e independent signal samples, becomes

$$D_e(n,n_e) = D_0(n)L_f(n_e) = \frac{D_0(1)L_i(n)L_f(n_e)}{n} \tag{4.5.7}$$

In dB format, we can write

$$D_e(n,n_e)_{dB} = D_0(1)_{dB} + L_i(n)_{dB} + (1/n_e)L_f(1)_{dB} - 10\log n \tag{4.5.8}$$

In the example calculated above, if there were $n_e = 2$ independent signal samples during the n pulses, the fluctuation loss in dB would be cut in half, and the required energy ratio would be reduced by 4.2 dB. For Case 2 signals, with $n_e = 24$, the loss would be only 0.4 dB and $D_2(24) = +3.0$ dB.

4.5.3 Other Target Models

Using the empirical expression for fluctuation loss, we are now in a position to solve for detectability factor of the generalized distributed target. For the general case the fluctuation loss is

$$L_f(Kn_e)_{dB} = (1/Kn_e)L_f(1)_{dB} \qquad (4.5.9)$$

The factor K represents half the number of independent Gaussian components added together to form a target signal, while n_e represents the number of independent signals integrated during n pulses. The steady target and all Swerling models are special cases of this generalized model, often referred to as a *Chi-square* model.

Case 0, Steady Target:	$n_e \to \infty$,	$K \to \infty$
Case 1 Target:	$n_e = 1$,	$K = 1$
Case 2 Target:	$n_e = n$,	$K = 1$
Case 3 Target:	$n_e = 2$,	$K = 2$

4.5.4 Diversity Gain

The theory of fluctuation loss provides a quantitative approach to determining the beneficial effects of diversity on target detection. The reduction in fluctuation loss may be considered as a *diversity gain* for a system taking samples over intervals in time or frequency. The diversity gain can be defined as

$$G_d(n_e) \equiv \frac{L_f(1)}{L_f(n_e)} = [L_f(1)]^{1-\frac{1}{n_e}} \qquad (4.5.10)$$

Four diversity cases may be distinguished:

(a) **Time diversity**, in which independent samples are received at intervals equal to the correlation time of the target;

(b) **Frequency agility**, in which independent samples are obtained rapidly, by changing transmitter frequency;

(c) **Frequency diversity**, in which parallel frequency channels are used to obtain independent samples; and

(d) **Combined diversity**, in which both time and frequency effects are used to increase the number or samples.

Time diversity requires that the integration interval exceed the *correlation time* of the target, t_c. The number of samples available is

$$n_e = 1 + t_o/t_c \qquad (4.5.11)$$

For rigid targets, the correlation time may be estimated as

$$t_c = \lambda/2\omega_a L_x \qquad (4.5.12)$$

where λ is radar wavelength, ω_a is the rate of rotation of the target about the line of sight, and L_x is the target dimension normal to the line of site and the axis of rotation. In a typical aircraft case, ω_a may be a few hundredths of one radian per second, and L_x may be a few tens of meters. Even at microwave frequencies (e.g., $\lambda = 0.33$ m), the typical target correlation time is tens of milliseconds, and there is not much time diversity effect unless long dwells are used for detection ($t_o \gg t_c$). If integration can be carried out over several scans of the antenna, then time diversity becomes more practical. However, as targets move between scans, it becomes difficult to perform the integration in narrow range cells, and the cumulative detection process is preferred.

Frequency agility radar can approach the Case 2 condition, if the agile bandwidth of the transmitter is great enough and if pulse-to-pulse frequency change is introduced. The number of independent target samples available in a bandwidth Δf is

$$n_e = 1 + \Delta f / f_c \qquad (4.5.13)$$

where the **correlation frequency** of the target is

$$f_c = c / 2L_r \qquad (4.5.14)$$

The velocity of light is c, and L_r is the radial length of the target. For a typical aircraft, L_r is 15 to 30 m, and the resulting $f_c = 10$ to 5 MHz. To obtain a Case 2 target in our previous example, with $n = 24$, the agile bandwidth would have to be 115 to 230 MHz, with the pulses equally distributed over that band. This is quite possible in microwave radar, provided there is no requirement for MTI or doppler processing to reject clutter. If MTI is required, the agility may be on a burst-to-burst basis, with reduction in the number of independent samples ($n_e = n/m$, where m-pulse bursts are used).

Frequency diversity radar offers a solution to the problem of large fluctuation loss, without compromising MTI or doppler performance. A number n_e of channels (often two) are operated in parallel, spaced by at least f_c within the radar band. Pulses are received in each of N_e channels, so the integration gain is increased as fluctuation loss is decreased.

Combined diversity obtains samples in both time and frequency, so that the total number of samples becomes

$$n_e = (1 + t_o / t_c)(1 + \Delta f / f_c) \qquad (4.5.15)$$

It should be noted that diversity gain is only possible if the *nondiverse* system would experience a fluctuation loss. From Figure 4.5.2, if the value of P_d is less than 0.33, diversity is not advantageous. On the other hand, if high probabilities of detection are required, some form of diversity is almost essential.

4.5.5 Binary Integration on Fluctuating Targets

In the prior discussion of binary integration, the occurrences of first threshold crossings were assumed to be independent. This is certainly true for a steady target in noise, but not for fluctuating targets. The question then arises, how can we determine the detectability factor $D_b(n)$ or detection probability P_d for the fluctuating target? A straightforward approach to determine $D_b(n)$ is to perform the calculation for the steady target, and then increase the snr by the fluctuation loss for the required value of P_d. However, as in the case of video integration, there is no direct approach to determining P_d from a known average snr, since the fluctuation and integration losses both depend on the final value of P_d. A solution is to perform the calculation for video integration, and to assume that the additional integration loss from the binary process is the same as that shown in Figure 4.4.2.

4.5.6 Cumulative Detection on Fluctuating Targets

In Section 4.4, we noted that the cumulative detection process was relatively ineffective, when compared to video or binary integration. In the case of fluctuating targets, however, the scan-to-scan build up of cumulative detection probability provides a significant benefit in reduced fluctuation loss, without the necessity of integration within a moving range cell. For early warning purposes, single-scan probabilities of detection as low as 0.33 can lead to cumulative probability $P_c >$ 0.9 within six scans, and even lower values of P_d can be useful. Fluctuation loss is then eliminated or actually converted into a gain factor.

4.6 Constant-False-Alarm-Rate Detection

If we were to establish the false-alarm rate (and by definition the P_{fa}), of a radar based upon a single parameter, e.g., the average noise level in the IF receiver, then any outside source of interference such as ECM, clutter, and noise above this threshold would trigger an excessive number of false alarms. The result would be a vastly degraded and virtually useless radar, especially in the case of automatic detection systems.

CFAR, for *constant false-alarm-rate*, is a technique used in most radars to deal with this problem, and has the formal definition [4.2]:

> A property of threshold or gain control devices specially designed
> to suppress false alarms caused by noise, clutter, or ECM of
> varying levels.

The use of some form of CFAR is essential if the output data are fed directly to communications channels or automatic data processors. If a human operator is used for detection and track initiation, that person's visual and mental processes will often provide the equivalent of a CFAR function to a limited degree, but even here circuitry to avoid display saturation may be required.

The basic CFAR process is to form an estimate of the noise and interference level *in the cell where the target detection is being carried out* and to set the detection threshold based on that estimate, rather than at some predetermined constant level. There are two basic approaches to performing this estimate:

(a) *Cell-Averaging CFAR*, in which the level is estimated by averaging over adjacent reference cells in range, doppler, angle, or some combination of radar coordinates [4.18];

(b) *Time-Averaging CFAR*, in which the level is estimated by averaging the output of the detection cell itself over several scans. The estimate may involve the mean level, other statistical properties of the interference, or the procedure may be designed to be independent of these statistics (*nonparametric CFAR*) [4.18, 4.19].

Time-averaging CFAR is especially useful when used in conjunction with digital processing and storage techniques to establish thresholds in ground clutter environments, but this technique is not adequate for the dynamic environment presented by ECM, and it has the additional potential disadvantage of suppressing slowly moving targets. The use of other forms of CFAR such as *two-parameter* [4.20] and nonparametric generally incurs severe losses that limit their applicability.

Averaging over reference cells is often done in range, because that coordinate has better resolution than angles, in most radars. A typical cell-averaging CFAR, with analog implementation, is shown in Figure 4.6.1. The envelope-detected outputs of m adjacent range cells are available simultaneously in a tapped delay line, and the center tap represents the detection cell. The m reference taps are averaged to form an estimate w of the local noise and interference within the radar beam, and the ratio of the detection cell amplitude x^s to this average is used as the video output. In effect, the threshold level, E_t is scaled to the estimate of local noise, rather than to an *a priori* value.

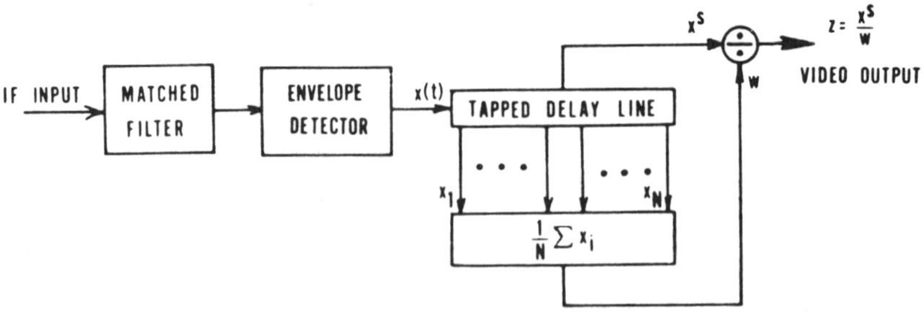

Figure 4.6.1 Range cell averaging CFAR [4.17].

4.7 Evaluation of Detectability Factor

The calculation of detectability factor involves intimate knowledge of the signal processing steps leading up to a detection decision, and also of the target statistics and diversity modes of the radar. In many cases, the complexity of modern systems challenges the ability of the evaluator to make accurate predictions based on theory. Simulation may be required to obtain an accurate value.

When it comes to field testing of the radar, to confirm the results of simulation, it often proves difficult to isolate the different factors and to identify the measured quantities. The over-all detection performance can be measured in system tests on known targets (see Section 3.3), but the factors which may contribute to departure from predicted results often remain obscure. While it is difficult to give general rules for evaluation on the system in the field, Chapter 13 contains guidelines for the types of tests, and the environmental factors which must be controlled (or at least known) if the test results are to be valid. Especially important is the planning and conduct of tests in such a way that the measured detection probability has statistical validity. If probabilities near 90% in detection probability, and near 10^{-6} in false-alarm probability, are to be reliable, there must be significant amounts of data gathered under the same conditions, rather than scattered tests over wide ranges of the many variables.

4.8 References

[4.1]   D. C. Schleher, *Introduction to Electronic Warfare*, Artech House, 1980.

[4.2]   IEEE Standard Dictionary of Electrical and Electronic Terms, *ANSI/IEEE Std. 100-1988*, 1988.

[4.3]   R. S. Raven, *Techniques for Signal and Noise Analysis*, Chapter 5 in *Airborne Radar*, (D. J. Povejsil, R. S. Raven and P. Waterman, eds.), D. Van Nostrand 1961.

[4.4]   A. M. Neville and J. B. Kennedy, *Basic Statistical Methods for Engineers and Scientists*, International Textbook Company, 1964.

[4.5]   S. O. Rice, Mathematical analysis of random noise, *Bell System Technical Journal* **23**, No. 3 July 1944, pp. 282-332, and **24**, No. 1, January 1945, pp. 46-156.

[4.6]   D. K. Barton, "Simple procedures for radar detection calculation," *IEEE Trans.* **AES-5**, No. 5, September 1969, pp. 837-846; reprinted in D. K. Barton (ed.), *Radars,* Vol. 2, *The Radar Equation*, Artech House, 1974.

[4.7]   J. V. DiFranco and W. L. Rubin, *Radar Detection*, Prentice-Hall, 1968; Artech House, 1980.

[4.8]   J. I. Marcum, "A statistical theory of target detection by pulsed radar," *Rand Research Memo. RM-754*, December 1947, with Appendix, *RM-753*, July 1948; reprinted in *IRE Trans.* **IT-6**, No. 2, April 1960.

[4.9]   W. A. Skillman, *Radar Calculations Using the TI-59 Programmable Calculator*, Artech House, 1983.

[4.10]   W. A. Skillman, *Radar Calculations Using Personal Computer*, Artech House, 1984.

[4.11]   D. K. Barton, *Modern Radar System Analysis,* Artech House, 1988.

[4.12]   M. Schwartz, "A coincidence procedure for signal detection," *IRE Trans.* **IT-2**, No. 4, December 1956, pp. 135-139.

[4.13]   B. H. Cantrell and G. V. Trunk, "Angular accuracy of a scanning radar employing a two-pole filter," *IEEE Trans.* **AES-9,** No. 5, September 1973, pp. 649-653.

[4.14]   L. V. Blake, "The effective number of pulses per beamwidth for a scanning radar," *Proc. IRE* **41**, No. 6, June 1953, pp. 770-774.

[4.15]   P. Swerling, "Probability of detection for fluctuating targets," *Rand Research Memo. RM-1217*, March 17 1954; reprinted in *IRE Trans.* **IT-6**, No. 2, April 1960.

[4.16]   D. P. Meyer and H. A. Mayer, *Radar Target Detection*, Academic Press, 1973.

[4.17]   V. Gregers Hansen, "Constant false alarm rate processing in search radars," *Radar - Present and Future*, IEEE Conf. Pub. No. 105, October 1973, pp. 325-332.

[4.18]   A. Farina and F. A. Studer, "A Review of CFAR Detection Techniques in Radar Systems," *Microwave Journal*, September 1986, pp.115-128.

[4.19]   G. V. Trunk and S. F. George, "Detection of Targets in Non-Gaussian Sea Clutter," *IEEE Trans.* **AES-6**, No. 5, September 1970, pp. 620-628.

[4.20]   M. Sakino, *et al.*, "Suppression of Weibull-distributed Clutter Using a Cell-averaging Tag/CFAR Receiver," *IEEE Trans.* **AES-14**, No. 5, September 1978, pp. 823-826.

Chapter 5

TARGETS AND INTERFERENCE

The primary function of any radar is to detect a "target," but the design target depends on the purpose of the radar. Air defense radars are designed to detect airborne targets such as aircraft and missiles, but there are many radars designed for special missions. Thus, for a ground-mapping radar the target is the surface of a portion of the earth itself or a celestial body like the moon or a planet. Tactical radars may be designed to detect tanks, ships, or motor shells. For weather radars, rain and clouds are targets, not clutter.

Although the fundamentals of radar target characteristics discussed in this chapter are applicable to a wide variety of objects, emphasis is placed on the description of airborne targets and of the characteristics of sources of unwanted interference that a monostatic radar must deal with in detecting them.

The performance of a radar system in detection and measurement can be predicted with confidence only when the actual target characteristics agree with their predicted values. Since a wide variety of targets may be encountered in use of radar systems, we must devise target descriptions of sufficient latitude to accommodate wide variation in characteristics of individual targets. The importance of the signal-to-noise power ratio and energy ratio has been shown in the previous two chapters. Knowledge of the *radar cross section*, or *backscattering coefficient*, of the target is essential in any such calculation. Although there are few cases in which this coefficient is a constant, it will generally be found to vary considerably for each target as the aspect angle changes, as internal motions of the target change its shape, and as radar frequency is varied. These changes force us to use statistical methods to describe the radar target. Similarly, the interfering signals backscattered from clutter (objects other than the desired target) will be described statistically, and as functions of radar-clutter path geometry, radar wavelength, and type of clutter.

In measuring target position and velocity, there are factors other then echo amplitude which are also important. The echo signal may consist of many components of energy scattered from points distributed over the surface of the target.

The amplitude and phase of each component will vary with time, aspect angle, and radar frequency, and the interaction of these components will affect the radar measurement process. This chapter will summarize these target characteristics.

5.1 Definition of Radar Cross Section

The first important characteristic of any radar target is the measure of its ability to reflect energy to the radar receiving antenna. The parameter used to describe this ability is the *radar cross section* or RCS of the target, also termed the *effective echoing area* or the *backscattering* (or forward-scattering, bistatic-scattering) *coefficient*. This is defined [5.1, p. 716] as

> 4π times the ratio of the power per unit solid angle scattered in a specified direction to the power per unit area in a plane wave incident on the scatterer from a specified direction... Three cases are distinguished (1) *monostatic* or backscattering RCS when the incident or pertinent scattering directions are coincident but opposite in sense, (2) *forward-scattering* RCS when the two directions and senses are the same, and (3) *bistatic* RCS when the two directions are different.

Thus, if the target were to scatter power uniformly over all angles, its RCS would be equal to the area from which power was extracted from the incident wave. Since a sphere has this ability to scatter isotopically, it is convenient to interpret the RCS in terms of an equivalent sphere.

5.1.1 Equivalent Sphere

A sphere with radius a that is large compared to the radar wavelength will intercept power contained in an area πa^2 of the incident wave, and will scatter this power uniformly over 4π steradians of solid angle. The RCS of the sphere, using the preceding definition, is therefore equal to its projected area πa^2 (if we neglect the forward-scattering RCS, to be discussed later). That portion of the surface which actually returns power to a monostatic radar is located close to the point where the wave first strikes the sphere, where the surface lies parallel to the wavefront. A physical interpretation of the RCS, for sufficiently large objects, is then: the RCS of a given target is the projected area of a conducting sphere which would produce the same signal as the target, if placed in the same position relative to the radar. For small targets, the equivalent sphere has a radius so small that a resonance phenomenon sets in, and this physical interpretation becomes inadequate.

The RCS of a sphere is exactly one-fourth the total surface area, and it can be shown that the *average* cross section of any large object which consists of continuous, curved surfaces will be one-fourth of its total surface area. This relationship can be used in estimating the RCS of elongated or irregular bodies, provided that the irregularities are not in the form of sharp edges or structures which are resonant at the radar wavelength. It should be kept in mind, however, that the average RCS may represent largely the contribution of one or more narrow lobes of large peak amplitude, leaving very low values over most aspect angles.

5.1.2 Equivalent Antenna

Another physical interpretation of RCS, which can be derived from the equivalent sphere, relates it to the *flat-plate area* A_p, which would provide an equal return signal if located at the target position and oriented normal to the radar beam for monostatic radar (or along the bisector of the two paths in bistatic radar). Since no portion of the sphere is actually flat, we shall take that portion which lies within a distance $\lambda/4\pi$ of the wavefront as it arrives at the sphere (Figure 5.1.1). This is a circular disk of radius $a_1 = \sqrt{a\lambda/2\pi}$, the area of which $A_p = \pi\, a_1^{\,2} = a\lambda/2$. If the nearly flat surface is considered as an antenna, the power intercepted will be reradiated toward the radar with gain $G = 4\pi\, A_p/\lambda^2$. The resulting RCS is given by the product of gain times interception area, or

$$\sigma = A_p G = (a\lambda/2)(4\pi a\lambda/2\lambda^2) = \pi a^2 \tag{5.1.1}$$

It can be shown that the area lying within the distance $\lambda/4\pi$ of the wavefront when it first makes contact with any smooth target will give the approximate flat-plate area A_p for that object, from which RCS can be calculated by using (5.1.1).

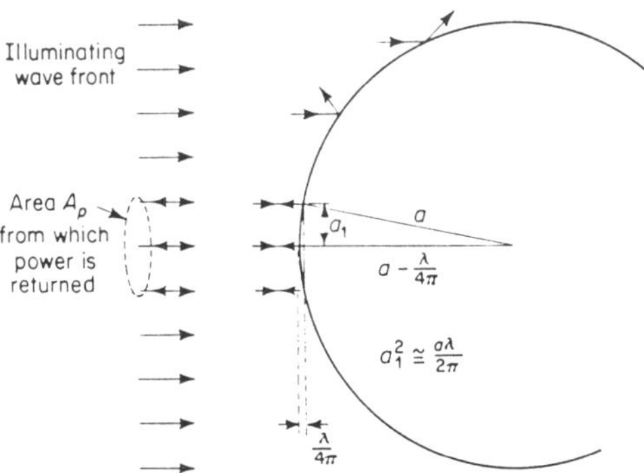

Figure 5.1.1 Effective flat-plate area of sphere [5.22].
© Artech House, reprinted by permission.

The analysis based on equivalent flat-plate area demonstrates the use of antenna theory to describe the re-radiation properties of targets. Whether the power which illuminates the surfaces of the target originates in the radar or at some other point makes no difference in the resulting radiation pattern, so long as the proper phase relationships are established. Thus, a rectangular flat plate, with area A, oriented normal to the radar beam can be equated to a uniformly illuminated aperture with constant phase, producing a radiation lobe directed at the radar. As the aspect angle changes by θ, the main RCS lobe will move in space by 2θ. Thus, this lobe and the sidelobes will be half as wide in the RCS pattern as in the antenna pattern for the same size aperture. The RCS for this normal incidence is

$$\sigma(0) = AG = 4\pi A^2/\lambda^2 \tag{5.1.2}$$

and the RCS pattern will follow the $(\sin^2 x)/x^2$ pattern of a rectangular antenna

$$\sigma(\theta) = \sigma(0) \left[\frac{\sin(kL \sin\theta)}{kL \sin\theta} \cos\theta \right]^2, \quad (k = 2\pi/\lambda) \tag{5.1.3}$$

This pattern is shown in Figure 5.1.2, with a main-lobe width $2\Delta_0 = 0.44\,\lambda/L$ between its half power points, $2\Delta_0 = \lambda/L$ between nulls, and sidelobe widths $\Delta_0 = \lambda/2L$ near broadside. For other shaped plates, the equivalent antenna patterns can be used, with the angle scale doubled for reflection.

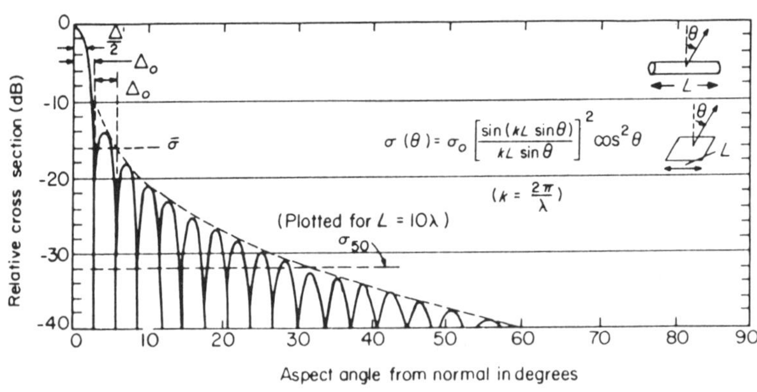

Figure 5.1.2 RCS lobe pattern of flat plate or cylinder [5.22].
© Artech House, reprinted by permission.

5.2 RCS of Simple Objects

5.2.1 Peak RCS and Number of Lobes

Equations describing the peak RCS, lobe widths, and total numbers of lobes presented by the target during a complete rotation are given in Table 5.1. Except for the *dipole*, these equations are subject to the limitation that target dimensions or radii of curvature must be larger than the wavelength. In some cases, more complex objects may be represented by combinations of these shapes, approximating the actual surface, to obtain an approximation of RCS characteristics.

A cylinder, for example, or a flat plate, may be represented by the lobe structure of Figure 5.1.2, uniformly distributed around the longitudinal axis and dependent on the length L, plus an end-lobe structure caused by the flat ends and dependent on the diameter D. In the region between side and end aspect, these two patterns will interfere with each other, producing irregular lobes which depend on the ratio of length to diameter.

5.2.2 Wavelength Dependence of RCS

The peak RCS values shown in Table 5.1 depend in different ways on the radar wavelength λ. The RCS of flat plates and corner reflectors (which appear as flat plates oriented normal to any radar aspect angle) vary as λ^{-2}. The RCS of a cylinder varies as λ^{-1}, and those of large doubly curved objects are independent of wavelength. Dipoles and pointed objects have RCS varying as λ^2. It has also been shown [5.2, pp. 178-179] that curved edges have RCS varying as λ, while discontinuities in radii of curvature produce RCS values varying as higher powers of λ.

Table 5.1   RCS of Simple Bodies

Object	σ_{max}	σ_{min}	Number of lobes	Major lobe width
Sphere	πa^2	πa^2	1	2π
Ellipsoid, $(k = a/b)$	πa^2	b^2/k^2	2	$\approx b/a$
Cylinder	$2\pi aL^2/\lambda$	null	$8L/\lambda$	λ/L
Flat plate	$4\pi A^2$	null	$8L/\lambda$	λ/L
Dipole	$0.88\,\lambda^2$	null	2	$\pi/2$
Infinite cone (half-angle α)	$\lambda^2\tan^4\alpha/16\pi$	null		
Convex surface	$\pi a_1 a_2$			
Square corner reflector	$12\pi a^4/\lambda^2$		4	$\pi/4$
Triangular corner reflector	$4\pi a^4/3\lambda^2$		4	$\pi/4$

The corner reflector, listed in the table, provides an interesting illustration of how RCS *can be much larger than the physical area of a target*. The action of the three orthogonal surfaces in the reflector is such as to reflect incident energy back toward the sources, regardless of aspect angle (Figure 5.2.1). In effect, the corner reflector is a flat plate oriented always normal to the incident ray, but with varying size, depending on how much of the ray experiences the triple bounce. In certain aspects, double or single bounce may replace triple bounce, giving the sharp lobes of the figure, but over most angles there is a broad, triple-bounce lobe. At the center of this lobe, the RCS (for a triangular corner) will approximate $4a^4\lambda^2$. Corner reflectors, mounted on navigation buoys, typically have $a = 0.5$ m, and at X-band ($\lambda = 0.03$ m) will give RCS = 280 m^2. Yet, the reflector target fits within a sphere having a diameter of only one meter.

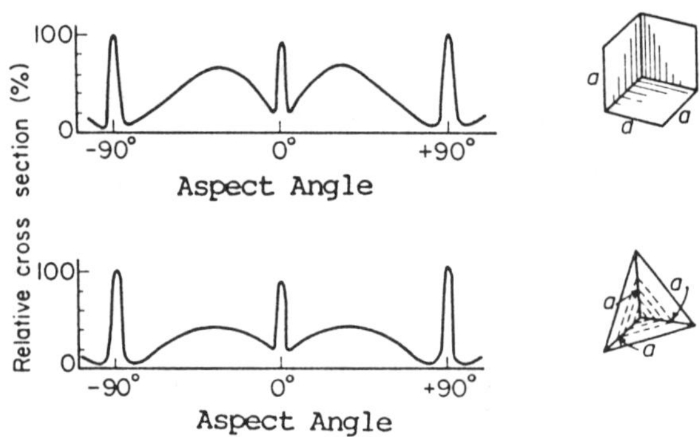

Figure 5.2.1 RCS patterns of square and triangular corner reflectors.

At the other extreme is the typical missile nose cone, the diameter of which may also be 1 m, length 2 m, and cone angle about 30°. The tip RCS, corresponding to the infinite cone in Table 5.1, is about 10^{-6} m^2 at X-band. With a rounded base, having discontinuities in curvature as it merges with the cone, RCS near $0.1a\lambda$ may appear (0.0015 m^2 at X-band), while with a flat base, much higher values will be seen, even near nose aspect. Only at aspects near normal to the conical surface, and to the rear, will there be RCS close to or above the physical area of the target (Figure 5.2.2).

Resonant phenomena also introduce *wavelength dependence*, even in the sphere which we have previously considered as having constant RCS. Figure 5.2.3 shows the RCS dependence of the conducting sphere on the ratio of circumference to wavelength. Three regions can be identified:

(a) *Optical region* (large spheres, $2\pi a/\lambda > 10$), where the RCS closely approaches the projected area;

(b) *Resonant region* ($0.5 < 2\pi a/\lambda < 10$), where large oscillations in RCS occur, reaching a peak of $3.7\pi a^2$ when $2\pi a = \lambda$; and

(c) *Rayleigh region* (small spheres, $2\pi a/\lambda < 0.5$), where the RCS drops rapidly.

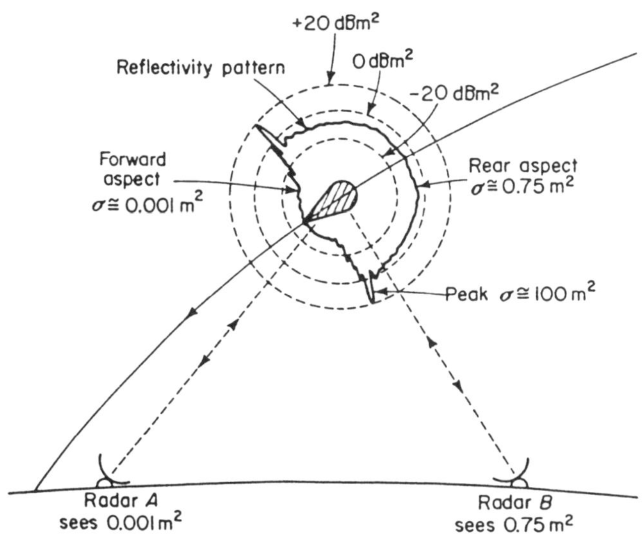

Figure 5.2.2 RCS pattern of typical reentry vehicle [5.22].
© Artech House, reprinted by permission.

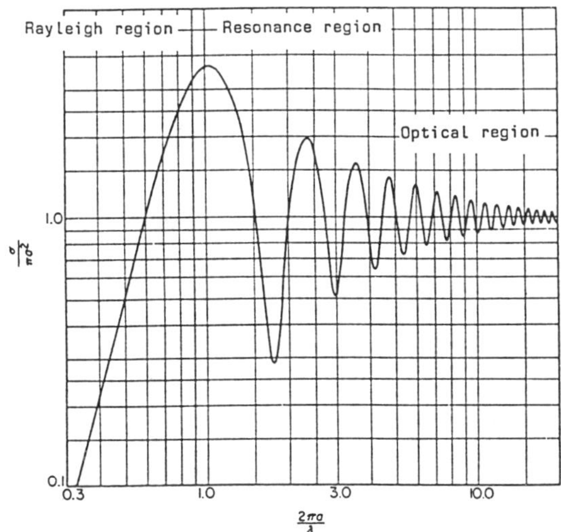

Figure 5.2.3 Normalized RCS of conducting sphere [5.22].
© Artech House, reprinted by permission.

Because of the predictability of their RCS, metallic spheres are often used in radar calibration. A typical calibration sphere, designed to be carried aloft on a small weather balloon, has a radius of 7.6 cm and an optical RCS of 0.018 m^2, or -17.4 dBm^2. At C-band, λ = 5.3 cm, the ratio $2\pi a/\lambda$ = 9, and the RCS will be within about 15% (or 0.7 dB) of its optical value. For accurate calibration, the exact radar wavelength would have to be known, and Figure 5.2.3 used to correct the optical RCS at C-band or longer wavelengths. At X-band and shorter wavelengths no correction would be required.

As an example of RCS of a small sphere, we can consider a test target sometimes used with sensitive CW radars. An air rifle pellet (a = 2.2 mm) fired near the radar will have enough velocity to pass the clutter rejection filter. At X-band, λ = 0.03 m, its RCS, will be 6×10^{-6} m^2. An individual raindrop will have an even smaller RCS value, in most cases, but huge numbers of them will lie within one radar resolution cell.

5.3 RCS of Complex Targets

In the case of simple objects, we have seen that the RCS has no single relationship to the physical area of the target. Most targets of interest present an even more complicated relationship, so that attempts to write equations based on physical dimensions are seldom successful. Prediction techniques based on elaborate computer programs can lead to useful results when theories of physical optics and the geometrical and physical theories of diffraction are applied [5.2, Chapter 5]. In most cases, however, full scale or small models of the actual targets are measured [5.3] to accurately determine RCS, or to validate the results of prediction programs.

A small aircraft target (L = 15 m), viewed by X-band radar (λ = 0.03 m), will have $8L/\lambda$ = 4000 lobes in a greater circle out passing through the nose and tail. Over all solid angles, there will be roughly $8(L/\lambda)^2 = 2 \times 10^6$ lobes, and a measurement program would have to take at least four data points per lobe for an accurate description of the pattern. Such measurements would have to be made at many frequencies, and with the polarizations of interest, leading to a huge data base, for only a single radar band. Even if measured and stored on tape, such a data base would be of limited value to most of those involved in radar performance analysis, unless it could be summarized in statistical form. This has been done on many targets of interest, leading to simple characterizations, such as mean RCS and probability density function (pdf), over different sectors of aspect angle. Tables of typical aircraft and ship cross sections are presented in Appendix D.

5.3.1 Swerling Target Models

In his classic paper on target modeling [5.4], Peter Swerling established four statistical models, distinguished by their probability density functions (pdf).

Cases 1 and 2: Exponential pdf of RCS, Rayleigh pdf of signal amplitude.

$$dP_\sigma = (1/\overline{\sigma})\exp(-\sigma/\overline{\sigma})\,d\sigma \tag{5.3.1}$$

$$dP_e = (E_s/S)\exp(-E_s^2/2S)\,dE_s \tag{5.3.2}$$

where $S = \overline{E_s^2}$ is the average signal power.

Cases 3 and 4: Chi-square, four-degree-of-freedom pdf of signal amplitude.

$$dP_\sigma = (4\sigma/\overline{\sigma}^2)\exp(-2\sigma/\overline{\sigma})\,dE \tag{5.3.3}$$

Cases 1 and 3 refer to slow fluctuation, in which the signal sample remains constant over integration of n pulses during observation time t_o, while Cases 2 and 4 have independent signal samples on each of n pulses. The detection performance corresponding to these four cases was discussed in Chapter 4. In that chapter, the more generalized chi-square distribution with $2K$ degrees of freedom was introduced. The Case 1 or Case 2 target has $K = 1$, while Case 3 or Case 4 has $K = 2$.

Physically, the Case 1 or Case 2 (Rayleigh) model corresponds to a target in which many scattering points are added with random phases. The amplitude pdf has the same form as that of the random noise envelope. When plotted in terms of the decibel value of RCS, the pdf of Figure 5.3.1 results. Values more than 10 dB above the mean are seldom seen, while fades of 10 dB or more below the mean occur with about 10% probability. *Virtually all targets in the real world are very close to this model,* since many scattering sources are involved. Figure 5.3.2 shows measured data [5.5] on two types of aircraft at L-, S-, and X-band.

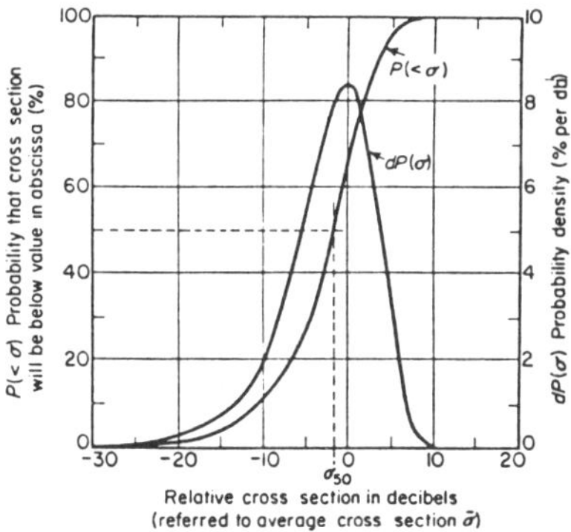

Figure 5.3.1 Log plot of Rayleigh distribution (Swerling Cases 1 and 2) [5.22]. © Artech House, reprinted by permission.

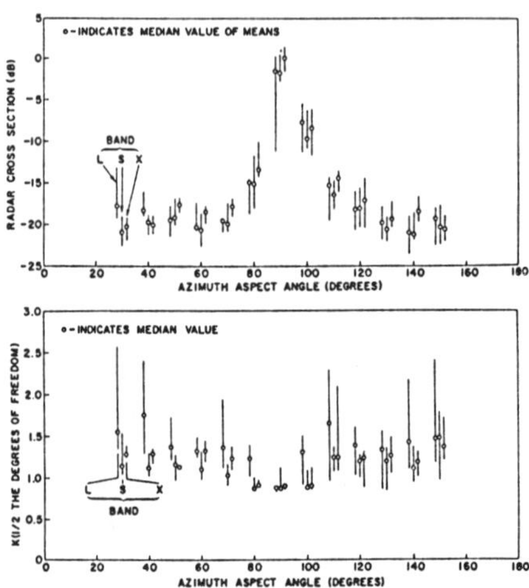

Figure 5.3.2(a) Mean RCS and K for jet fighter in three radar bands [5.5].

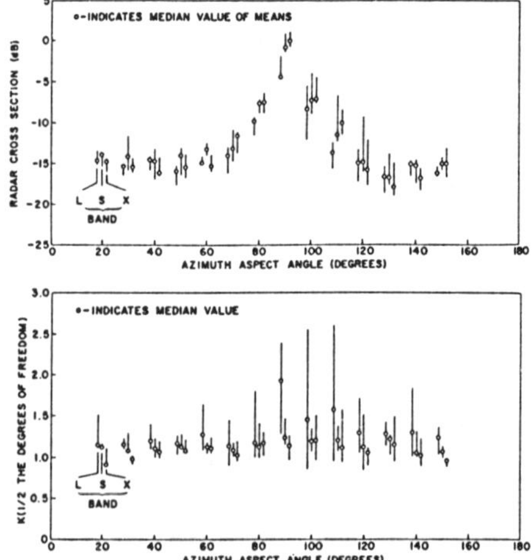

Figure 5.3.2(b)   Mean RCS and K for propeller aircraft
in three radar bands [5.5].

5.3.2 Target Spectrum and Correlation Time

The *rate at which RCS changes* is important in both detection and measurement applications. This rate, and the corresponding target correlation time t_c, will
determine radar performance. When signals are integrated over an observation
time t_o, the resulting output pdf will depend on whether only one sample ($t_c \gg t_o$)
or several independent samples ($t_c < t_o$) have been observed, and fading of the

output will be most pronounced with only a single sample (Case 1). When a cumulative probability of detection is to be determined, the usual assumption is that the successive trials at intervals t_s are independent ($t_c < t_s$). In sequentially sampled measurement processes, the best performance is obtained if the target remains constant during the measurement ($t_c \gg t_o$).

The *fluctuation rate* of the target is the rate at which RCS lobes pass across the line of sight to the radar. For a target having scatterers that are uniformly distributed over the span L_x across the line of sight, the spectrum of the received signal will be rectangular with a width

$$f_{max} = 2\omega_a L_x / \lambda \tag{5.3.4}$$

where ω_a is the rate of change of aspect angle. The correlation time can be expressed as

$$t_c = 1/f_{max} = \lambda/2\omega_a L_x \tag{5.3.5}$$

5.4 Spatial Distribution of Cross Section

Radar measurement of target position and velocity is based on the assumption that some point on the target may be defined as the position reference. Targets which are small with respect to the radar wavelength present no problem in this regard, since they are seen as small point sources in all coordinates. Larger targets, however, show significant shifts in all coordinates, including doppler frequency, as a function of changing aspect angles. As with target amplitude or RCS, these shifting apparent positions are best modeled statistically [5.8].

5.4.1 Target Glint

Glint is defined [5.1, p. 391] as

> the inherent random component of error in measurement of position or doppler frequency of a complex target due to interference of the reflections from different elements of the target.

Although the interference among signal components which leads to glint will also cause amplitude fluctuations and a related *scintillation error* in some tracking systems, the two effects are quite different and must be considered independently of glint.

It is sometimes erroneously assumed that glint merely represents the wandering of the dominant reflection (or radar center of gravity) over the extent of the target, but in fact the glint may lead to a radar pointing angle far beyond the physical span of the target.

An equivalent glint standard deviation for uniform scatters over a target span L is given by

$$\sigma_g = L/3 \tag{5.4.1}$$

where L is either cross-range width L_α, for angle glint, or radial length L_r, for range measurement. For most aircraft, if L is the wingspan or fuselage length, σ_g will be $L/4$ to $L/6$.

It is convenient to illustrate glint effects by considering the two-element target, in which two spheres or other point sources are rigidly connected with spacing L, and are rotated about the line of sight. Let the signal amplitude of source 1 be unity, and that of source 2 be k. If the axis connecting the sources makes an angle α with the line of sight, the received signal voltage will be

$$E_s = \sqrt{1 + k^2 + 2k \cos \phi} \tag{5.4.2}$$

where $\phi = (4\pi L/\lambda) \sin \alpha$ is the phase angle between the two signal components. The **glint** error with respect to the center point of the target pair, and normalized to the projected target span $L \cos \alpha$ can be written

$$\sigma = (1/2) \frac{1 - k^2}{1 + k^2 + 2k \cos \phi} \tag{5.4.3}$$

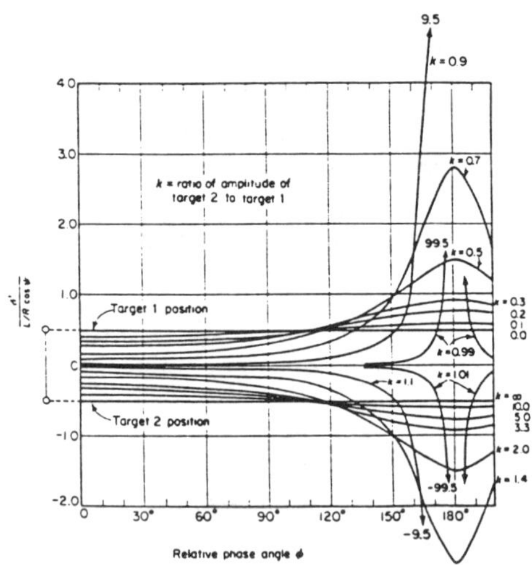

Figure 5.4.1 Glint for the two-element target [5.22].
© Artech House, reprinted by permission.

This is plotted in Figure 5.4.1 as a function of phase angle ϕ for different values of k. The extreme values of glint can be found for $\phi = 0°$ and $180°$. For $k \to 1$, the tracking point lies at the center of the pair, $\delta \approx 0$, for all phase angles except $\phi = (1 + k)/2(1 - k)$. For small k, the tracking point will oscillate sinusoidally about the position of the stronger source, $\delta \approx -k \cos \phi$. *An important result is that the average tracking point will lie at the stronger source,* with deviations occurring in both directions from this source. As the second target becomes the stronger, the average tracking point shifts abruptly to it, and the interference from the first target causes deviations on both sides of the second.

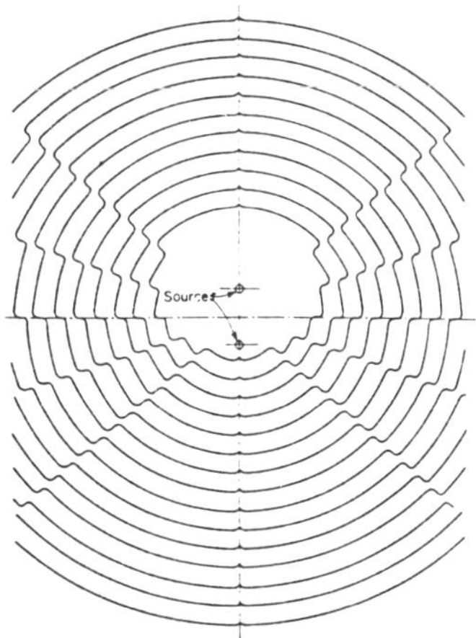

Figure 5.4.2 Phase front radiated by the two-element target.
(Courtesy of D. D. Howard, Naval Research Laboratory)

5.4.2 Angle, Range, and Doppler Glint

In angle measurement or tracking, the glint error δ will be normalized to the angle L_x/R subtended by the target span at range R. The deviations described in the previous section can be attributed to the ripples in the phase fronts arriving at the radar antenna [5.9], as illustrated in Figure 5.4.2. Abrupt discontinuities in these phase fronts, and corresponding peaks of glint, will occur when the target signal goes through deep fades, between lobes in the RCS pattern. The basic rms glint error in radians will then be $\sigma_{\theta_g} \approx L_x/3R$. This value, of course, will be modified by the processing steps and dynamic response of the tracker, to the extent that the circuits may not follow the sharp peaks of glint when the target fades. The value also depends, to some extent, on the beamwidth λ/D of the radar antenna, relative to the subtended angle L_x/R of the target. As the target extent approaches dimensions of the radar beamwidth, the edges of the target will no longer receive full illumination or weighting by the antenna pattern, and their contributions to glint will be reduced. The condition for target extent approaching beamwidth is equivalent to the antenna width approaching that of the RCS lobe arriving at the radar, providing smoothing of the phase front over the antenna. This effect also applies to doppler glint. The condition for this glint reduction also corresponds to peak errors approaching the beamwidth, if the antenna is permitted to follow the error signal, and loss of target may result in spite of the reduced value of normalized glint.

In range, the glint error has the same form, with peaks also occurring during signal fades. The rms glint error will be $\sigma_{rg} \approx L_r/3$, where L_r is measured radially, along the radar beam. When the range resolution cell of the radar waveform and signal processing approaches the radial length of the target, glint noise error may be reduced in a way similar to the beamwidth effect in angle. In doppler, the effect is the same, but the spectral span of the target is $f_{max} = 2\omega_a L_x/\lambda$, as given in (5.3.4). The rms glint error is $\sigma_{fg} \approx 2\omega_a L_x/3\lambda$.

5.5 Radar Clutter

Radar clutter was defined in Section 2.2.3 as unwanted echoes, typically from the ground, sea, rain or other precipitation, chaff, birds, insects, and aurora [5.1, p. 151]. In some cases, land vehicles may be added to this list. This is the definition of clutter for a typical air surveillance radar, but the distinction between clutter and target in a radar system depends on the purpose of the radar. Clutter may be divided into sources distributed over a surface (land or sea), within a volume (weather or chaff), or concentrated at discrete points (structures, birds, or vehicles). The methods of analyzing the radar response to these three types of clutter will differ.

Whatever the definition for a particular radar application, a number of parameters must be specified if the clutter is to be included in the analysis of radar performance. These are:
- Clutter RCS or density of RCS (reflectivity) over the surface or volume;
- Numbers of discrete sources;
- Spatial extent and distribution of sources;
- Velocity extent and distribution (spectrum);
- Wavelength dependence of RCS;
- Amplitude distribution (pdf);
- Spatial correlation of amplitudes; and
- Polarization properties.

The propagation factor F_c along the path from radar to clutter will play an important role in establishing the clutter power received by the radar. In many cases, the factor F_c^4 is embedded in the measurement of RCS or its density, and must be removed if the paths to be considered differ from those in the measurement program.

To include distributed clutter in a radar analysis, the volume V_c or surface area A_c within the radar resolution cell is found (Section 2.2.3) and multiplied by the reflectivity (or RCS density) η_v or σ^0 to obtain the clutter RCS in each cell. For discrete sources, the RCS is taken directly from a model. The pattern-propagation factor F_c^4 for the clutter is then included in the radar equation to find clutter power or C/S ratio. The radar antenna pattern and resolution considerations will be discussed in Chapters 6 and 7, and here we shall be concerned with the reflectivity, the clutter propagation factor, and the other parameters listed above.

5.5.1 Surface Clutter

In Section 2.2.3 surface reflectivity was: described by the *constant-γ* model;

$$\sigma^0 = \gamma \sin \psi \qquad (5.5.1)$$

where ψ is the grazing angle at the surface and γ is a parameter describing the scattering effectiveness of the surface. This model is in excellent agreement with most measurements, at angles high enough that $F_c \approx 1$, but not too close to 90°. At low grazing angles, the measurements fall below the model because of the propagation factor. Near zenith, the measurements rise because of quasispecular reflections from the surface facets.

The propagation factor for paths between the radar antenna at height h_r and the clutter sources on the undulating, random surface of the earth can be written, for low grazing angles, as [5.10, p. 219]:

$$F_c \approx \psi/\psi_c = \psi(4\pi\sigma_h/\lambda) = R_1/R \qquad (5.5.2)$$

where σ_h is the rms surface height deviation from the average height, R is range, and R_1 is the range at which ψ_c is reached. Thus, below the critical grazing angle $\psi_c = \lambda/4\,\pi\sigma_h$, or beyond range R_1, the measured value $\sigma^0 F_c^4$ should vary as ψ^5 rather than ψ (using the small-angle assumption, $\sin \psi \approx \psi$). This relationship may be modified by diffraction for large values of σ_h.

Sea Clutter

When the preceding model is applied to sea clutter, averaging all wind directions, it was found (Section 2.2.3) that γ depends on the Beaufort wind scale K_B and radar wavelength, such that

$$10 \log \gamma = 6K_B - 10 \log \lambda - 64 \qquad (5.5.3)$$

where wavelength is in meters. This relationship is plotted as solid lines in Figure 2.2.8, for a medium sea condition (Beaufort scale 4 to 5 wind). The rms surface deviation σ_h for this wind state depends on the time and space over which the wind has acted on the sea, but for steady-state conditions we can make the approximations:

$$\sigma_h \approx K_B^3/300 \qquad (5.5.4)$$

$$\psi_c \approx 24\lambda/K_B^3 \qquad (5.5.5)$$

$$\beta_0 \approx 0.04 \quad \text{radians}$$

The reason for the near constant value of rms slope is that the wavelength and wave height both increase with wind speed.

The range extent of sea clutter is determined primarily by the radar antenna height, but also by atmospheric conditions, which can cause *supperrefraction* and *ducting*. The horizon range can be expressed as a function of height h_r and effective earth's radius ka.

$$R_h = \sqrt{2kah_r} \approx 4130\sqrt{h_r} \qquad (5.5.6)$$

where $a = 6.5 \times 10^6$ m and the approximation applies to $k = 4/3$, with h_r and R_h given in meters. Even before the horizon is reached, the declining value of $\sigma^0 F_c^4$ will usually bring the clutter level below the noise level.

The velocity spectrum of sea clutter [5.11, pp. 243-247] with a wind velocity v_w will be approximately Gaussian, with a standard deviation $\sigma_v \approx v_w/8$ and a mean (downwind) between $v_w/8$ and $v_w/4$. After envelope detection, the standard deviation will be greater by $\sqrt{2}$. The amplitude distribution will be approximately Rayleigh for grazing angles above ψ_c, and also at low angles for large resolution cells, in which many wave tops contribute to the clutter power. As resolution increases, the distribution departs from Rayleigh, corresponding to large regions in which clutter is very low, punctuated by sharp peaks at the larger wave tops.

Land Clutter

The reflectivity of land clutter is much more difficult to characterize than that of the sea. The constant γ model is still followed at the higher grazing angles, with values of γ between - 10 and - 15 dB widely applicable to land covered by crops, bushes, and trees. Desert, grassland, and marshy terrain is more likely to have γ near - 20 dB, while urban or mountainous regions may have $\gamma \rightarrow$ - 5 dB. These values are almost independent of wavelength and polarization, and apply to modeling of the mean clutter reflectivity.

At low grazing angles, applying to ground-based radars, propagation considerations become dominant. One study [5.13] showed that practically all of the variations in measured reflectivity of vegetation-covered terrain could be attributed to the propagation factor, so that use of a constant $\sigma^0 = $ - 30 dB could closely reproduce the measured clutter mean and pdf, at least in the microwave region. In fact, the definition or calculation of a grazing angle becomes quite indefinite for a ground-based radar looking out over typical terrain. Another difficulty is that large regions of the surface tend to be shadowed completely, and the measured data are necessarily restricted to regions in which clutter is visible to the radar. The true mean reflectivity is then the measured mean multiplied by the fraction of terrain over which the data have been measured, often less than 10%.

A simple modeling approach [5.14] is to assume a homogeneous surface with given σ_h and $\gamma \approx$ - 14 dB, and to assume the radar antenna to be located on a local $2\sigma_h$ point in the terrain, so that its height above the mean is $h_r' = h_r + 2\sigma_h$. A single curve for σ^0 then results, which may be modified for very flat and for mountainous terrain, and for the propagation factor at different wavelengths (see Figure 2.2.7). In range, there are three distinct clutter regions, depending on the propagation mode. For $R < R_1$, $F_c \approx 1$. Between R_1 and R_h, reflection-interference phenomena dominate the path, and F_c is given by Eq. (5.5.2). Beyond the horizon, a choice is made between smooth-sphere and knife-edge diffraction (see Chapter 8), either one leading to a rapid reduction in F_c with range. The effective reflectivity as a function of range, for different antenna heights, may then be plotted as in Figure 5.5.1. The horizon range is computed from Eq. (5.5.6), and propagation beyond the horizon is computed by using smooth-sphere or knife-edge diffraction approximations. This captures the average effects of propagation, as well as the average reduction in grazing angle with increased range. Observed variation in land clutter reflectivity with wavelength is explained primarily by the propagation factor.

The spatial extent of land clutter depends not only on antenna height, siting, and propagation conditions, but on terrain roughness. Mountains rising to 5000 m altitude will normally be visible at ranges near 300 km, but, with atmospheric ducting, can appear at 500 to 600 km range. Clutter from normal, rolling terrain ($\sigma_h = 15$ m) will be visible some 15 km beyond the radar's horizon range, but will be extended to hundreds of km under ducting conditions. The frequency spectrum of land clutter is narrow, having an approximately Gaussian shape with $\sigma_v < 0.5$ m/s for vegetation in windy conditions.

A major analytical problem with land clutter is the determination of its amplitude distribution. Knowledge of the mean clutter is not adequate for prediction of its effect on the radar, unless the pdf is known.

5.5.2 Volume Clutter

Rain and Snow

The reflectivity of volume clutter is described by a factor η_v, expressed in m^2/m^3. This factor can be expressed, for rain at $\lambda > 0.02$ m, as

$$\eta_v = 5.7 \times 10^{-14} r^{1.6}/\lambda^4 \tag{5.5.7}$$

where r is the rainfall rate in mm/h. For dry snow at these wavelengths,

$$\eta_v = 1.2 \times 10^{-13} r^2/\lambda^4 \tag{5.5.8}$$

where r is the rate in terms of water content of the snow. For shorter wavelengths [5.16], the scattering properties depart from the pure Rayleigh RCS prediction, as shown previously in Figure 2.2.9. This reflectivity is for linearly polarized radar, or circularly polarized systems using opposite sense for transmission and reception. For same-sense circular polarization, cancellation up to about 20 dB can be expected.

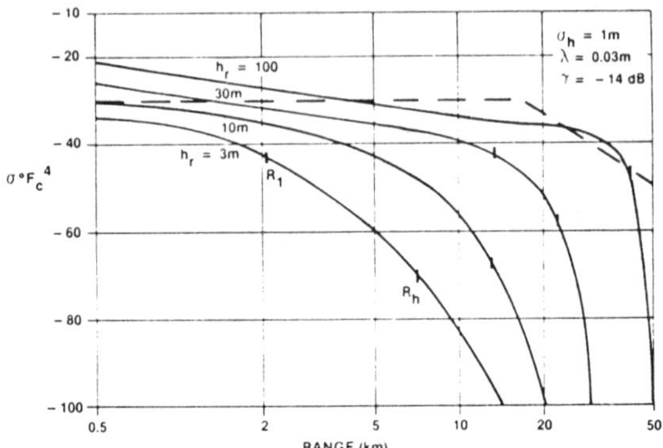

(a) X-band clutter reflectivity *versus* range, for different radar heights over level terrain.

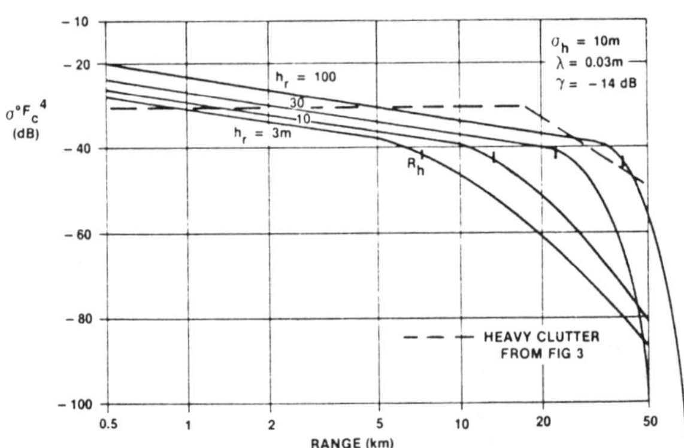

(b) X-band clutter reflectivity *versus* range, for different radar antenna heights over rolling or hilly terrain.

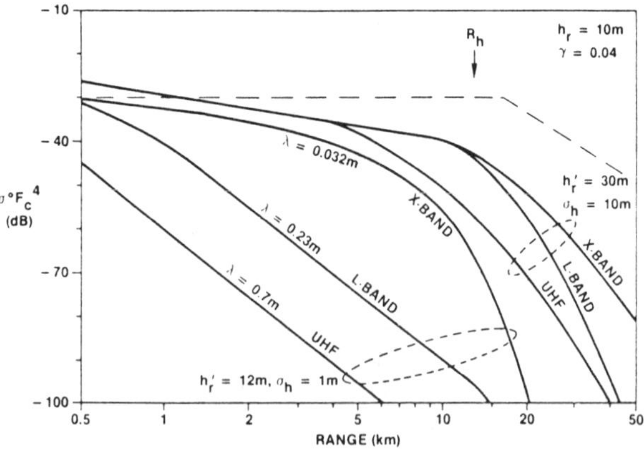

(c) Comparison for different bands over rough and smooth terrain.

Figure 5.5.1 Predicted average land reflectivity vs. grazing angle for different values of antenna height and surface roughness.

Chaff

The RCS of an individual dipole of chaff is approximately $0.88\lambda^2$, when viewed broadside. When averaged over all aspect angles, this drops to $0.15\lambda^2$. Early data on chaff gave a total RCS, as a function of its weight W in kg,

$$\sigma = 660 W_{kg}/f_{GHz} = 22,000\lambda W \,(\text{m}^2) \tag{5.5.9}$$

(with λ in m), for chaff dipoles cut to resonance in a specific band. Recent data [5.15, p. 187] indicate that more modern chaff can achieve this RCS over a two-octave band, with λ in Eq. (5.5.9) taken as the geometric mean value (Figure 5.5.2). For $\lambda > 0.3$ m, resonant dipoles usually give way to long streamers called rope, achieving somewhat lower RCS for its weight, but being less subject to breakage. Chaff normally falls with a motion which randomizes the orientation of individual dipoles, and hence is not very sensitive to the radar polarization.

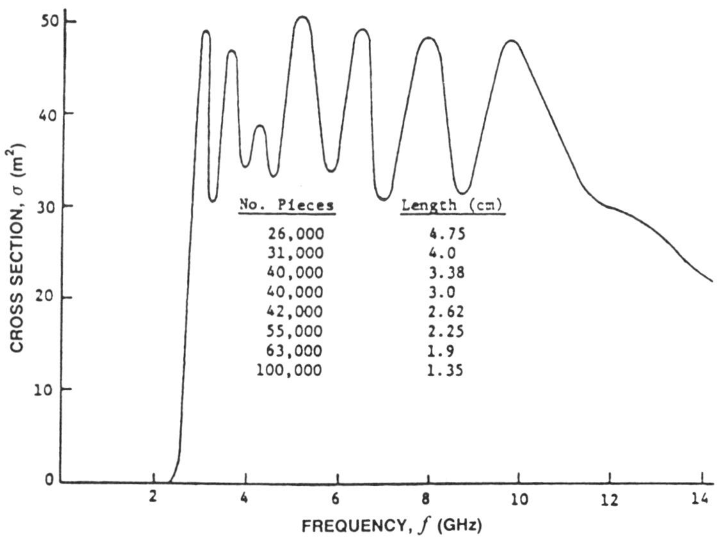

Figure 5.5.2 Radar cross section of a chaff package [5.15].

The density of a chaff cloud is thus entirely dependent on the weight of the chaff dispensed into a given volume. For example, in order to achieve the same reflectivity as 3 mm/h rain at X-band $\eta_v = 3 \times 10^{-7}$ m²/m³, or 300 m²/km³, chaff must be dispensed at a rate of 0.45 kg/km³. With uniform distribution, there would be one dipole per 17 m cube, and an average mass density of 450 g/km³. This compares with 3 mm/h rain, which typically has 2×10^8 g/km³.

Spatial Extent of Volume Clutter

The spatial extent of rain is limited by its top altitude, usually 3 to 7 km, but extending up to 10 km in thunderstorms. Widespread rain, below 5 km altitude, can extend to the horizon in all directions; about 300 km range. Heavier rainfall will be more limited horizontally, but may extend to greater altitudes. Chaff is limited by the operational altitude of the dispensing aircraft, normally to about 12 km (for which the horizon range is 450 km). The horizontal extent of chaff depends on the routes flown by the aircraft, the time since dispensing, and winds.

Velocity Spectrum of Volume Clutter

The velocity spectra of rain and chaff translate to a doppler spread that may be detected by coherent radars, and are dependent on the wind conditions, and in particular on the *wind shear* in altitude. For volume clutter filling the elevation beam, and subject to a wind shear k_{sh} in m/s per km altitude, Nathanson [5.11, p. 207] gives the standard deviation of the velocity spectrum as

$$\sigma_v = 0.3 k_{sh} R \theta_e \cos \alpha \qquad (5.5.10)$$

where R is in km, α is the azimuth angle between the beam and the wind direction, and the elevation beamwidth θ_e is in radians (the constant in this equation has been adjusted for our use of the one-way 3 dB beamwidth). Values of k_{sh} between 2 and 4 m/s per km are common, with a total velocity change up to 40 m/s between the surface and the highest altitude clutter. The mean clutter velocity, with constant wind shear, will be equal to the wind radial velocity at the center of the beam.

The velocity spectrum of volume clutter competing with a particular target doppler frequency will depend not only on the wind and beamwidth, but also on the presence of range ambiguities in the radar response.

The vertical velocity component of rain, near the surface, reaches several m/s, but is not usually a significant contributor to the spectral spread, even for beams at high elevation. Chaff has a very small vertical velocity, given in [5.15, p. 183] as near 0.3 m/s at sea level. This implies a rate near 0.6 m/s at 12 km altitude, and a fall time of eight hours to the surface from that altitude.

Amplitude Distribution of Volume Clutter

The amplitude distribution of volume clutter is normally assumed to be Rayleigh. However, both rain and chaff have local variations, leading to peaks well above those which would be predicted from the Rayleigh distribution alone.

5.5.3 Discrete Clutter Sources

Of the several types of *discrete clutter*, buildings (including towers and water tanks) present the largest RCS, and birds are the most extensive in space and velocity. Table 5.2 shows several different models of discrete land clutter, along with a recommended model for general use. These large, rigid sources can be regarded as having very small spatial extent (typically only one or a few meters) and spectra of essentially zero width. *They establish the dynamic range at the input to the receivers of most systems,* and place special requirements on methods of controlling false alarms.

Birds, although much smaller in RCS, present serious problems for radars which must detect small targets or targets at low altitude. The bird RCS for mid-microwave bands can be represented by a log-normal pdf, with a median of - 30 dBm^2 and a standard deviation of 6 dB. This implies that only 0.13% of the bird population will exceed - 12 dBm^2. Variation with radar wavelength and type of bird is discussed in [5.21]. Migrating birds will have an exponential distribution in altitude, with a scale height near 700 m. Most local birds will be at lower altitudes. The most damaging feature of the bird population, for modern doppler radar systems, is its velocity distribution. The birds will move at the wind velocity, ± 15 m/s air speed. Bird RCS is not sensitive to polarization, nor are the larger bird RCS values sensitive to wavelength for $\lambda < 0.3$ m. Small birds will become resonant near that wavelength, and will be in the Rayleigh region at longer wavelengths. Large birds will resonate in the UHF bands, but all will be in the Rayleigh at VHF and below.

Table 5.2 Models of Discrete Point Clutter

			Mitre		
REFERENCE MODEL	[5.17]	[5.18]	[5.19]	[5.20]	
PARAMETER	RRE	WARD	RURAL	METRO	SUGGESTED
Density (per km^2): $\sigma_c = 10^2 m^2$	3.5	1.8			2.0
$= 10^3$	0.8	.36	.02	2.0	0.5
$= 10^4$	0.15	.18	.002	.2	0.2
Resulting $\overline{\sigma}^0$ dB	-26.0	-26.0	-44.0	-24.0	-26.0
Number of points per 1° beam, 0-5 km: $\sigma_c = 10^2$ m^2	.7	.35			0.4
$= 10^3$	.16	.07	.004	.4	0.1
$= 10^4$	.03	0.35	.0004	.04	0.04

5.6 Jamming

5.6.1 Noise Jamming

The most common form of jamming, noise jamming, was described in Section 2.2.4 in terms of effective radiated power (ERP) and bandwidth.

Typical ECM transmitters operate with average powers between 100 W and a few kW. The ERP is increased by using directional antennas, to the extent this is possible on the type of vehicle employed. For example, the SOJ can use fairly large antennas (e.g., $D = 0.5$ m), if the beamwidths are consistent with the coverage required. For simultaneous jamming of several radars, the azimuth beamwidth must be on the order of $60°$, and this requirement combined with the limited vertical aperture size normally limits the antenna gain to from 10 to 15 dB. The corresponding ERP levels for SOJs are typically 30 to 100 kW. For the SSJ, antenna gains nearer 5 dB are used, resulting in ERP levels of from 1 to 10 kW. These values are usually taken as the effective values after reduction for noise quality.

A very effective version is the continuous noise jammer which radiates random noise with bandwidth B_j that is wider than the radar receiver bandwidth B_n. The effect is to raise the total spectral density of the background noise in the receiver from N_0 to $N_0 + J_0$, where the received jamming spectral density is:

$$J_0 = \frac{P_j G_j G_r \lambda^2 F_j^2}{(4\pi)^2 B_j R_j^2 L_{aj}} \tag{5.6.1}$$

Here, $P_j G_j$ represents the ERP of the jammer, $P_j G_j / B_j$ is the effective radiated power density at the frequency to which the radar is tuned, F_j^2 is the one-way pattern-propagation factor from the jammer into the radar antenna, R_j is the range to the jammer, and L_{aj} is the one-way atmospheric attenuation.

For the case of a stand-off jammer, which maintains a constant range from the radar, it is convenient to consider the jammer as one more component of noise temperature to be added to T_s at the receiver input:

$$T_j = \frac{P_j G_j G_r \lambda^2 F_j^2}{(4\pi)^2 k B_j R_j^2 L_{aj}} \tag{5.6.2}$$

If the radar range equations (3.1.21) to (3.1.23) or the Blake chart is modified to replace T_s with $(T_s + T_j)$, the target detection range of the radar can be calculated directly. In many cases, if the main lobe of the radar antenna looks directly at the jammer, $F_j^2 \approx 1$, the resulting T_j is so great that the calculated radar range is negligible. However, in search sectors where only antenna sidelobes see the jammer, $F_j^2 \ll 1$, T_j may be comparable to T_s and useful detection range is still obtained. For two or more jammers, individual values T_{ji} may be calculated, and the total input temperature will be the sum $T_s + T_{ji} + ... + T_{jn}$. All receiver processing considerations remain the same for this jamming as for receiver noise, so long as B_j covers the bandwidth B_n of the radar waveform.

Several noise jammer examples are given in Appendix E.

5.6.2 Deception Jamming

Section 2.2.4 introduced the concept of the deceptive jammer. In some cases, rather than emitting continuous noise to force an increase in the radar detection threshold, the deceptive jammer emits signals which simulate the radar transmission, producing false echoes in the radar receiver. These false echoes may be generated at range delays and with doppler shifts other than those of the real target. If the jamming is sufficiently strong to penetrate the sidelobes of the radar receiving antennas, i.e., has a high J/S, the false echoes may appear when the antenna is looking at angles far removed from that of the jamming source. The received energy for each pulse of this type of jamming is given by Eq. (5.6.1) with $B_j = 1/\tau_j$, where τ_j is the radar pulsewidth replicated by the jammer. After processing in the receiver, the false echo outputs will have a J/N power ratio equal to $J_0/N_0 L_m$, and are otherwise indistinguishable from a target echo, except for imperfections in the replicated waveform and possible angle-of-arrival information implicit in the pattern-propagation factor F_j^2. Methods of recognizing and rejecting such jamming pulses will be discussed in Chapter 12.

5.7 References

[5.1]   IEEE Standard Dictionary of Electrical and Electronic Terms, *ANSI/IEE Std. 100-1984, 1984.*

[5.2]   E. F. Knott, J. F. Shaeffer, and M. T. Tuley, *Radar Cross Section*, Artech House, 1985.

[5.3]   N. C. Currie (ed.), *Techniques of Radar Reflectivity Measurement*, Artech House, 1984.

[5.4]   P. Swerling, Probability of detection for fluctuating targets, *RAND Corp. Research Memo. RM-1217*, March 17, 1954; reprinted in *IRE Trans.* **IT-6**, No. 2, April 1960.

[5.5]   J. D. Wilson, "Probability of detecting aircraft targets," *IEEE Trans.* **AES-8**, No. 6, November 1982, pp. 757-761.

[5.6]   W. Weinstock, Target cross section models for radar systems analysis, doctoral dissertation, University of Pennsylvania, Philadelphia, 1964.

[5.7]   D. K. Barton and H. R. Ward, *Handbook of Radar Measurement,* Prentice-Hall, 1969; Artech House, 1984.

[5.8] R. V. Ostrovityanov and F. A. Basalov, *Statistical Theory of Extended Radar Targets,* Soviet Radio Publishing House, 1982; Artech House, 1985.

[5.9] D. D. Howard, "Radar target angular scintillation in tracking and guidance based on echo signal phase-front distortion," *Proc. NEC* **15**, 1959, pp. 840-849.

[5.10] M. W. Long, *Radar Reflectivity of Land and Sea,* Artech House, 1983.

[5.11] F. E. Nathanson, *Radar Design Principles*, McGraw-Hill, 1969.

[5.12] G. V. Trunk, "Radar properties of non-Rayleigh sea clutter," *IEEE Trans.* **AES-8**, No. 2, March 1972, pp. 196-204.

[5.13] S. Ayasli, "Propagation effects on radar ground clutter," *Proc. IEEE National Radar Conference*, 1985.

[5.14] D. K. Barton, "Land clutter models for radar design and analysis," *Proc. IEEE* **73**, No. 2, February 1985, pp. 198-204.

[5.15] D. C. Schleher, *Introduction to Electronic Warfare*, Artech House, 1986.

[5.16] R. K. Crane, "Microwave scattering parameters for New England rain," *MIT Lincoln Laboratory Tech. Rep 426*, October 3, 1966.

[5.17] A. K. Edgar, E. J. Dodsworth, and M. P. Warden, "The design of a modern surveillance radar," *IEEE Conf. Pub. No. 105, Radar-73,* October 1983, pp. 3-13.

[5.18] H. R. Ward, "A model environment for search radar evaluation," *IEEE Eascon Record*, 1971, pp. 164-171.

[5.19] W. J. McEvoy, "Clutter measurements program: Operations in Western Massachusetts," *Mitre Corp. Rep. MTR-2074*, March 1972, DDC Doc. AD742297.

[5.20] W. J. McEvoy, "Clutter measurements program: Operations in the metropolitan Boston area," *Mitre Corp. Rep. MTR-2085*, March 1972, DDC Doc. AD742298.

[5.21] G. E. Pollon, "Distributions of radar angles," *IEEE Trans.* **AES-8**, No. 6, November 1972, pp. 721-727.

[5.22] D. K. Barton, *Modern Radar System Analysis*, Artech House, 1988.

Chapter 6

RADAR ANTENNAS

Of necessity, much material concerning radar antennas has already been covered in this handbook. In Chapter 1, the basic function of the radar antenna was discussed, along with beamwidth and gain relationships to antenna area and wavelength. Definitions of key antenna characteristics such as *antenna pattern, illumination,* and *reciprocity* were also given, and some of the many different ways that antennas may be classified were listed (Table 1.2). Chapter 2 included a discussion of *multipath lobing* and other propagation effects that can occur when a radar radiates near the earth's surface, and gave examples of antenna scan patterns for volume search. In Chapter 3 the concepts of *power density, effective aperture,* and *antenna loss* were explored, and a derivation of the *solid angle of search* was presented. The $\csc^2$ shaping of an antenna pattern for large volume search was also described. Monopulse antennas are discussed in Chapter 10.

From this material, we may correctly conclude that the antenna is an extremely important subsystem of the radar, in that it serves as the link or interface to the outside world. A review of radar history reveals that there have been and exist today a bewildering variety of radar antenna designs. Further, many different designs have been adopted by different users (or countries) to serve what are essentially identical applications. This point illustrates a key factor common to many fields of engineering: there is no single "correct" solution. Antenna designs that become part of operational radar systems often reflect preferences based upon the technical competence of the radar designers in related radar subsystems such as transmitters, receivers, and signal processors, and, in some cases, historical precedent weighs heavily in their decisions regarding the antenna design selected for a particular role.

Independent of these factors, however, and regardless of the specific purpose for which a radar is designed, the resultant antenna generally represents the designer's best effort to maximize the gain term in the radar *effective radiated power* expression $P_t G_t$. Previous chapters have shown that antenna gain is directly proportional to antenna area, but even here there are important limitations, and the antenna design ultimately adopted for use represents a compromise imposed by considerations of allowable size, weight, environment, cost, and complexity.

In this chapter we will examine the topic of *antenna response* more fully by giving a more complete description of the antenna pattern, including ways available to alter its shape for improved performance. In addition, we will examine several important classes of antenna designs, including *reflector* antennas, *phased arrays*, and *multi-beam* antennas. We conclude with a discussion of several typical *ultra-low sidelobe* antenna designs and their associated performance capabilities and limitations.

6.1 Antenna Fundamentals

The earliest radar antennas, and some of the most recent, use arrays of *half-wave dipole* elements (Figure 6.1.1). The pattern of a dipole is a toroid surrounding the dipole axis, and the gain in any direction normal to this axis is approximately $\pi/2$. When the dipole is mounted $\lambda/4$ in front of a *ground plane,* the rear half of the toroid is reflected into the front hemisphere, reinforcing the forward lobe, narrowing its width in the θ coordinate, and increasing its gain to $\approx \pi$.

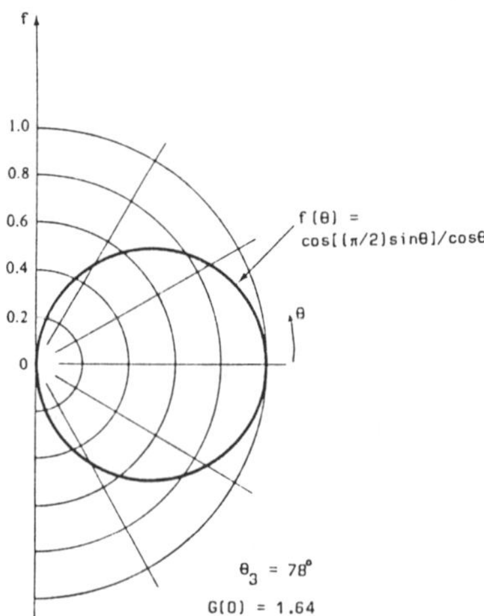

Figure 6.1.1 Radiation pattern of half-wave dipole [6.1].

It is convenient to consider any antenna as an array of $T = n_a \times n_e$ elements, each $\lambda/2$ square with gain π. Such elements can be thought of as small dipoles, spaced $\lambda/2$ apart in both directions and mounted in front of a continuous ground plane. Elements other then dipoles may be used, or the antenna may actually be a continuous reflector or lens illuminated by a horn feed, in which case the elements will merely be small areas on the surface, $\lambda/2$ square. If the elements are illuminated uniformly and in phase, the antenna gain will be

$$G_0 \equiv G(0,0) = \pi T \qquad (6.1.1)$$

where the z axis ($\theta = 0$) is normal to the ground plane. For a rectangular aperture of area $A = w \times h$, there will be $n_a = 2w/\lambda$ columns and $n_e = 2h/\lambda$ rows of elements, giving $T = 4A/\lambda^2$ total elements. The corresponding gain is

$$G_0 = \pi T = 4\pi A/\lambda^2 \qquad (6.1.2)$$

The **beam pattern** in this case will be described by the product of two $(\sin x)/x$ functions:

$$f(\theta, \phi) = \frac{\sin[(\pi w/\lambda)\sin\theta\cos\phi]}{(\pi w/\lambda)\sin\theta\cos\phi} \frac{\sin[(\pi h/\lambda)\sin\theta\sin\phi]}{(\pi h/\lambda)\sin\theta\sin\phi} \qquad (6.1.3)$$

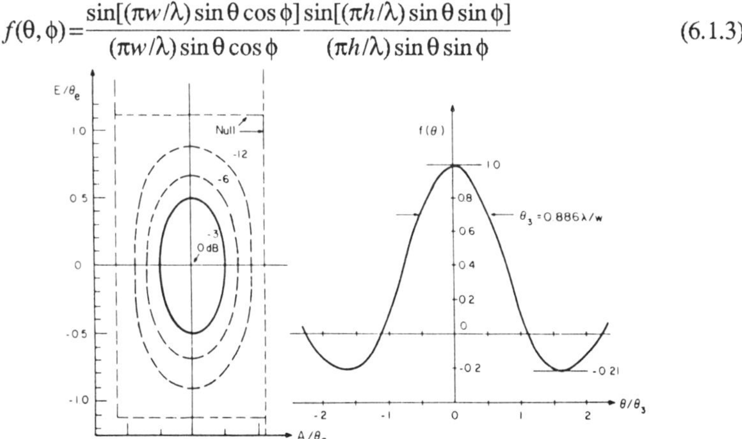

Figure 6.1.2 Antenna pattern plots.
(a) Contours in two angular coordinates. (b) Principal-plane voltage pattern [6.4]. © Artech House, reprinted by permission.

In this case, not only is the antenna pattern separable from the range-doppler response, but the azimuth and elevation patterns are separable from each other. This response is shown in Figure 6.1.2, with the angle scale normalized to the 3 dB beamwidth. The first nulls, beside the main lobe, will occur at $\sin A = \sin\theta\cos\phi = \lambda/w$ and at $\sin E = \sin\theta\sin\phi = \lambda/h$, and other nulls will occur at integral multiples of these angles. Solving Eq. (6.1.3) for the half-power beamwidths, we find

$$\sin\theta_a \approx \theta_a = 0.886\lambda/w \quad \sin\theta_e \approx \theta_e = 0.886\lambda/h \text{ (uniform illumination)} \quad (6.1.4)$$

Considering the azimuth coordinate only (elevation will be similar with h replacing w), and making the small-angle assumption $\sin\theta \approx \theta$, the pair of first sidelobes, at $\theta = \pm 1.4\,\lambda/w$, $\theta/\theta_a = 1.6$, will have an amplitude -0.21, or -13.6 dB from the main-lobe peak. The sidelobe peak amplitudes will fall off inversely with $(\pi w/\lambda)\sin\theta$, and will not fall below -30 dB until the tenth sidelobe at $\theta = 10.5\lambda/w = 11.9\theta_a$. The relationship between *number of elements* and *beamwidth* is

$$T = 3.14/\theta_a\theta_e \text{ (in radians)} = 10,300/\theta_a\theta_e \text{ (in degrees)} \quad (6.1.5)$$

The gain and the beamwidth are related by

$$G_0 = 9.84/\theta_a\theta_e \text{ (in radians)} = 32,300/\theta_a\theta_e \text{ (in degrees)} \quad (6.1.6)$$

The constant 9.84 corresponds to an adjustment factor $L_n = 1.28 = 1.1$ dB in the gain-beamwidth relationship $G = 4\pi/\theta_a\theta_e L_n$.

We will discuss later in this section how these rather fundamental relationships among gain, beamwidth, and number of elements (or elementary areas) change as the aperture illumination is tapered for lower sidelobes, or as the aperture itself is made elliptical to match the illumination provided by a horn feed.

For ground-based antenna systems we often speak of antenna patterns in terms of azimuth and elevation coordinates. These can be related to the spherical coordinate system best used to make antenna calculations, as shown in Figure 6.1.3. The antenna aperture is located in the x-y plane, and the broadside beam will be directed along z. The angle θ is the deviation from broadside, while ϕ indicates the direction of this deviation. Placing the aperture in the vertical plane (Figure 6.1.4), with its width w along with the x axis and height h in the y axis, and z pointing north, we can see that the azimuth angle $A = \theta\cos\phi$, while elevation $E = \theta\sin\phi$.

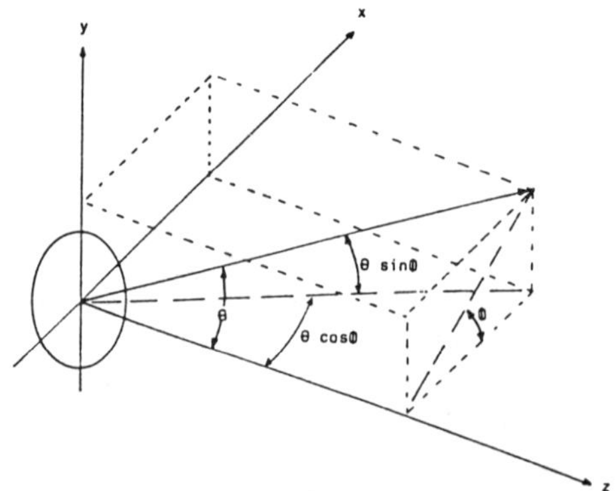

Figure 6.1.3 Antenna pattern coordinates [6.4].
© Artech House, reprinted by permission.

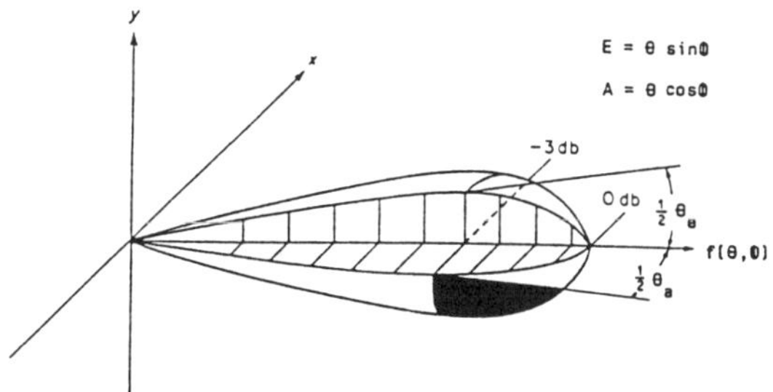

Figure 6.1.4 Coordinates for horizontal beam [6.4].
© Artech House, reprinted by permission.

A typical plot of the *power pattern in decibels* for a pencil-beam tracking
antenna is shown in Figure 6.1.5. Several typical antenna patterns are shown in
Appendix F.

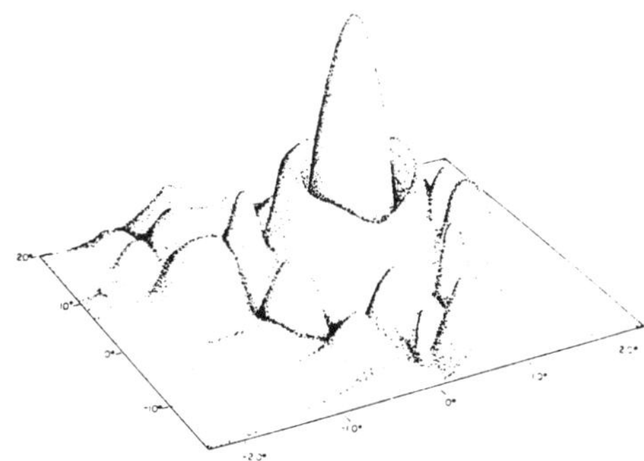

Figure 6.1.5 Power gain for two angular coordinates [6.2].

In general, the voltage gain of an antenna is a complex quantity, with magnitude and phase angle measured relative to the on-axis voltage. Many types of
antennas, however, produce real patterns with phase angles essentially zero over
the main lobe and oscillating between 180° and zero over the sidelobes. Gain is
also a function of *polarization*, $f_t(\theta,\phi)$ being defined for the designed transmitting
polarization and $f_r(\theta,\phi)$ for reception of that polarization. Polarization diversity
systems require separate functions for each polarization used.

Reciprocity of Antenna Patterns

An important concept in antenna theory is the principle of reciprocity. Simply stated, the patterns of a given antenna for receiving and transmitting will be the same. We may measure the pattern by radiating from the antenna and measuring field strength as a function of angle, at any convenient long range from the antenna, or by radiating from a small source at that range and measuring the received voltage at the antenna terminal, as a function of the same offset angle from the source. In the first case, we may consider that the antenna aperture has been *illuminated* by the transmitter through its feed system, causing it to radiate a beam into space. If the aperture is composed of discrete elements, these are said to be *excited* by the transmitter through the feed system. In the second case, the aperture is placed in a field of uniform intensity, created by the small remote source, and components of this field are *weighted* by the feed system in forming the received beam pattern. In either case, the amplitude and phase of signal components corresponding to different portions of the aperture are controlled by the feed system. For simplicity, we will refer to this process as *illumination* for all types of antennas in both transmitting and receiving modes. Uniform illumination, with phase matched to a plane wave normal to the desired beam direction, gives maximum gain in that direction.

Reciprocity is violated in actual antennas only when a ferromagnetic device such as a circulator or *ferrite phase shifter* is included in the circuit, or when a nonlinear device such as a gas-tube *duplexer* is included.

6.2 Four-Coordinate Radar Response

Radar operates in four-dimensional space: two angles (often specified as azimuth and elevation), range (measured as time delay), and frequency (including the doppler shift caused by target radial velocity). A target is said to be *resolved* if its signal is separated at the radar output from those of other targets in *at least one of these coordinates*. For example, a surveillance radar may scan its beam in azimuth and elevation, detecting a target in a particular range gate and doppler filter. A second target signal may be resolved if it lies in a different azimuth or elevation beam position, or in a different range gate or doppler filter. A useful criterion for resolution is that the position of the desired target should be measurable by the radar with only small errors caused by the other target.

Figure 6.2.1 shows that the relative phase and amplitude of the second target affect the resolution process. However, for two targets of equal amplitude and arbitrary constant phase, resolution is normally possible when they are separated by approximately the half-power beamwidth or the half-power width of the processed pulse or doppler filter.

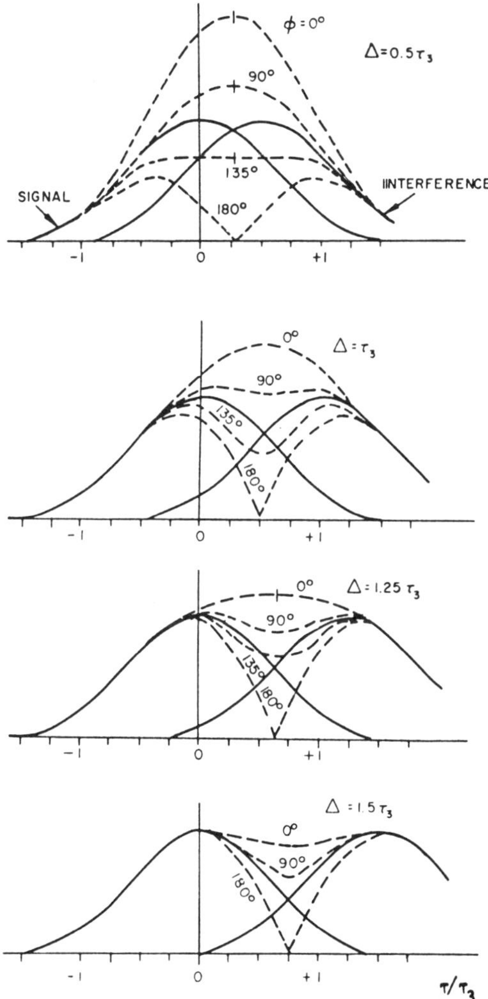

Figure 6.2.1 Resolution of signals from closely spaced target returns in time [6.4]. © Artech House, reprinted by permission.

Resolution is thus determined by the relative response of the radar to targets separated from the target to which the radar is directed, or to targets separated from the target to which the radar is directed, or matched. The antenna and receiver, at a given time, are configured to produce a maximum response on a target at particular angles, range delay, and frequency. The radar will be designed to respond with reduced gain to targets at other locations. This *response function* can be expressed as a surface in a five-dimensional coordinate system: relative output voltage, as a function of the target displacement from the **matched point** in the four radar coordinates: $\chi(\theta, \phi, t_d, f_d)$. The parameters of the radar range equation determine this voltage at the matched point, at which χ is normalized to unity.

Separable Angle Response

In most cases, the angles response of the radar can be considered independent of the time delay and doppler response:

$$\chi(\theta, \phi, t_d, f_d) = \chi(\theta, \phi)\chi(t_d, f_d) \tag{6.2.1}$$

The angle response $\chi(\theta, \phi)$ is simply the two-way voltage gain pattern of the antenna, which has been represented in the radar equation by a pattern component $f^2(\theta, \phi)$ of the pattern-propagation factor F^2. In cases where the transmitting and receiving antennas have different patterns, the two-way voltage gain pattern will be the product of the two one-way patterns:

$$\chi(\theta, \phi) = f_t(\theta, \phi)f_r(\theta, \phi) \tag{6.2.2}$$

It is these one-way voltage gain patterns, or their squares, that represent power gain patterns, which will normally be specified or measured to characterize the resolution properties of the antenna system. The power gain pattern, $G(\theta, \phi) = G(0,0) f^2(\theta, \phi) = G_m f^2(\theta, \phi)$, is often plotted on a decibel scale to permit sidelobe levels to be more readily seen.

6.3 Antenna Sidelobes

Ideally, the sidelobe levels of an antenna would be determined by the *design illumination function* described in Appendix F. However, in practice, the sidelobes will be higher because of errors in generating this illumination. In reflector antennas, these errors also result from spillover, electrical currents on the edge of the reflector, blockage of the aperture by the feed structure, feed alignment, and the shape of the reflector surface. In array antennas, there will be errors in the phase and amplitude of the illumination of individual antenna elements, resulting from both the feed network and the phase shifters, as well as edge current effects and mechanical deviations in location of the elements.

While it has long been customary to specify the sidelobe characteristics of antennas in terms of the *peak lobes* (usually the first pair beside the mainlobe), it is often the *average sidelobe levels* (rms voltage) over an angular region away from the main lobe that will be important (Figure 6.3.1). This *far sidelobe* level can be calculated as the average gain of an additive pattern established by the aperture when illuminated by the error from the intended illumination function. For example, if the phase errors of small elements ($\lambda/2$) of the aperture are independent, and have an rms value σ_ϕ radians ($\sigma_\phi \ll 1$), a broad, random error pattern will be produced for which the power gain, averaged over the entire sphere, is $G_s = \sigma_\phi^2$ relative to isotropic gain. In the forward hemisphere, the average level will be about twice this value. Near broadside to the array, where the gain of each element is maximum ($\approx \pi$), the gain relative to the mainlobe gain G_m is given by

$$G_s/G_m \;=\; \pi\sigma_\phi^2/G_0\eta_a \;=\; \sigma_\phi^2\lambda^2/4A\eta_a \tag{6.3.1}$$

If the phase errors are not random over the array, but correlated by row, column, or group (e.g., subarray), the average sidelobe level remains unchanged, but particular lobes will be much higher than the average.

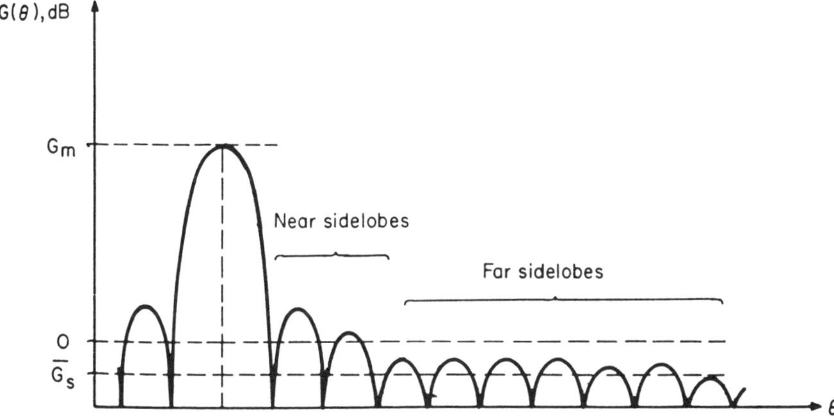

Figure 6.3.1 Typical antenna pattern showing peak and average sidelobe levels [6.4]. © Artech House, reprinted by permission.

6.4 Tapered Aperture Illumination

We have seen in Eq. (6.1.2) that uniform illumination provides the maximum possible gain for a given antenna aperture. Unfortunately, the high sidelobes provided by the uniformly illuminated antenna are often not acceptable in a real radar application. This problem is dealt with by *tapering* the illumination as a function of distances x and y from the center of the aperture. Tapering is accomplished by applying a weighting to the illumination function which alters the antenna pattern and hence the *effective* area A_r of aperture. Figure 6.4.1 shows the difference between an antenna pattern provided by uniform illumination and one (for the same antenna) provided by a cosine-weighted illumination function. By applying the cosine taper, dashed lines, three changes have resulted:

(a) The sidelobe levels have been reduced;

(b) The mainbeam (and 3 dB beamwidth) has broadened; and

(c) The peak antenna gain has been reduced.

The gain of the tapered aperture can be expressed as $G_m = G_o\eta_a$, where η_a is the aperture efficiency.

In a rectangular aperture with separate illuminations (one for the x-dimension and another for the y-dimension) the aperture efficiency n_a will be

$$\eta_a \;=\; G_m/G_o \;=\; G_m\lambda^2/4\pi A \;=\; \eta_x\eta_y \tag{6.4.1}$$

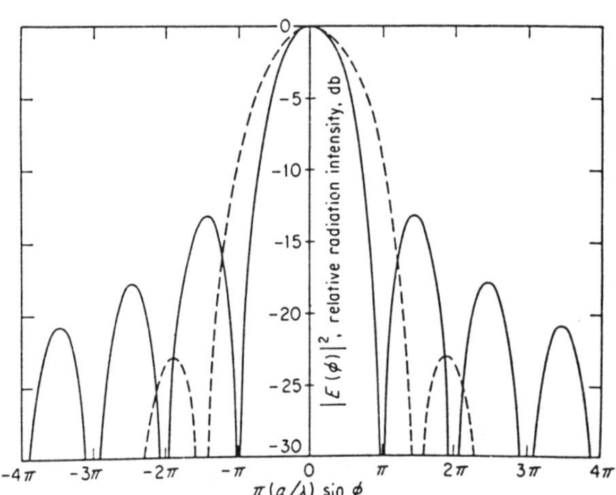

Figure 6.4.1 Uniform vs. cosine illumination [6.3].

Figure 6.4.2 shows how beamwidth and efficiency vary with the level of the first sidelobe for several common tapering functions. Here $\eta_x = 1.0$ and $K_\theta = 0.9$ apply to the unweighted reference or (uniform) illumination function. For example, an aperture with cosine taper in both coordinates will be described by the following parameters:

$$\eta_a \;=\; \eta_x\eta_y \;=\; (0.8)^2 \;=\; 0.64, \quad \text{or} \quad -1.9\,\text{dB}$$

$$\theta_a \;=\; \theta_e \;=\; K_\theta\lambda/w \;=\; 1.19\lambda/w \;(\text{rad}) \;=\; 51\lambda/w \;(\text{deg})$$

$$G_s/G_m, \quad \text{first sidelobes}, \;=\; -23\,\text{dB}$$

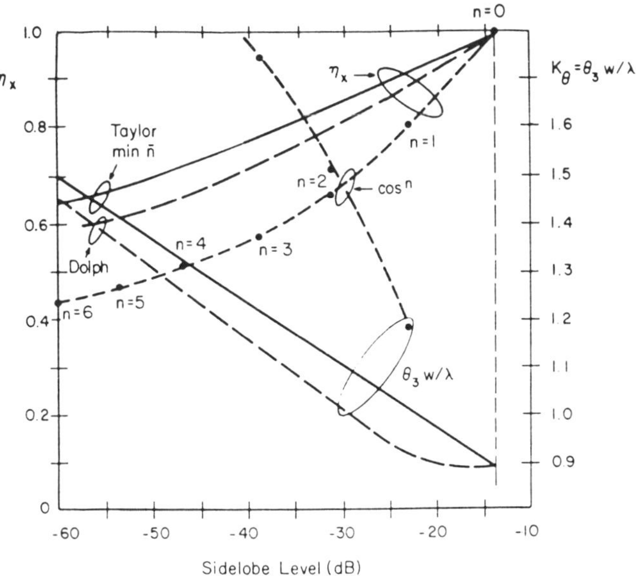

Figure 6.4.2 Beamwidth and aperture efficiency vs. level of first sidelobe [6.4]. © Artech House, reprinted by permission.

Elliptical and Circular Apertures

When low-sidelobe tapers are used for $g(x)$ and $g(y)$, the combined $g(x,y)$ illumination amplitudes near the corners of the aperture are very small, and those elements or portions of the aperture can be omitted with little effect on performance. The result is an elliptical aperture with gain and pattern that are essentially the same as those of the rectangle having the same width w and height h. The area of the ellipse and the number of elements are $\pi/4$ times those of the rectangle, *and hence its aperture efficiency* η_a *will be greater by about* $4/\pi = 1.05$ dB.

6.5 Reflector Antennas

A paraboloid of revolution, illuminated by a feed at the focal point, reflects a plane wave along its axis (Figure 6.5.1). If we measure the electric field strength in a plane just beyond the feed, it will appear as if it had been radiated from a planar array with the illumination $g(x,y)$ of the reflector, but with three important differences:

(a) That portion of the aperture shadowed by the feed and its support structure will not contribute to the field, and will produce a broad pattern of *blockage lobes* in the forward hemisphere;

(b) The power intercepted by the feed structure will be scattered over a wide angle, mostly in the rear hemisphere occupied by the reflector; and

(c) Some of the power radiated by the feed will miss the reflector and produce a second, direct *spillover* pattern in the rear hemisphere.

The *far-field* antenna pattern will thus consist of the sum of the intended pattern plus the three unintended components: *blockage*, *feed scattering*, and *spillover*. In practice, the illumination taper will establish the level of the principal sidelobes only, blockage will control the level in most of the forward hemisphere, while feed scattering and spillover will control the level in the rear hemisphere (which may also have components caused by reflector edge effects — diffraction and electrical currents circulating around the edges onto the rear surface). The lobes caused by direct feed spillover can be calculated readily from the pattern of the feed horn, but the other components usually require measurement.

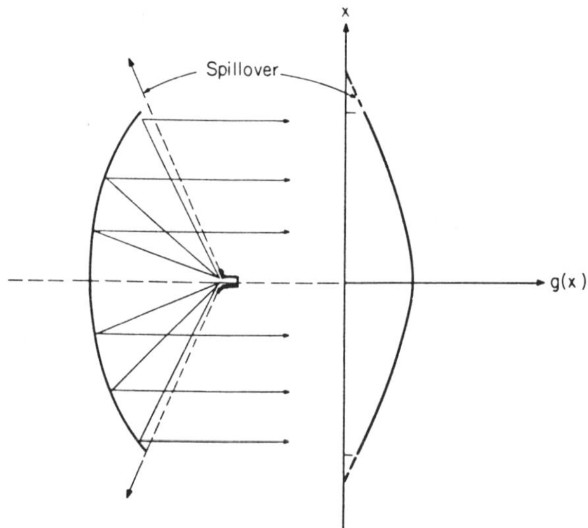

Figure 6.5.1 Parabolic reflector and equivalent planar aperture illumination [6.4]. © Artech House, reprinted by permission.

6.5.1 Cassegrain Reflector Antennas

The *Cassegrain* antenna (Figure 6.5.2) is the most commonly used of the folded-optics type of reflector system. The *hyperboloidal* subreflector of diameter *d* intercepts the rays coming from the main paraboloidal reflector, focusing them on the feed. Waveguide runs from the feed to the transmitter and receiver can be made shorter than in front-fed designs, and the supporting structure for the sub-reflector can be lighter, with less blockage. The subreflector is typically allowed to have a diameter about 1/10 that of the main reflector.

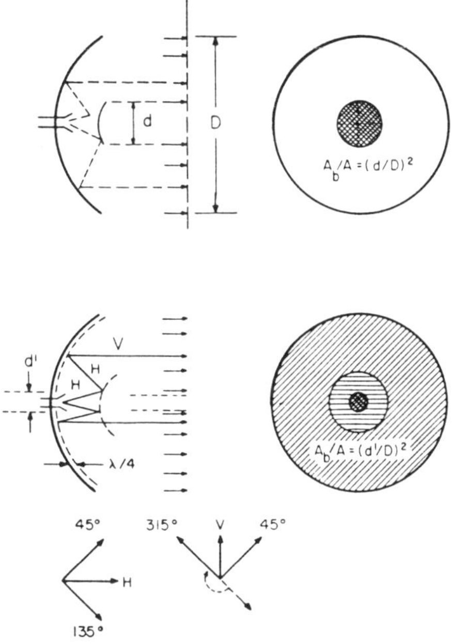

Figure 6.5.2 Cassegrain reflector configurations.
(a) Solid subreflector system. (b) Polarization-twist Cassegrain system [6.4].
© Artech House, reprinted by permission.

The subreflector can be increased in size, without blockage, if the *polarization-twist* Cassegrain configuration is used (Figure 6.5.2b). In this case, the subreflector is composed of thin wires embedded in a plastic surface of hyperboloidal shape. In a typical tracking antenna, these wires will be horizontal, and the feed will be horizontally polarized. The horizontally polarized illumination will reach the surface of the main reflector, where it is rotated to vertical by a grid of wires oriented at 45° angle and located $\lambda/4$ in front of the solid reflector surface. The radiation into space is then vertically polarized, *passing through the subreflector without blockage*. The return signal goes through the process in reverse, appearing horizontally polarized at the feed.

Efficiency of Reflector Antennas

The efficiency of the reflector antenna system may be written as the product of factors expressing effects of the illumination function, blockage, spillover, and surface tolerances:

$$\eta_a = \eta_i \eta_b \eta_s \eta_t \tag{6.5.1}$$

6.6 Phased Arrays

The concept of an *array* antenna, an aperture consisting of a collection of individual elements, was introduced at the beginning of this chapter. In fact, the earliest radars used simple arrays, but they were, for the most part, relatively low frequency (UHF and VHF), used fixed dipole elements, were mechanically steered, and operated as an ensemble with no control over individual elements. Antennas of this type are still in use, e.g., the *Yagi* array, but the development of techniques and devices such as the ferrite and digital phase shifter, switches with extremely fast switching rates, and high-speed computers, has transformed the original concept of array antennas and ushered in the age of the multi-function *phased array radar*.

The phased array is an electromagnetically steerable antenna in which the phase shift of each individual element, row, or column of elements, is precisely controlled. The resultant phase gradient across the aperture effectively steers the beam over a broad region about the broadside direction, and by manipulation of both row and column, the beam may be pointed to any desired direction within the antenna coverage volume. This unprecedented flexibility permits the phased array radar to perform multiple tasks previously requiring, at the least, separate antennas, and in many cases completely separate radars.

Today a sophisticated phased array radar under computer control may perform volume search, IFF, multiple target track, and target illumination for semi-active homing missiles. Such a radar is capable of altering its beamwidth, frequency, and waveform to optimize radar performance for each of these separate functions and in accordance with the demands of the natural or man-made hostile environment in which they occur. These requirements, together with the intricate time-sharing and radar power management capabilities needed to successfully integrate them all within the radar system's operating cycle, have lead to a new and critical dependence on software for the radar's operational reliability.

Table 6.1 shows the major ways in which phased arrays may be classified. In addition to these, it should be mentioned that the *shape* of the array is yet another way in which phased arrays may be classified. The *linear* array is literally a single row (or column) of elements enabling the antenna to electronically scan in one dimension only. A *planar array* contains elements in two dimensions constituting a plane. It is possible to design *conformal* arrays to fit a three-dimensional shape such as a pyramid, cube, or a solid of revolution such as the fuselage of an aircraft.

Table 6.1 Classification of Phased Arrays

Scan coordinates[1]	Elevation only (mechanical in azimuth) Azimuth and elevation
Steering method	Phase scan Frequency scan Switched beam-forming matrix
Feed method	Constrained (corporate) Space (optical)
Amplifier location	At feed input At rows or columns At subarrays At individual element
Amplifier function	Receiving only Transmitting and receiving

[1] Azimuth-elevation coordinates may be replaced by direction cosines, yaw-pitch, or any other selected pair of angles.

6.6.1 Principle of Operation

The principle of operation of a phased array antenna can best be explained by using a simple linear array as an example. Figure 6.6.1 represents such an array consisting of n_a elements, of which four are shown. The spacing between elements is some distance d. Now let us apply current i to each element, but vary the phase at each element by a constant value $\Delta\phi$. Thus at element 1, the reference element, the phase is 0, at element 2 the phase is $\Delta\phi$, at element 3 the phase is $2\Delta\phi$, and so on.

The value of the electric field radiated by the array is maximum at some angle θ, such that at that angle the signals from each element are in phase with each other, and the whole ensemble forms the plane wave phase front indicated in the figure. Since the phase front is perpendicular to the direction of propagation θ, we see that at time t_1, the distance $a = d \sin \theta$. The number of wavelengths represented by a is then simply $(d/\lambda) \sin \theta$ and the phase angle at t_1 is then (number of wavelengths) $\times$ (2π rad/wavelength) or $(d/\lambda \sin \theta) \times (2\pi) = (2\pi \, d/\lambda) \sin \theta$. The maximum field occurs therefore when

$$\Delta\phi = (2\pi d/\lambda)\sin\theta \qquad\qquad (6.6.1)$$

The phase difference is applied by a *phase shifter,* and is normally applied *modulo* 2π, so that the phase shifter need only be controllable over that range.

Using the analogy of time, Table 6.2 shows the phase coherence for each of the four elements of Figure 6.6.1.

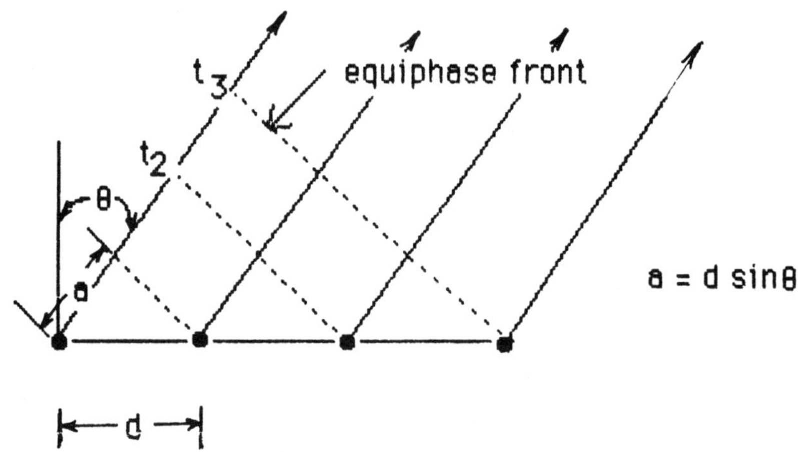

Figure 6.6.1 Phase front propagation.

TABLE 6.2 Phase Settings of Array Elements

Time	Element 1	Element 2	Element 3	Element 4
t_0	0	$\Delta\phi$	$2\Delta\phi$	$3\Delta\phi$
t_1	$\Delta\phi$	$2\Delta\phi$	$3\Delta\phi$	$4\Delta\phi$
t_2	$2\Delta\phi$	$3\Delta\phi$	$4\Delta\phi$	$5\Delta\phi$
t_3	$3\Delta\phi$	$4\Delta\phi$	$5\Delta\phi$	$6\Delta\phi$

6.6.2 Antenna Pattern

In Figure 6.6.2, assume that the linear array shown consists of n isotropic elements equally spaced distance d apart, and radiating uniformly with equal phase and amplitude. If the gain at broadside ($\theta = 0$) is considered equal to unity, then the ***array pattern factor*** takes the form of Eq. (6.1.3) for one dimension

$$f_a(\theta) = \frac{\sin n\pi(d/\lambda)\sin\theta}{n\,\sin[\pi(d/\lambda)\sin\theta]} \tag{6.6.2}$$

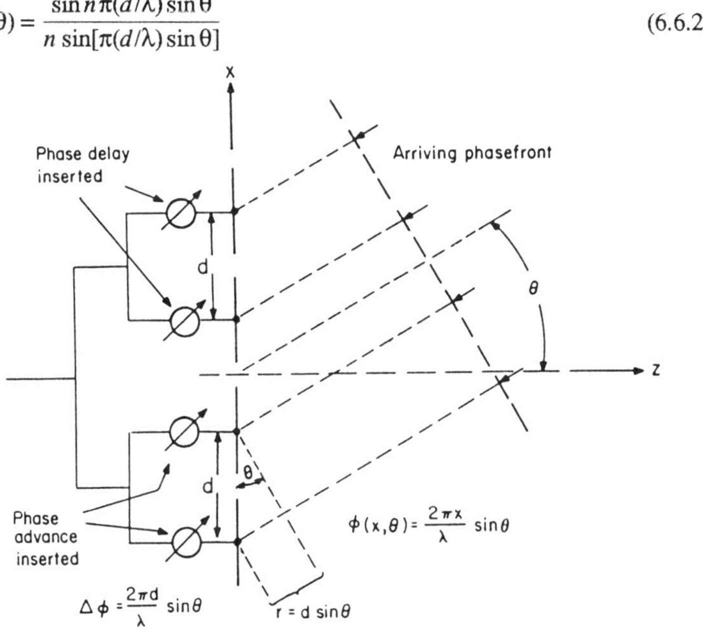

Figure 6.6.2 Phase relationships for beam steering [6.4].
© Artech House, reprinted by permission.

This pattern is shown for $n = 10$ elements in Figure 6.6.3. The figure shows the presence of ***grating lobes,*** which are repetitions of the main lobe structure that occur when the separation between elements is such that $\pi(d/\lambda)\sin\theta = 0$, π, 2π, ..., $n\pi$. Because grating lobes create ambiguities which are indistinguishable from the true mainlobe response (i.e., can create false targets), elements should be spaced less than

$$d = \lambda/(1 + |\sin\theta|) \tag{6.6.3}$$

For example, if the maximum scan angle $\theta_m = 60°$, $d < \lambda/1.86 = 0.65\,\lambda$ is required to suppress grating lobes. The use of *directive* elements rather than isotropic elements allows some additional control over grating lobes and permits a closer spacing between elements.

The complete pattern factor is the product of the array factor and the element factor

$$f(\theta) = f_a(\theta)f_\theta(\theta) \tag{6.6.4}$$

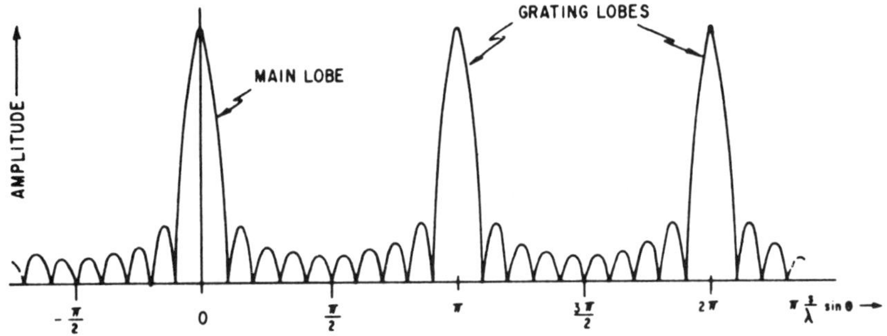

Figure 6.6.3 Array factor with $n = 10$ elements [6.3].

Since the radiated lobe is the product of the array pattern and element pattern, and the latter is essentially zero at 90°, spacings of 0.6 to 0.65 λ are generally permissible without significant grating lobe problems. Arrangement of the elements in a triangular grid helps suppress grating lobes without raising the number of elements in the array too close to the value $4A/\lambda^2$, which would result from half-wavelength spacing in both coordinates.

Ideally, $f_e(\theta) = \cos\theta$, so that the array gain is proportional to the aperture area projected normal to the beam direction. The short dipole would have $f(\theta) = \cos\theta$, while the half-wave dipole would produce an even narrow pattern. Hence, radiating elements based on the dipole will normally have additional structure associated with each element to spread the pattern over the desired scan angle. In order to obtain proper matching and to control mutual coupling between elements as the beam scans, it is often necessary to configure the radiating elements such that

$$f_e^2(\theta) = \cos^{3/2}\theta \tag{6.6.5}$$

A compromise element matching can give a relationship approaching $\cos\theta$, but with a fixed loss of about one dB relative to the usual $G_e(0) = \pi \approx 5$ dB.

For wide-angle scanning arrays, the number of elements can be estimated as

$$T = n_a \times n_e = (1.4w/\lambda)(2.4h/\lambda) = 2A/\lambda^2 \qquad (6.6.6)$$

Expressing T in terms of the product of the beamwidths, for a tapered illumination with $\theta_3 = K_\theta \lambda/w \approx 1.25$, we find

$$T = 2K_\theta^2/\theta_a\theta_e \approx 3.14/\theta_a\theta_e \text{ (in radians)} = 10,300/\theta_a\theta_e \text{ (in degrees)} \qquad (6.6.7)$$

which is the same result as given in Eq. (6.1.5) for uniform illumination with $\lambda/2$ spacing.

If scanning is restricted to more limited sectors ($\theta_m < 20°$), directional elements may be used; subarrays fed from a single phase shifter, large horns, Yagis, or other end-fed directional elements. Control of grating lobes is made difficult by both the closer spacing of the lobes, at intervals such that $\sin\theta_i - \sin\theta_0 = i\lambda/d$, and the effects of control quantization over the smaller number of elements.

Thinned Arrays

To reduce the number of radiating and phase-shifting elements (and the number of amplifiers, in an active array), dummy elements are sometimes used at random locations in the array. A fraction of active elements F_a can be distributed in a random way so as to create the desired average taper of illumination, while avoiding the generation of large sidelobes in particular directions.

The total number of active elements required to establish given beamwidths in the thinned array is proportioned to F_a:

$$T_f \approx 3.14F_a/\theta_a\theta_e \text{ (in radians)} = 10,000F_a/\theta_a\theta_e \text{ (in degrees)} \qquad (6.6.8)$$

The narrow beam formed by a thinned array is advantageous in resolving closely spaced targets, achieving high tracking accuracy, and reducing the clutter within the main beam. There are, however, three major disadvantages attached to the use of thinning:

(a) The gain is reduced, when compared with the fully filled, uniformly illuminated array;

(b) The presence of inactive elements causes an increase in the average sidelobe level; and

(c) In the case of a search radar, or a tracker in the acquisition mode, the reduced gain will apply for both transmitting and receiving.

6.6.3 Phase Error Effects

Electronic scanning at a constant frequency is performed by using digitally controlled phase shifters at each element (or each row, for one-dimensional scanning). Equation 6.6.1 applies with ϕ calculated for each element position x,y. The digital-to-phase conversion process, carried out to m-bit precision, introduces a phase error for which the standard deviation is

$$\sigma_\phi = \frac{2\pi}{2^m\sqrt{12}} \tag{6.6.9}$$

This *phase quantization error* causes three distinct effects on the array radar system:

(a) *Quantization loss:*

$$L_q = 1 + \frac{\pi^2}{3 \times 2^{2m}} \tag{6.6.10}$$

(b) *Increased sidelobes:* The peak sidelobes for scanning in a **cardinal plane**, or for a linear array, are

$$G_i/G_m = (1/i2^m)^2 \tag{6.6.11}$$

where i is the sidelobe number.

The average sidelobes for independent phase errors are

$$\overline{G}_s/G_m = \frac{L_q - 1}{T\eta_a} = \frac{\pi^2}{3 \times 2^{2m}T\eta_a} \tag{6.6.12}$$

(c) *Beam-steering error:* The peak angle error for scanning in a cardinal plane, or for a linear array, is

$$\varepsilon_\theta = \frac{2.3\lambda}{2^m w \cos\theta}\sqrt{G_d(\theta)/G_0} \approx (\theta_3/2^m)\sqrt{G_d(\theta)/G_0} \tag{6.6.13}$$

where θ is the scan angle and $G_d(\theta)$ is the normalized pattern derivative, or monopulse pattern, for this angle [6.3]. The average angle error for independent phase error is

$$\sigma_\theta = \frac{\theta_3\sigma_\phi}{1.61\sqrt{T}} \tag{6.6.14}$$

Practical phase shifter and feed designs introduce additional random errors, often equal to the value of σ_ϕ caused by quantization. The loss in dB is then coupled, as are the sidelobe powers and the variance of the pointing error.

6.6.4 Phase Shifters

The element phase can be controlled by diode switching of RF paths or by variation in waveguide phase velocity through ferromagnetic action. The control input in either case is digital, but, with the ferrite, a D/A conversion process is used to control current or residual magnetization (through control of $\int i\,dt$). Neither approach is inherently superior in all respects, as both have their advantages and disadvantages (see Table 6.3).

Table 6.3 Phase Shifter Characteristics

Characteristic	*Diode*	*Ferrite*
Controllable element	Switched paths	Adjustable phase velocity
Control input	Inherently digital	Inherently analog, via D/A converter
Switching speed	Fast	Slower than diodes
Power limitation	Surface density	Volume density
Power rating	Lower	Higher
Reciprocity	Reciprocal	Nonreciprocal, or reciprocal with higher loss
Loss	Higher	Lower
Control current	Continuous	Latching or continuous
Cost	Lower	Higher
Size	Smaller	Larger
Weight	Lower	Higher

While the qualitative characteristics listed for a particular type of phase shifter will apply generally, specialized designs can overcome a disadvantageous characteristic, often at the expense of another. Still, the ferrite phase shifter has been preferred in high-power radars without element amplifiers, while diodes are generally used in lower power systems, receiving arrays, and active amplifier arrays (where they appear at the inputs to the final power amplifiers and the outputs of the low-noise receiving amplifiers).

6.6.5 Frequency Scanning

The scanning of a planar array to an angle θ in the plane containing the x axis requires inserting of equal phase increments $\Delta\phi$ between elements, as given by Eq. (6.6.1). A simple way of controlling phase shift is to locate feed slots with spacing s along a waveguide (Figure 6.6.4), and to steer the beam by changing radar frequency. The ***cut-off wavelength*** of a rectangular waveguide is $\lambda_c = 2a$, where a is the larger dimension of the guide. Shorter wavelengths will propagate in the guide, but as $\lambda \to \lambda_c$ the phase velocity increases, following the relationship

$$c/c_g = \lambda/\lambda_g = \sqrt{1 - (\lambda/\lambda_c)^2} \qquad (6.6.15)$$

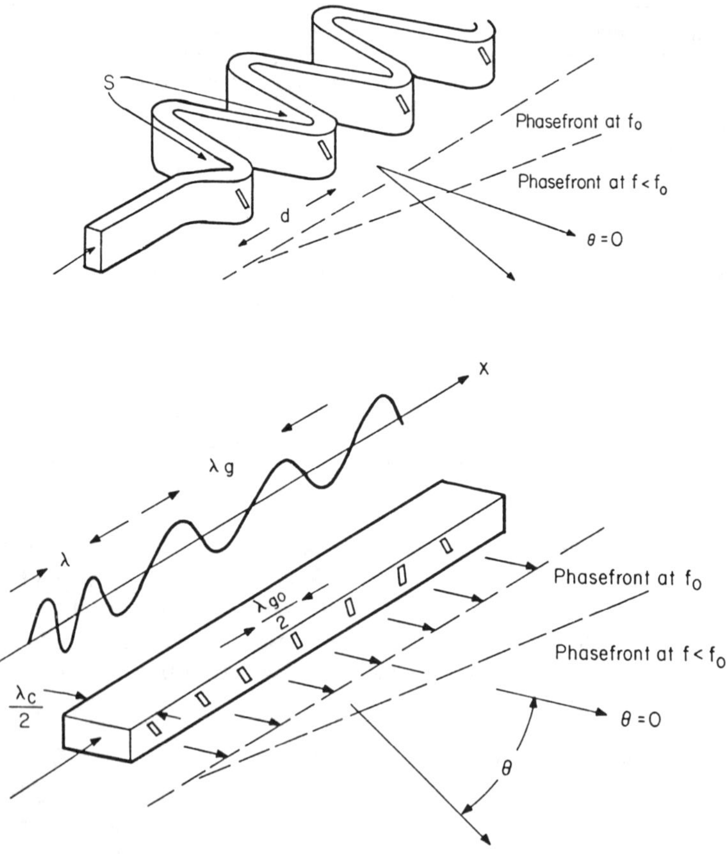

Figure 6.6.4 Frequency scanning waveguide antennas.

The phase increment between slots spaced a distance s along the guide will be

$$\Delta\phi = -2\pi s/\lambda_g \tag{6.6.16}$$

where the negative sign implies a phase delay in the waveguide. Alternate slots will normally be given an additional phase shift π by their location or angle relative to the waveguide wall, so that the incremental phase delay will be

$$\Delta\phi = (2\pi s/\lambda_g) - \pi \qquad (6.6.17)$$

To place the beam at broadside, for a frequency $f_0 = c/\lambda_0$, we set $\Delta\phi = 0$, and the slot spacing is

$$d = \lambda_{g_0}/2 = a/\sqrt{(2a/\lambda_0)^2 - 1}$$

Then, at other frequencies, the beam will squint by an angle

$$\theta = \sin^{-1}[(\lambda/\lambda_g) - (\lambda/\lambda_{g_0})] \qquad (6.6.18)$$

In order to scan over wider angles (e.g., 45° within the usual 5% tuning band), the waveguide feed is formed into serpentine or toroidal loops (Figure 6.6.4b) to give $s = n\lambda_{g_0} \gg d$, n is an integer for identical slots, or an integer $+ 1/2$ for out-of-phase slots. The spacing d of the radiating elements is selected to give the desired scan angle without grating lobes. The frequency scan equation now becomes

$$\theta = \sin^{-1}\{[s/d][(\lambda/\lambda_g) - (\lambda/\lambda_{g_0})]\} \qquad (6.6.19)$$

As the beam scans from broadside (a line perpendicular to the array face) its width varies according to

$$\theta_3(\theta) = \theta_3(0)\sec\theta \qquad (6.6.20)$$

6.6.6 Array Feed Systems

Two basic methods exist for coupling transmitter power to (and collecting receiver power from) the elements of an array:

(a) The *constrained* or *corporate feed*, in which the signal is confined to waveguide stripline, or coaxial structures with discrete branches connecting to the elements; and

(b) the *space feed*, in which the signal is radiated from a horn and propagates as a wave through air or a continuous dielectric medium to reach the elements. The general characteristics of the two types are shown in Table 6.4.

Constrained feeds are classified as either *series* or *parallel* networks. The series feed is simpler, and may be implemented as a slotted waveguide or with a number of two-port couplers in series, each diverting the power required for its associated element or group of elements (Figure 6.6.5) defocusing (if center-fed) by varying the frequency. Parallel feed networks (Figure 6.6.6) split the input power into two or more fractions, and continue to subdivide each of these until the levels for each element are reached. Either type of feed may be designed for equal path lengths to each element, providing a very wide bandwidth.

TABLE 6.4 Feed System Characteristics

Constrained (Corporate)	Space (Optical)
Closed circuits	Radiation through space
Coupler control of illumination	Illumination determined by horns in focal plane
No spillover	No divider loss
No rear-face reflections	No divider mismatch
Complex coupler networks	No networks unless many horns are coupled
Flat panels possible	Conical volume required
Monopulse networks very complex	Monopulse horn clusters easily implemented
Multiple-beam networks very complex	Multiple-beam feeds easily implemented

A type of feed often used, for economy and ease of fabrication, is the slotted waveguide. This is similar to the frequency scanning structure discussed earlier in this section, and indeed some squint (in an end-fed line) or defocusing (in a center-fed line) is inevitable with this approach. The individual element phase shifters can compensate for this at a specific frequency, but the instantaneous signal bandwidth may be affected. It is the ability to control phase and amplitude very precisely by milling of the waveguide slots that makes this feed very attractive.

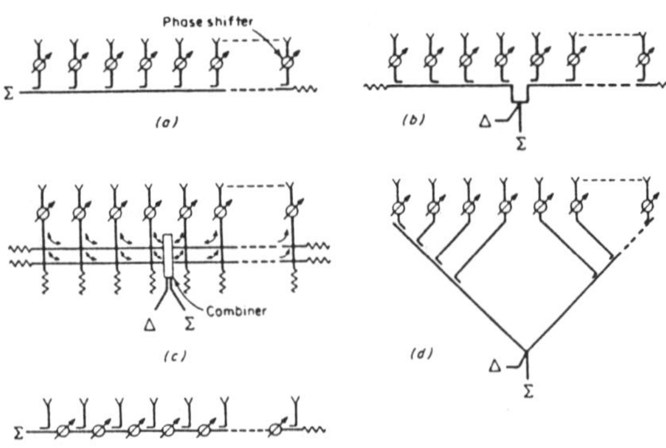

Figure 6.6.5 Series feed networks [6.5]. (Copyright 1970 by McGraw-Hill Book Co. Reprinted with permission.)

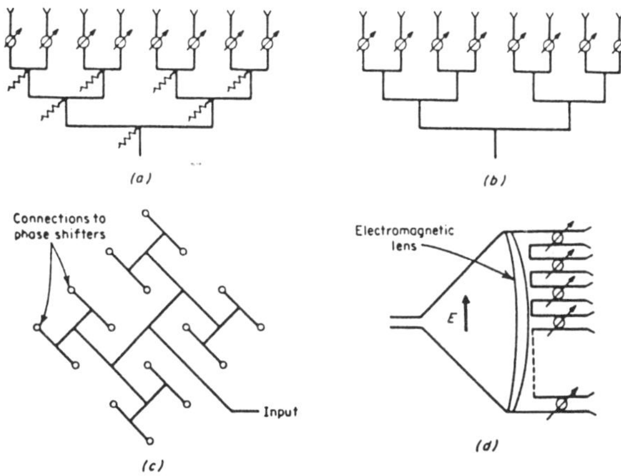

Figure 6.6.6 Parallel feed networks [6.5]. (Copyright 1970 by McGraw-Hill Book Co. Reprinted with permission.)

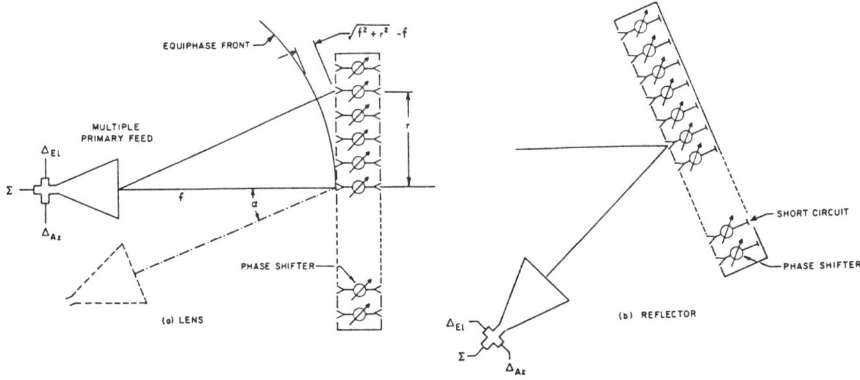

Figure 6.6.7 Space-fed arrays [6.5]. (Copyright 1970 by McGraw-Hill Book Co. Reprinted with permission.)

The space-fed array may be implemented either as a reflector or lens (Figure 6.6.7). The lens may be a planar array of phase shifting elements, or a passive lens of the Rotman type (Figure 6.6.8). The Rotman lens is a one-dimensional divider, and so to produce narrow beams in both dimensions it is necessary to stack a number of lenses to form the two-dimensional beam matrix (Figure 6.6.9). The space-fed planar array is a two-dimensional system that can form a pencil beam. Additional beams may be formed simultaneously, within a limited angle sector, by adding more feed horns in the focal plane of the array, in which case the entire beam cluster is scanned together by the phase shifters.

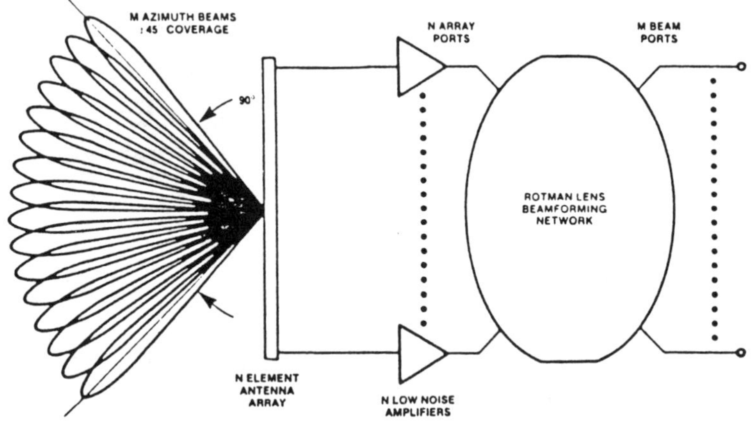

Figure 6.6.8 Array with Rotman lens feed [6.6].

6.6.7 Amplifier Arrays

Regardless of the feed arrangement, it is possible to place the final transmitter amplifier and the first receiving amplifier at the array element. In constrained feed systems, these amplifiers may be placed, instead, at any level in the dividing network: at subarrays, or at row or column level, as well as the individual elements. An advantage of placing amplifiers at the element (or at rows in one-dimensional scanning array) is that the phase shifter may be placed on the feed side of the amplifiers, reducing its power rating and avoiding loss in output power and receiving noise factor. Only the final duplexer (a switch or circulator device) and a short cable or guide to the radiating element will then appear in the loss budget for L_t and L_r. In a typical solid-state modular array, separate feed networks are used for transmitting and receiving providing uniform illumination for efficient transmitting, while tapering the receive illumination for low sidelobes. A common phase shifter at each element is connected to the amplifiers through a circulator.

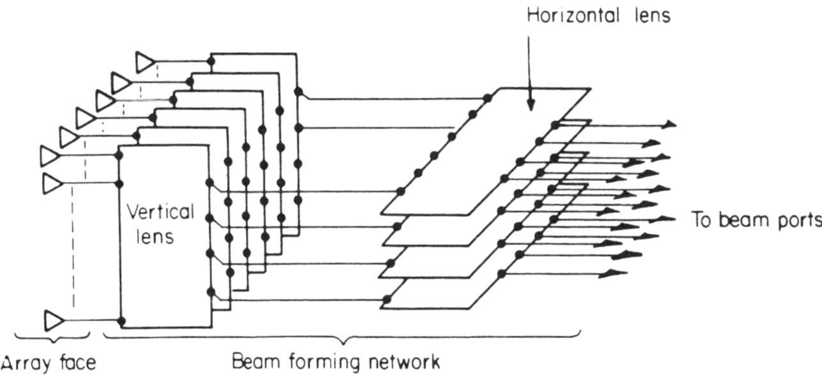

Figure 6.6.9 Two-dimensional Rotman lens feed [6.7].

6.7 Ultra-Low-Sidelobe Antennas

6.7.1 Definitions

There are no established standards defining low-sidelobe, very-low-sidelobe, and ultra-low-sidelobe antennas, but Table 6.5 may be used as a guide.

Table 6.5 Definitions of Sidelobe Levels

Antenna Description	*Sidelobe Levels*		
	dB below G_m		dB below isotropic
	Peak	Average	Average
Normal	> - 25	> -30	> -3
Low-sidelobe	-25 to -35	-35 to -45	-3 to -10
Very-low-sidelobe	-35 to -45	-45 to -55	-10 to -20
Ultra-low-sidelobe	< -45	< - 55	< - 20

A very-high-gain antenna is bound to have low average sidelobes over most solid angles, since the average sidelobe level outside the main beam must be less than $(1 - \eta_a)/G_m$ relative to the main beam, or $1 - \eta_a$ relative to isotropic. Hence, to fall within a given description, all three sidelobe criteria of Table 6.5 must be met.

Random phase and amplitude error tolerances are difficult to meet in electronically scanned arrays. An early example of an ultra-low-sidelobe array is the AWACS array, which uses slotted waveguide radiating lines (Figure 6.7.1). This and the subsequent array design for the AN/TPS-70 3D radar have azimuth sidelobes so low that the manufacturer cannot disclose the numbers, for security reasons, but both rotate mechanically in azimuth. The phase and amplitude errors introduced by the elevation network or phase shifters produce much larger peak sidelobes in the elevation plane. It is only in the azimuth plane and slant planes that ULSA performance from these antennas can be expected.

Figure 6.7.1 AWACS waveguide ULSA.
(Photo courtesy of Westinghouse Electric Co.)

6.7.2 Scanning ULSA Systems

Arrays that use phase shifters at each element, for two-coordinate scanning, are more difficult to design for ULSA performance. Random phase and amplitude errors can originate in the feed network, the phase shifters, or the radiating elements themselves. A major share of the total error must be assigned to the phase shifters, imposing very tight tolerances on the feed and the radiators. Consider an error budget to achieve -20 dBi average sidelobes, for which eight equal allocations are made for the following error sources: phase shift insertion phase, phase quantization, phase response to control, feed network phase, phase shifter insertion amplitude, phase shifter amplitude change, feed amplitude, and radiator amplitude. The four phase errors will each be 0.02 radian = 1.1° rms, and the amplitude errors will be 2% = 0.17 dB. A six-bit phase shifter will not quite meet the requirement.

The requirement for peak sidelobes at least 45 dB down from the main lobe is also difficult to meet, unless the array is very large.

When a *corporate feed* is used to illuminate an array $n_a \times n_e = 64 \times 64$ elements, the receiver will normally connect to the elevation network, from which will branch 64 networks. The 1.1° allowance for random errors in the row networks is difficult to meet, but the tolerance on the elevation feed is eight times tighter, since it contributes errors which are correlated across each row. In addition, when the array is steered in one of the cardinal planes, the peak sidelobes will be increased by the periodic phase quantization error, Eq. (6.6.10). It is the practical impossibility of meeting requirements for low correlated errors that leads the designer to orient the array with the first feed in the elevation plane, where sidelobe requirements can normally be relaxed.

In a *space feed*, (Figure 6.7.2) the phase error problem is significantly relaxed because correlated errors are avoided by the system geometry. There is no elevation column feed to introduce errors correlated over rows of elements. In addition, the phase command to a given element will be the sum of three terms: a row command, a column command, and a focusing correction.

The illumination function produced by a simple horn in a space-fed array system does not produce ULSA performance for two reasons: the approximate cosine shape of the horn function does not match the complex curvature of the Taylor or Dolph function (Figure 6.7.3), and the spillover may radiate into space or generate reflections from the structure surrounding the array elements. More elaborate feeds (e.g., **corrugated horns** or focal-plane arrays of horns) are required to generate the required illumination.

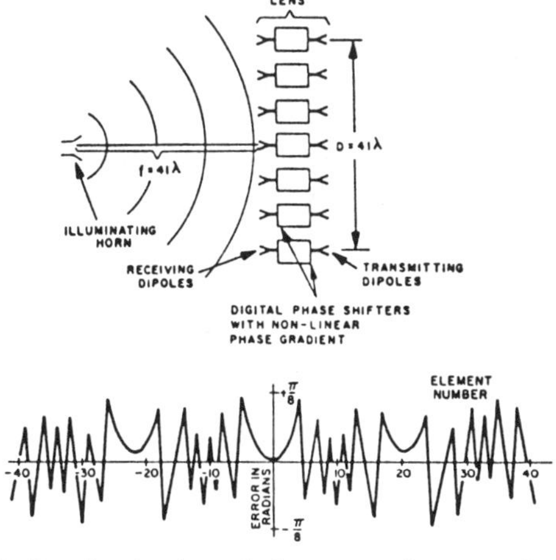

Figure 6.7.2 Randomization of phase error in space-fed array [6.8].

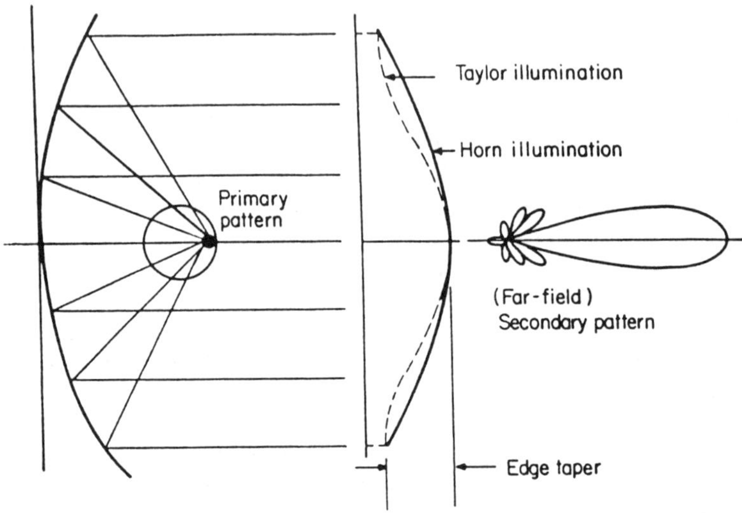

Figure 6.7.3 Horn illumination compared to Taylor function [6.10].

Figure 6.7.4 Line-fed parabolic cylinder antenna.
(Photo courtesy of Marconi Radar Systems, Ltd.)

6.7.3 Ultra-Low-Sidelobe Reflector Antennas

The conventional reflector antenna system using a horn feed and doubly curved paraboloid, is designed to give a reasonable compromise between efficiency and sidelobe level. Reflector antennas often achieve between 50 and 60% efficiency, with first sidelobes about -25 dB from the main-lobe peak, and spillover lobes near -35 dB.

Two design approaches have been used to approach ULSA performance.

(a) *Parabolic cylinder with line feed.* In this design, a constrained line feed is used to produce the low-sidelobe illumination function (usually in the azimuth plane). The feed illuminates a parabolic cylinder, which is made somewhat wider than the feed to minimize spillover (Figure 6.7.4). The parabola is fed from a position offset in elevation to avoid blockage.

(b) *Doubly curved reflector with complex feed.* In this approach a precision reflector surface is illuminated by an oversized feed. The feed must be offset to avoid blockage, and must be designed to approximate the 45 dB Taylor or similar illumination function. An example is shown in Figure 6.7.5. Spillover lobes near 120° are about -15 dBi or -53 dB from the main lobe. The first sidelobes are replaced, in this design, by shoulders on the main lobe at about -40 dB, and the first separated sidelobes are at -42 dB. Another example is the Gregorian configuration, Figure 6.7.6, with first sidelobes again near -40 dB and average sidelobes below -30 dBi, -78 dB with respect to the mainlobe. Designers of satellite communications antennas have gone beyond their radar counterparts in reflector ULSA systems.

6.8 Multiple-Beam Antennas

6.8.1 Stacked-Beam Systems

In order to obtain adequate time-on-target for MTI or doppler processing, high-performance 3D radars must cover the entire elevation search sector in parallel with a *stacked beam* configuration, rather than scanning sequentially with a single pencil beam. In the 1945 to 1975 period, such radars would use an array of feed horns in front of the doubly curved reflector (Figure 6.8.1). In the lower portion of the elevation sector, each horn provides a single beam, illuminating the entire reflector. At higher elevations, outputs of two or more horns are added to produce broader beams, approximating the $\csc^2$ receiving gain requirement. The transmitter couples through duplexers into each beam, producing a single fan beam with $\csc^2$ extension. A basic problem with this class of antenna is the blockage area of the extended feed structure, which causes large sidelobes, particularly in azimuth.

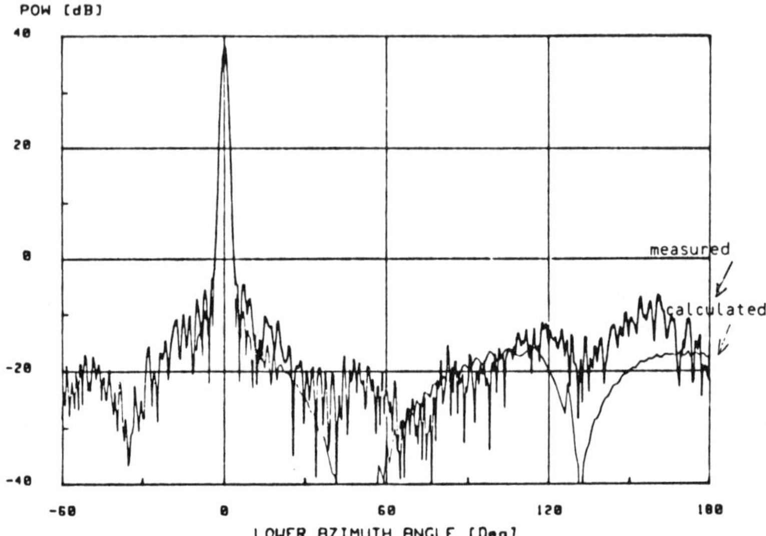

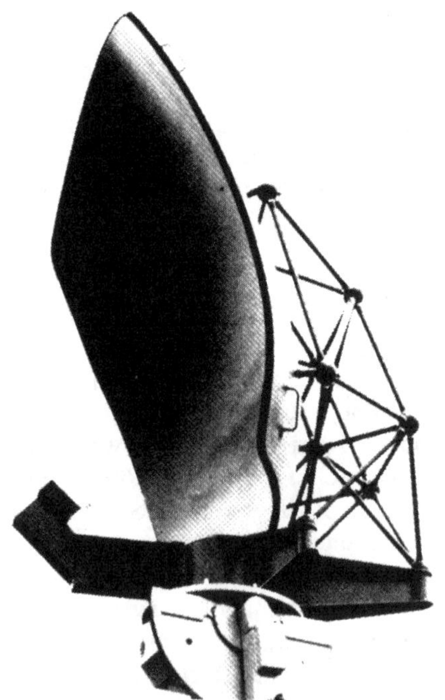

Figure 6.7.5 ULSA reflector antenna [6.9].
(a) Azimuth beam pattern. (b) Antenna on test rotor.
(Courtesy of L. M. Ericcson Co.)

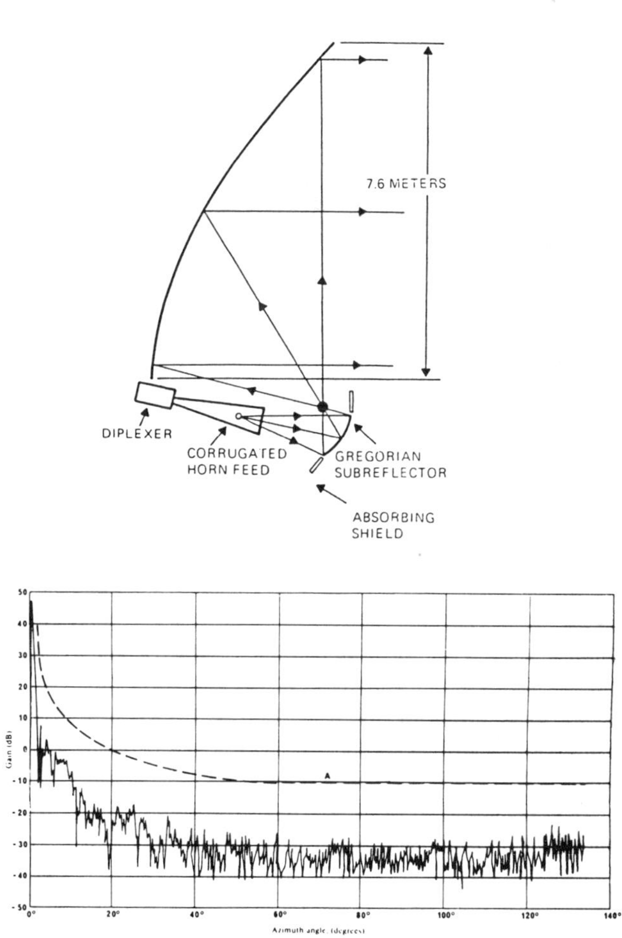

Figure 6.7.6 ULSA reflector antenna using Gregorian configuration [6.10].

Modern stacked-beam antennas (Figure 6.8.2) use planar arrays, in which n_e rows of n_a elements each are fed from an elevation beamforming matrix. The antenna is still scanned mechanically in azimuth. The row feeds and radiating elements can consist of slotted waveguides, or corporate feed networks with radiating dipoles. The elevation beam-forming matrix can be at RF, using a Rotman lens (Figure 6.6.9), at IF, or in the form of digital processing (Figure 6.8.3). Since only the single elevation matrix is used, the elevation sidelobes will not be very low unless great precision is maintained in coupling to the n_e rows of radiating elements. On the other hand, the one matrix can be fabricated with considerable care, since the one matrix will not be a major expense.

Figure 6.8.1 AN/TPS-43 stacked-beam antenna.
(Photo courtesy of Westinghouse Electric Co.)

In some systems, and necessarily in those using IF or digital beam forming, amplifiers are used prior to the matrix. Placing amplifiers before an RF matrix makes it easier to design the matrix, since moderate losses can be tolerated without affecting the system noise temperature. However, the gain and phase stability of these amplifiers is critical in establishing the elevation sidelobe levels. In the digital beamformer, these amplifiers consist of the entire receiver, providing outputs through phase detectors and A/D converters. This makes it more difficult to maintain the phase and gain stability required for low elevation sidelobes. In addition, whenever amplifiers are used, the failure of one row amplifier will introduce large elevation sidelobes, even though it may have small effect on the gain of each elevation beam channel.

Figure 6.8.2 AN/TPS-70 stacked-beam array.
(Photo courtesy of Westinghouse Electric Co.)

6.8.2 Digital Beam Forming

The ultimate in flexibility of receiving arrays is provided by digital beam-forming systems (Figure 6.8.3). Each of T elements passes its signal through a low-noise receiver to $2T$ phase detectors into A/D converters, which operate at a rate equal to the signal bandwidth B on the I and Q baseband outputs. These outputs, having a digital word rate $2TB$, can be processed to yield $M \leq T$ orthogonal beams. These beams can be further combined with weighting to produce any number of overlapping beams, each with the full signal bandwidth and with pattern and sidelobe characteristics determined by T, by the number of bits in the digital words, and by errors in the element channels and digital weighting coefficients.

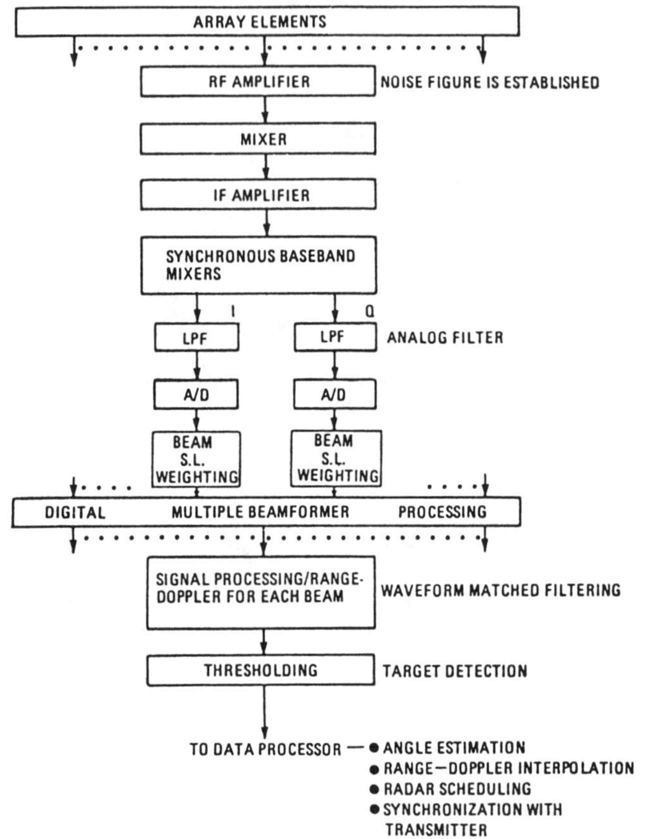

Figure 6.8.3 Digital beam-forming system [6.11].

6.9 References

[6.1] R. S. Elliott, *Antenna Theory and Design*, Prentice-Hall, 1981.

[6.2] D. D. Howard, Analysis of the 29-foot monopulse Cassegrainian antenna of the AN/FPQ-6 and AN/TPQ-18 precision tracking radars, *U.S. Naval Research Laboratory Memo.* 1776, June 1967, DDC Doc. AD 816772.

[6.3] M. I. Skolnik, *Introduction to Radar Systems*, McGraw-Hill, 1962.

[6.4] D. K. Barton, *Modern Radar System Analysis,* Artech House, 1988.

[6.5] T. C. Cheston and J. Frank, "Array Antennas," Chapter 11 in *Radar Handbook* (M. I. Skolnik, ed.), McGraw-Hill, 1970.

[6.6] L. Cardone, "Ultra-wideband microwave beamforming technique," *Microwave J.*, April 1985, pp. 121-131.

[6.7] D. Archer, "Lens-fed multiple beam arrays," *Microwave J.*, October 1975, pp. 37-42.

[6.8] P. J. Kahrilas, "HAPDAR - An operational phase array radar," *Proc. IEEE* **56**, No. 11, November 1968, pp. 1967-1975.

[6.9] E. Carlsson *et al.*, "Search radar reflector antenna with extremely low sidelobes," *Proc. Military Microwaves '82,* Microwave Exhibitions and Publishers, Ltd., Turnbridge Wells, England, October 1982, pp. 500-505.

[6.10] H. E. Schrank, "Low sidelobe reflector antennas," *IEEE APS Newsletter,* April 1985, pp. 5-16.

[6.11] A. E. Ruvin and L. Weinberg, "Digital multiple beamforming techniques for radar," *IEEE Eascon '78 Record*, pp. 152-163.

Chapter 7

WAVEFORMS AND SIGNAL PROCESSING

The concept of the radar waveform and some examples of the simpler types were discussed in Chapters 1 and 2. This chapter builds on the material outlined therein, with some additional attention given to the general factors that attend waveform design. Some of the more complex, but common, waveforms used in current radar systems are described here. Pulse compression waveforms, which represent an important subset of the matched filter concept, are presented first, followed by a treatment of the processing of pulse trains in MTI and pulsed doppler waveforms. Finally, a brief section is devoted to a more recent development, *impulse radar,* in which very short pulses are transmitted and processed in wideband receivers.

7.1 Waveform Design Considerations

In Chapter 2 we showed by way of the radar search equation (2.4.7) that target detection range was a function, among other factors, of the total energy on target during the radar's time-on-target. Neither the radar operating frequency f_t, nor the form of that energy appeared in the equation. The energy relationship of (2.4.7) can therefore be considered as a minimum design requirement for detection, but until the radar waveform and frequency are specified, virtually nothing can be known about either the kind of information that a radar provides, or its quality. The waveform (which includes the radar operating frequency) determines which target parameters the radar can measure, how accurately these can be measured, the radar's resolution capability, and the degree to which the radar can reject unwanted sources of interference such as noise and clutter.

Waveform design is the process of determining the preferred modulation characteristics of the transmitted signal and the receiver's response to the target echo after it has been modified by other reflections or sources of interference during its two-way path. In the early days of radar, the waveforms that could be generated practically were only of two simple types (see Chapter 1): continuous wave (CW) and regular pulse, and the radars developed as a consequence were limited in their ability to perform measurements of target parameters. In addition to angle, the early CW radars could measure radial velocity but not range, and the pulse radars could measure angle and range but could only estimate radial velocity by dividing successive range measurements by the time between measurements. During World War II, D. O. North developed the mathematical concept of the *matched filter* [7.1], which stimulated major advances in waveform design and signal processing, the first and subsequently highly successful example of which was the pulse compression waveform cited in Chapter 1. Today, the design, analysis, and evaluation of waveforms is based on the concept of the matched filter in conjuction with the work of P. M. Woodward, who in 1953 [7.2] developed the radar ambiguity function, a simple but powerful mathematical formula that allows the radar engineer to quantitatively describe the ability of any waveform to resolve multiple radar reflected signals from the reference target of interest. (See Appendix G.)

Although an infinite number of waveform designs are theoretically possible, there is no single waveform that is optimum for all radar applications. The particular waveform (or waveforms) that a radar employs is intimately tied to the functions and mission of that radar. Operating frequency f, plays an important role in determining the angle and velocity resolution of a radar, and is a major factor when acceptable performance in weather is required. For these and other practical reasons (see Appendix K), the operating frequency is decided upon first, and the requirements for the remainder of the waveform, i.e., the modulation applied to the carrier, are then established based upon the target parameter measurement requirements (type, accuracy, resolution, and data rate) and other expected characteristics of the operational environment. Examples of several different types of waveform modulation techniques are given in the sections which follow.

7.2 Pulse Compression

The concept of time compressing a received radar signal was independently developed by several individuals during the World War II time period. This concept permitted an increase in the range resolution capability (i.e., requiring a narrow pulse) of a radar without loss in detection range, which is generally associated with high-energy, long-duration pulses. Frequency-modulated pulse compression, referred to in Chapter 1, is most easily illustrated by assuming that the carrier of the transmitted signal whose amplitude is shown in Figure 7.2.1(a) is linearly swept, as depicted in Figure 7.2.1(b). If the receiver contains an ideal lossless filter with the time-delay vs. frequency characteristic of Figure 7.2.1(c), this would cause the first part of the received signal to be delayed relative to the latter parts of the signal. With the right amount of delay this would produce a narrower pulse at the filter output, Figure 7.2.1(d), with an increased peak amplitude. The linear time-delay characteristic of the filter acts to delay the high-frequency components of the frequency-swept input more than the low-frequency components, with the frequency components in between experiencing a proportional delay. In this example we can assume that the principle of conservation of energy applies, so that the increase in peak power of the time-compressed pulse τ_p would be proportional to the ratio of the widths of the filter input and output pulses, or

$$P_o/P_i \;=\; \tau/\tau_p \tag{7.2.1}$$

where

P_o = peak power of the compressed pulse
P_i = peak power of the input pulse
τ = uncompressed pulse width
τ_p = compressed pulse width.

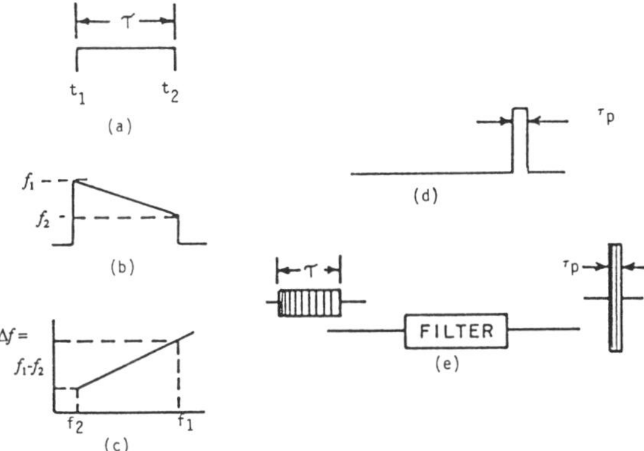

Figure 7.2.1 Idealized pulse-compression characteristics. (a) Wide-pulse envelope. (b) Carrier-frequency modulation. (c) Filter time delay characteristics. (d) Compressed-pulse envelope. (e) Input-output waveforms [7.3].

The fact that a long duration signal can be transformed into a much narrower duration signal by the process outlined above does not represent a contradiction, because the frequency modulated carrier contains the wideband information necessary to construct the narrower compressed pulse.

If the product of the transmitted pulse width τ and the frequency deviation $\Delta f = f_1$-f_2 is large, the linear progression of the carrier frequency between f_2 and f_1 should result in an approximately rectangular spectrum-amplitude distribution as shown in Figure 7.2.2. The linear frequency modulation results in a spectrum phase component (i.e., imaginary part of the complex spectrum) that is parabolic in shape, also shown in Figure 7.2.2. When the parabolic phase is removed from the input signal by the conjugate phase shift of the linear delay filter, optimum pulse compression occurs, and the filter output pulse is the Fourier transform of the rectangular amplitude spectrum. The shape of such a signal is given by

$$a_o(t) \;=\; \sqrt{\tau\Delta f}\,\frac{\sin \pi\Delta f t}{\pi\Delta f t} \tag{7.2.2}$$

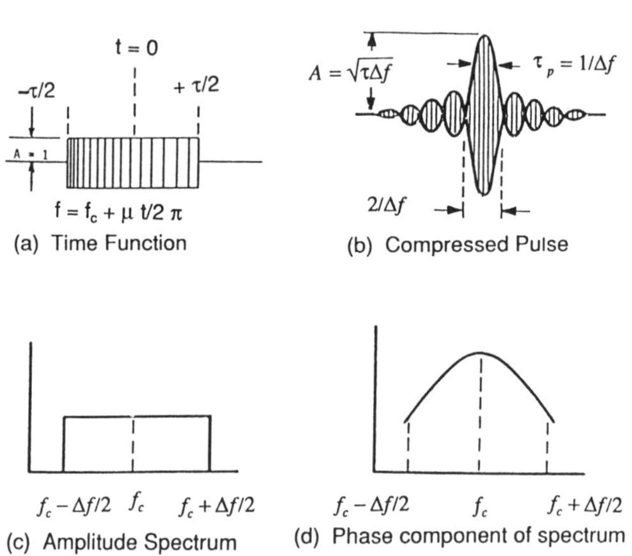

Figure 7.2.2 Linear FM waveform and spectrum details.

The width of this pulse is $\tau_p = 1/\Delta f$ measured 4 dB down from the peak, as indicated in Figure 7.2.2(b). Thus, the filtering process to compress the input signal changes the rectangular shape of the input signal to a $(\sin x)/x$ shape that exhibits the *range sidelobes* shown in Figure 7.2.2(b). This is not the most desirable pulse waveform shape for some radar applications, and additional filtering techniques are usually employed to reduce the range sidelobe levels. This is discussed in greater detail in Section 7.2.2.

7.2.1 Advantages of Pulse Compression Waveforms

The previous discussion described the use of pulse compression in terms of increasing the transmitted signal bandwidth in order to obtain, by means of matched filtering at the receiver, a receiver output signal with greater range resolution capability. Listed below are more formal statements regarding the principal advantages of pulse compression waveforms.

1. More efficient use of average power available at the radar transmitter and, in some cases, avoidance of peak power problems in the high power sections of the transmitter. If the transmitter employs relatively low-power solid-state components, pulse compression provides a method of generating a signal with a high "effective" peak power.

2. Increased system resolving capability, both in range and velocity, since flexibility is provided in independently choosing the transmitted pulse duration and the compressed pulse width. In the case of range resolution, the generation of extremely fast rise-time, high peak power signals is unnecessary when pulse compression techniques are used.

3. Reduction of vulnerability to certain types of interfering signals that do not have the same properties as the coded pulse-compression waveform. As an example, pulsed interference having the same bandwidth as the pulse compression signal will be smeared out and reduced in peak amplitude by the receiver matched filter. This can be a powerful ECCM feature.

4. Extraction of information from the signals present at the receiver input to obtain estimates of important parameters associated with the individual signals, such as range, velocity, and possibly acceleration. This aspect of radar matched filter processing is referred to as *parameter estimation*. The ability of any specific waveform to perform these types of functions is related to its ambiguity function characteristics. The waveform ambiguity function is briefly discussed in Appendix G.

7.2.2 Linear FM Pulse Compression

Linear FM pulse compression, outlined above, is the most common type of pulse compression waveform used in search and tracking radars. This section provides fuller details about this important radar waveform. The exact equation for this signal format is given by [7.1]

$$s(t) = \cos\left[2\pi f_o t + \frac{1}{2}\mu t^2\right], \; -\frac{\tau}{2} \le t \le \frac{\tau}{2} \tag{7.2.3}$$

where $\mu/2\pi = \Delta f/\tau$ is the linear FM modulation rate, and $f_0+(\Delta ft)/\tau$ is the waveform instantaneous frequency modulation function. The output of the filter matched to this signal is

$$f(t) = \sqrt{\tau\Delta f}\frac{\sin\left[(\pi f_d + \mu t/2)(\tau - |t|)\right]}{(2\pi f_d + \mu t)\tau/2}\cos(2\pi f_o + \pi f_d)t \tag{7.2.4}$$

Figure 7.2.3 shows the detected output waveform of the matched filter for different values of doppler shift normalized to the swept bandwidth Δf. This illustrates the delay-doppler shift ambiguity, as well as the loss in peak filter output and the broadening of the compressed pulse for large doppler shifts. These latter two effects are the result of the effective narrowing of the signal spectrum as the doppler shift causes it to "slide" out of the approximately rectangular passband of the matched filter. Figure 7.2.4 illustrates the details of the matched filter envelope as a function of time-bandwidth product (or compression ratio), illustrating that the compressed pulse is only an approximation to the $(\sin x)/x$ shape assumed in the discussion in Section 7.3. As shown in this figure, the approximation becomes better for larger pulse compression ratios.

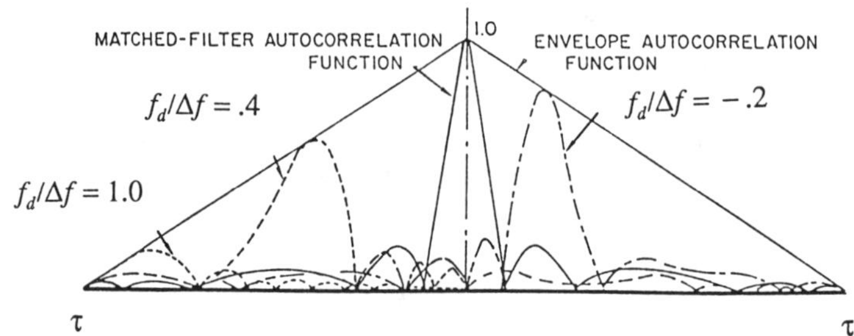

Figure 7.2.3 Matched-filter output waveforms for frequency shifted input [7.3].

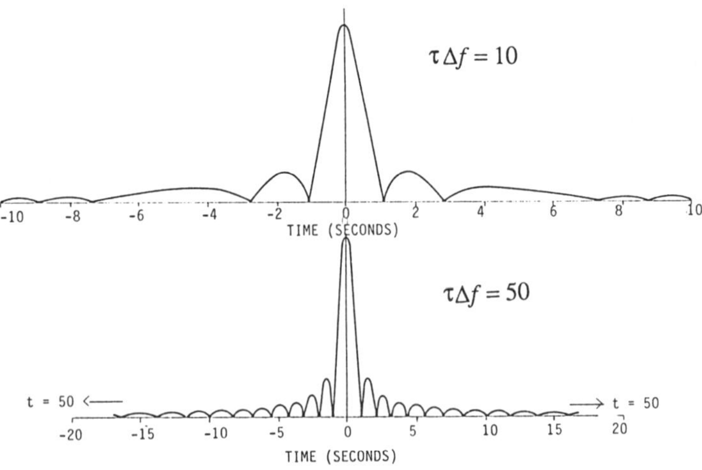

Figure 7.2.4 Linear FM matched-filter response for $\tau\Delta f = 10$ and $\tau\Delta f = 50$ (Δf normalized to unity) [7.3].

Figure 7.2.5 depicts the amplitude component of the signal spectrum for three compression ratios. These show that the spectrum magnitude tends to become more rectangular in shape as the time-bandwidth product increases. However, this function can never become exactly rectangular since the waveform given in (7.2.3) is time limited. Nevertheless, for many first-order analyses a rectangular spectrum magnitude is often assumed. The amplitude ripples in the spectrum result from the assumption of the infinitely fast rise-time of the rectangular time envelope. These ripples can limit the effectiveness of sidelobe reduction techniques for time-bandwidth products < 50. The magnitude of these ripples can be made smaller if the rise-time is more gradual [7.4]. Whether this can be achieved in practice depends on the design of the radar transmitter being consistent with maintaining a slower pulse rise time. In evaluating linear FM pulse compression it is important that both the compressed pulse and the spectrum magnitude be symmetrical and well-formed, as shown in Figures 7.2.4 and 7.2.5.

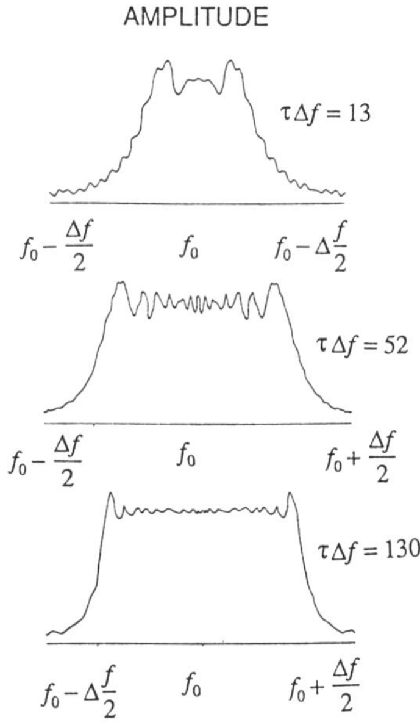

Figure 7.2.5 Linear FM spectra.

The linear FM waveform can be generated by either active or passive means. In the former, an accurate voltage controlled oscillator is linearly swept, as shown in Figure 7.2.6(a), or a phase controlled oscillator is driven by a square-law function, as depicted in Figure 7.2.6(b). With the phase-controlled oscillator the output frequency function is proportional to the derivative of the control voltage. Passive generation of the linear FM waveform is achieved by impulsing the linear FM matched filter as shown in Figure 7.2.7. The output is a stretched, linearly swept signal. The linear frequency modulation of the pulse expansion signal has the opposite direction from that of a signal that would be compressed by the matched filter. In order to use the same matched filter for both pulse expansion and pulse compression, a sideband inversion circuit must be used to reverse the direction of the FM sweep. If this is not done, then a separate compression filter will be needed. With either of these approaches to linear FM generation, the signal spectrum must conform to the examples shown in Figure 7.2.5.

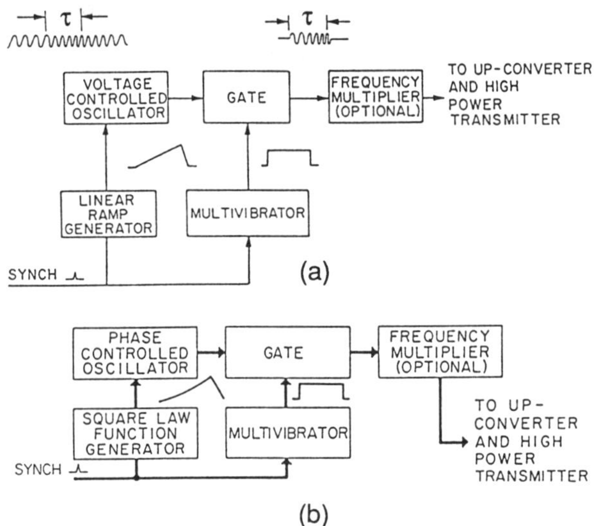

Figure 7.2.6 Techniques for active generation of linear FM signal [7.3].
(a) Frequency modulation. (b) Phase modulation.

If the time-bandwidth product of the signal is very large, the matched filter for pulse expansion and pulse compression can be designed using parallel linear delay line elements as illustrated in Figure 7.2.8. The advantage of this approach is that it reduces the size of the dispersive delay line design, which for a single-channel implementation, is directly proportional to $\tau\Delta f$. With the parallel channel approach, each channel has a time-bandwidth product of $\tau\Delta f/N^2$, and N channels result in an overall time-bandwidth product of $\tau\Delta f$. In early pulse-compression radars linear delay matched filters were built from lumped constant bridged-T networks. Current pulse compression radars generally use some form of acoustic wave device [7.5] or a digital filter equivalent. Radars that require several pulse compression modes to match the waveform energy content and resolution capability to the target environment often use software controlled microprocessors to generate and process the pulse compression signals.

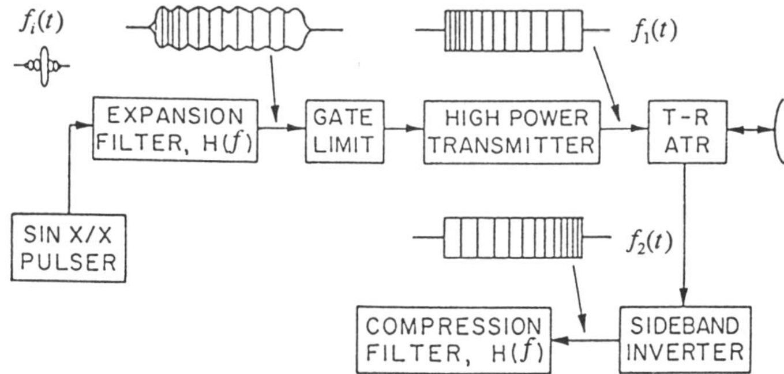

Figure 7.2.7 Matched-filter radar using sideband inversion [7.3].

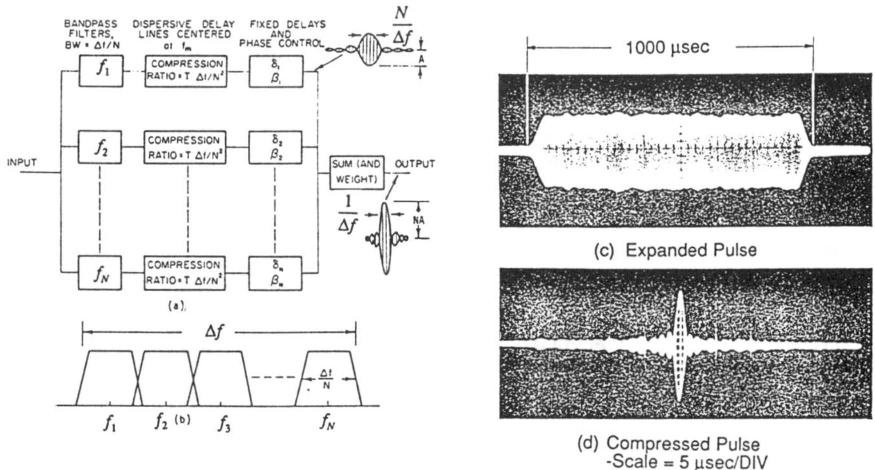

Figure 7.2.8 Parallel channel pulse compression processor for large time-bandwidth linear FM signals [7.3]. (a) Parallel channel block diagram. (b) Bandpass filter characteristics. (c) Expanded pulse. (d) Compressed pulse.

As noted previously, the $(\sin x)/x$ shape of the linear FM matched filter signal has time sidelobes that can mask smaller signals. These sidelobes can be reduced either by time-shaping the transmitted pulse envelope or by frequency-shaping (weighting) the amplitude response of the receiver. Shaping or weighting the transmitted envelope is generally not possible with high power radar transmitters, and frequency response weighting is the most common method of sidelobe reduction. This is illustrated by the block diagram of Figure 7.2.9. A number of different weighting functions can be used. One of the most common is the Hamming function, which is a special case of the $\cos^n$-on-pedestal weighting function given by

$$W(f) \;=\; k + (1-k)\cos^n\!\left(\frac{f-f_o}{2\Delta f}\right) \tag{7.2.5}$$

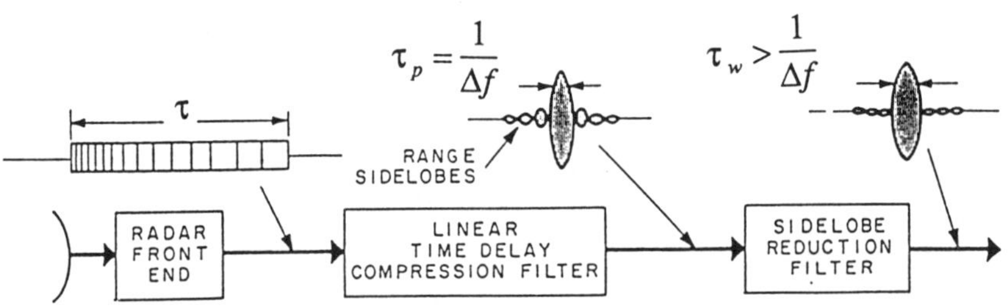

Figure 7.2.9 Range sidelobe reduction by weighting of receiving filter.

The Hamming function is obtained when $k = 0.08$ and $n = 2$. This is an "optimum" weighting function in the sense that it results in a minimum broadening of the compressed pulse, as well as minimum detection loss compared to the matched condition, when a -40 dB sidelobe level is specified. For many radars, a sidelobe level of -40 dB is adequate. Figure 7.2.10 shows the effect of the weighting function of (7.2.5) on the compressed pulse for different values of k and n.

Table 7.2.1 lists the performance data for several typical weighting functions. If the rate of fall-off of the time sidelobes distant from the compressed pulse is an important factor, then it should be noted that any weighting function with a pedestal results in a far sidelobe theoretical fall-off rate of $1/t$. During system evaluation, levels of the sidelobes closest to the compressed pulse and the rate of fall off of those farthest from the compressed pulse must be measured and compared to the values in the system specification.

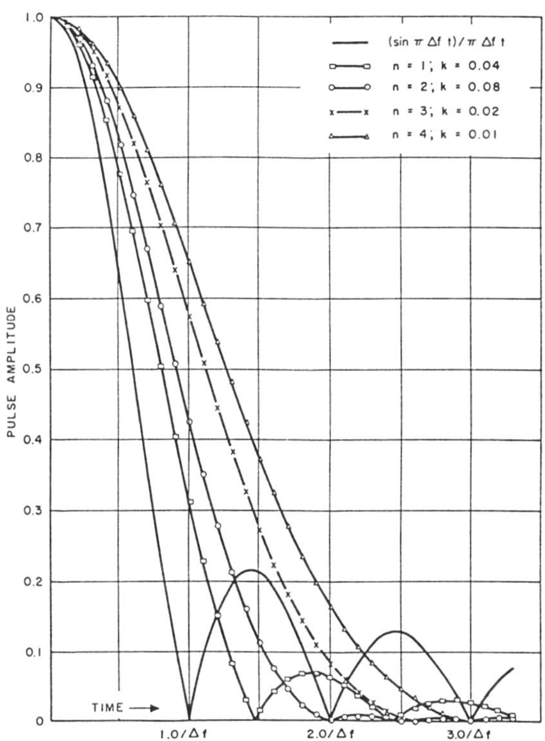

Figure 7.2.10 Waveforms for $\cos^n$ weighting on pedestal of height k.

Table 7.2.1 Weighting Function Data[1]

Weighting function	Peak side-lobe (dB)	Pulse widening	Mismatch loss (dB)	Far sidelobe fall-off rate
Dolph-Chebyshev	-40	1.35	1.5	1
Taylor, n=6	-40	1.41	+1.2	$1/t$
$k + (1-k) \cos^n$				
Hamming (k=0.08, n=2)	-42.8	1.47	+1.34	$1/t$
$\cos^2$ (k=0, n=2)	-32.2	1.62	1.76	$1/t^3$
$\cos^3$ (k=0, n=3)	-39.1	1.87	2.38	$1/t^4$
$n = 1, k = 0.04$	-23	1.31	0.82	$1/t$
$n = 2, k = 0.16$	-34	1.41	1.01	$1/t$
$n = 3, k = 0.02$	-40.8	1.79	2.23	$1/t$

[1] Note: All data are based on rectangular unweighted spectrum.
(From Cook and Bernfeld [7.3])

In practice, the distant sidelobes will be dominated by the time (range) sidelobes associated with the amplitude ripples in the linear FM spectrum depicted in Figure 7.1.5. These range sidelobes tend to peak in the interval $\pm\, \tau/2$ relative to the compressed pulse and have an approximate level $(20 \log\tau\Delta f + 3)$ dB below the compressed pulse peak [7.6]. Thus, the system specification with respect to compressed pulse range sidelobes must take these additional sidelobes into account.

When performing evaluation tests on linear FM pulse compression waveform generators and processors, it should be possible to obtain the type of test waveforms illustrated in Figure 7.2.11. Figure 7.2.11(a) shows the linear FM pulse expansion waveform and the compressed pulse obtained with an acoustic pulse expansion/pulse compression processor for $\tau\Delta f = 100$. Figure 7.2.11(b) shows the compressed pulse on an expanded time scale to illustrate the symmetry of the $(\sin x)/x$ waveform. The delay is a constant of the compression filter that is a reference for the generation of the receiver time-base (or ranging) sweep. In recent acoustic device pulse-compression processors the sidelobe reduction weighting response is an integral part of the compression delay line. In this case the pulse compression performance criteria can be represented in a single diagram, as shown in Figure 7.2.12. In the example shown, the unweighted compressed pulse width would be 100 ns, with $\tau\Delta f = 100$. The pulse widening caused by sidelobe reduction is given by $\alpha = 1.43$, and $\delta_w = 1.15$ dB is the loss in compressed pulse signal-to-noise ratio. In this figure the spectrum ripple (Fresnel) sidelobes are clearly depicted and separately specified from the nominal sidelobes reduced by the Taylor weighting function [7.7], which is a weighted cosine series added to a constant, or pedestal.

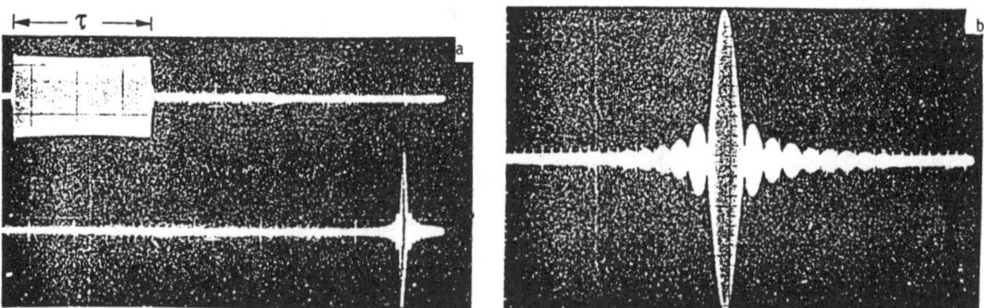

Figure 7.2.11 Typical pulse compression waveforms, $\tau\Delta f = 100$, obtained with an acoustic pulse compression/expansion processor.

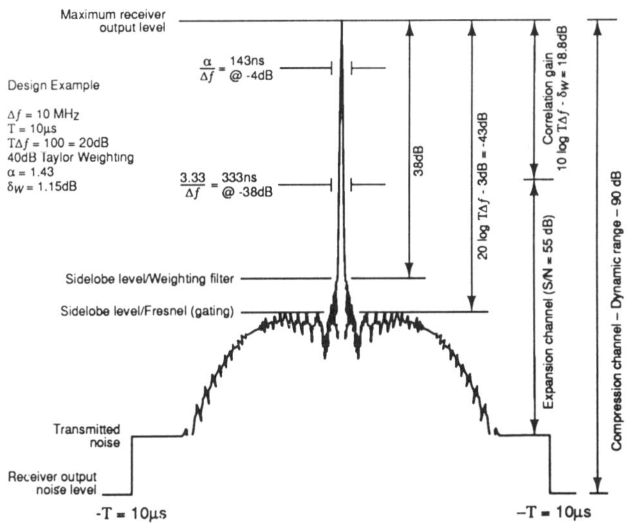

Figure 7.2.12 Example of a pulse compression receiver specification for a pulse compression filter with integral sidelobe reduction weighting (Courtesy Anderson Laboratories, Bloomfield, Connecticut).

7.2.3 Nonlinear FM Pulse Compression

The approximately rectangular linear FM spectrum is a special case of a general approximation that states that the magnitude of the signal spectrum of a rectangular envelope, frequency modulated signal is proportional to the inverse of the derivative of the frequency vs. time function.

This relationship suggests that matched filters for pulse compression signals with low time sidelobes (e.g., with weighted spectrum magnitudes) can be achieved with frequency modulation functions that are nonlinear vs time. Examples of these are described by Key, et al. [7.8] and by Cook [7.9]. The disadvantage of nonlinear FM pulse compression waveforms is that the time sidelobes increase rapidly as a function of doppler shift. However, in applications where the doppler shift is only a fractional percentage of the FM bandwidth, these signals can provide a more efficient method of producing low time-sidelobe compressed pulses by avoiding the mismatch loss associated with weighting the approximately rectangular spectra of linear FM waveforms.

7.2.4 Linear Step-FM

Linear step-FM [7.3] is an important waveform for two reasons. First, it approximates the linear FM pulse compression signal without requiring the use of linear dispersive delay lines. Second, it provides a conceptual bridge to a class of *polyphase pulse compression codes* that also approximate linear FM behavior, with the exception that the matched filter outputs of these polyphase codes have very low time sidelobes. These will be discussed in the next section.

The parameters of the linear step-FM waveform are shown in Figure 7.2.13. The frequency vs. time function of step-FM comprises N equal steps of $1/\delta$, where δ is the time duration of each frequency step. The frequency vs. time function in this figure is depicted as increasing with time; however, the waveform could just as well have decreasing FM steps. The total duration of the waveform is $N\delta$ and the total frequency excursion Δf is N/δ. Thus the time- bandwidth product of this waveform is N^2. The frequency steps in Figure 7.2.13 are coherent; that is, there is phase continuity at the time points where the frequency steps from f_n to f_{n+1}. The matched filter for this waveform is shown in Figure 7.2.14. It has N parallel channels, each centered in frequency to match one of the frequency steps. The overall matched-spectrum filter shown at the processor output has approximately the same shape as that for a linear FM spectrum. This filter, too, can depart from the ideal response without having a serious effect on the compressed pulse output. The time resolution of the output is $1/\Delta f = \delta/N$.

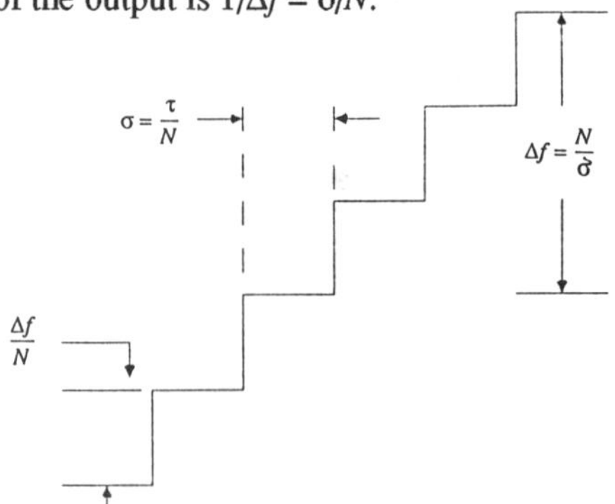

Figure 7.2.13 Frequency vs. time function of linear step-FM.

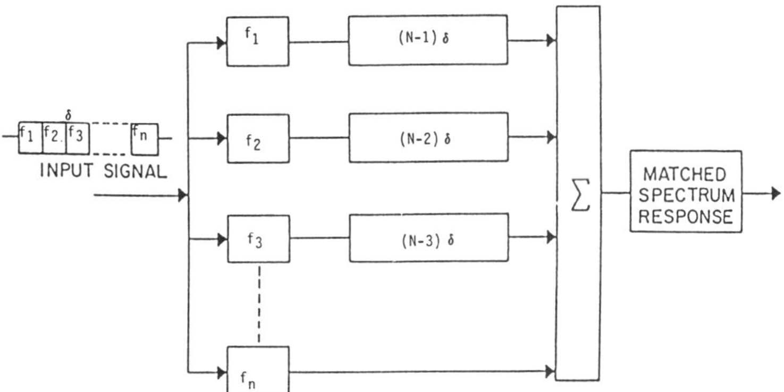

Figure 7.2.14 Matched-filter processor for linear step-FM.

The step-FM matched filter output has similarities to and differences from that of linear FM. The main feature of the step-FM output function is the ridgeline that indicates that the range/doppler coupling of step-FM is essentially identical to that of linear FM. The zero-doppler compressed pulse also has an approximate $(\sin x)/x$ structure that is very close to that of linear FM. However, the major departure from the linear FM compressed pulse waveform is the presence of "grating" time sidelobes that are created when the doppler shift is such that the received step-FM has a mismatch to the step-FM function assumed in the matched filter. The maximum mismatch, and the largest grating sidelobes, occur when

$$f_d = \frac{(2n-1)}{2N}\Delta f, \quad n = 1, 2, 3, ..., N/2 \tag{7.2.6}$$

The grating sidelobes have zero amplitude when

$$f_d = \frac{m}{N}\Delta f, \quad m = 1, 2, 3, ..., N \tag{7.2.7}$$

At these values of doppler shift there are only $(N\text{-}m)$ matches between the steps of the received waveform and the matched filter channels. Thus, the effective output bandwidth for these cases is $(N\text{-}m)/N\Delta f$ and the compressed pulse width is $\delta/(N\text{-}m)$. Examples of the linear step-FM matched filter output as a function of doppler shift are shown in Figure 7.2.15. The existence of the grating sidelobes reduces the amount of waveform energy in the main compressed pulse. Thus, the peak amplitude of the step-FM matched filter output does not fall off linearly as a function of doppler shift, as with linear FM. The comparative compressed pulse amplitudes vs. doppler shift for linear FM and step-FM vs. doppler shift are shown in Figure 7.2.16. It should be noted that sidelobe reduction weighting can be used with step-FM; however, this has relatively little effect on the grating sidelobes.

Thus, step-FM would only be used where the doppler shift (or doppler spread) is very small (i.e., $f_d \ll 1/\delta$). Step-FM can be used to create a linear FM waveform by filling in the FM steps with a sawtooth FM function. The matched filter for this signal would be either a continuous linear dispersive delay line or the parallel channel implementation shown in Figure 7.2.8.

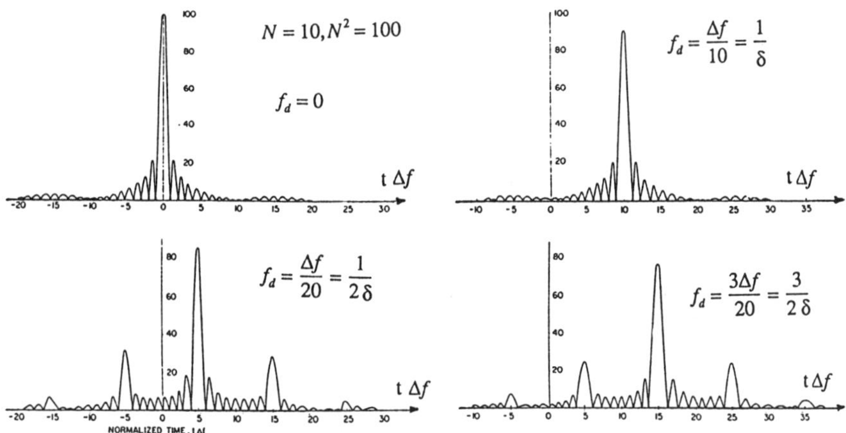

Figure 7.2.15 Linear step-FM matched-filter outputs
as a function of doppler shift [7.3].

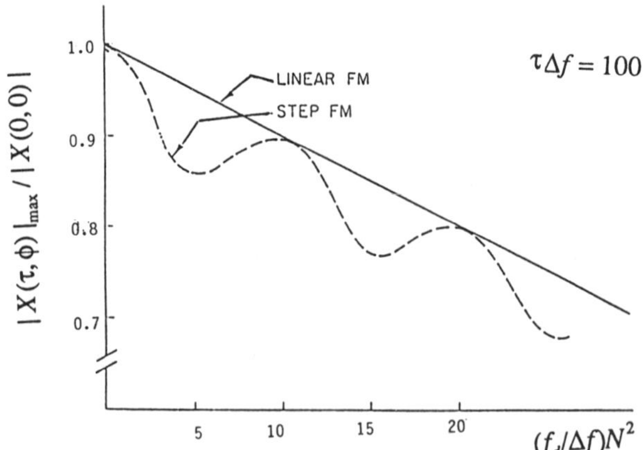

Figure 7.2.16 Comparative peak matched-filter outputs
for linear FM and linear step-FM vs. doppler shift.

7.2.5 Polyphase Coded Waveforms

The time (range) sidelobes associated with linear FM, step-FM, and other pulse compression waveforms can create problems in multiple target and/or clutter environments, particularly if there is a requirement for a large dynamic range of received target signals. Many types of waveforms have been considered to achieve very low time sidelobes. Some of these can be obtained only with a nonuniform transmitted signal amplitude over the uncompressed pulse width combined with a phase code function that is difficult to synthesize at the transmitter. An important class of uniform-amplitude, discrete polyphase codes has been discovered that does exhibit superior sidelobe behavior compared to linear FM and step-FM. Interestingly, these are phase-quantized approximations to step-FM and linear FM, and have a number of properties that are similar to these waveforms. The polyphase codes have the advantage of being generated and processed with digital techniques. Thus, they are applicable to modern computer-controlled radar systems. Historically, the first of this class of polyphase codes considered for radar application was the Frank code format. This waveform was originally developed by R. L. Frank for Loran-C. It was later seen to be equivalent to a pulse-compression, matched filter waveform.

Frank Code Properties

Frank codes [7.10] are composed of N sets of phase sequences that can be described by an $N \times N$ matrix in which the phase in the ith row and jth column is given by

$$\Phi_{i_j} = (2\pi/N)(i-1)(j-1) \qquad\qquad (7.2.8)$$

An example of a Frank code 4×4 matrix is shown in Table 7.2.2.

Table 7.2.2 Frank Code Matrix, $N^2=16$ (modulo 2π)

0	0	0	0
0	$\dfrac{8\pi}{16}$	$\dfrac{16\pi}{16}$	$\dfrac{24\pi}{16}$
0	$\dfrac{16\pi}{16}$	0	$\dfrac{16\pi}{16}$
0	$\dfrac{24\pi}{16}$	$\dfrac{16\pi}{16}$	$\dfrac{8\pi}{16}$

The elements of this matrix, on a row-by-row basis, are the phase shifts given to the carrier frequencies of N^2 contiguous pulses, each of duration τ_p. Thus, the total sequence duration $\tau = N^2 \tau_p$, and the compression ratio $\tau/\tau_c = N^2$. Eq. (7.2.8) represents a phase sequence that is a sampled approximation to a step-FM waveform, as illustrated in Figure 7.2.17. If the received waveform having a Frank-code sequence is processed through a tapped delay-line matched filter (or its digital equivalent), the envelopes of the compressed pulse outputs appear as shown in Figure 7.2.18. The width of the compressed pulse is τ_c, and the signal bandwidth $B = 1/\tau_c$. These compressed pulse signals have much lower time sidelobes than those of linear FM or step-FM. For large N, the Frank-code peak range sidelobes are approximately $20 \log(\pi N)$ dB below the maximum value of the compressed pulse.

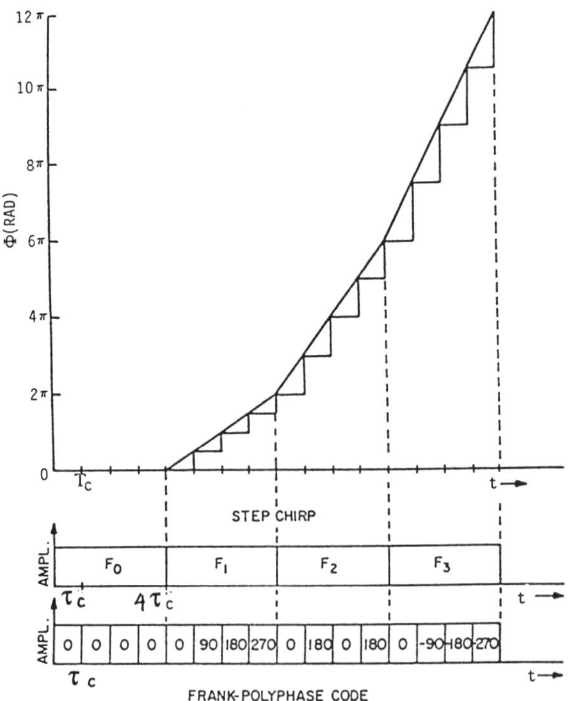

Figure 7.2.17 Step-chirp and Frank polyphase-code relationships [7.11].

The Frank-code spectra have a roughly $(\sin\pi f/B_f)/(\pi f/B_f)$ shape with asymmetric indentations, as illustrated by Figures 7.2.19(a) and 7.2.19(b). This asymmetry is much more pronounced when N is even.

The doppler shift behavior of the Frank-code matched filter output is shown in Figure 7.2.20. This is similar to that of step-FM, exhibiting large secondary peaks for $f_d = (2n-1)B/2N$ and zero for $f_d = nB/N$. In addition, the doppler-shifted outputs of Frank codes contain an *image signal* that grows larger as the doppler shift increases. When $f_d = B/2$, the matched filter output splits into two identical signals, as depicted in Figure 7.2.20(d). The matched filter output also has the same range/doppler coupling that is the trademark of linear FM and step-FM. The large secondary and image responses need not be a drawback to the use of Frank-code waveforms if the carrier frequency and/or doppler shift is not too large. Lewis and Kretchmer [7.12] state that these doppler-created sidelobes are not objectionable if $f_d/B \ll 0.005$. This translates to a target traveling at a Mach 5 radial velocity for a carrier frequency at L-band and $\tau_c = 0.5\ \mu s$. A more serious problem arises with Frank-code waveforms when there is band limiting at the radar front end. Such band limiting is typical, particularly in sampled digital implementations, to avoid spectrum foldover. In this case the asymmetry of the Frank-code spectrum causes the time sidelobes to increase if the band limiting results in a serious mismatch to the waveform spectrum.

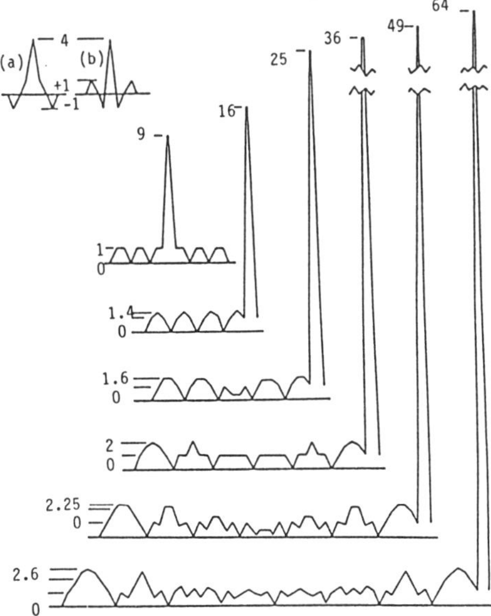

Figure 7.2.18 Frank polyphase code response functions.

P-Code Properties

Polyphase P-codes [7.11 - 7.13] are related to and are extensions of the Frank codes. There are four P-codes that have been described: the P1, P2, P3, and P4 codes. They have properties that are similar to linear FM, Step-FM, and the Frank codes. The P-codes are designed to offset some of the problems inherent in the Frank-codes and step-FM.

The P1 and P2 codes, similar to the Frank codes, can be represented by an $N \times N$ phase matrix. The phase values for the elements in the P1 matrix are given by

$$\Phi_{i_j} = -(\pi/N)[N - (2j - 1)][(j - 1)N + (i - 1)] \tag{7.2.9}$$

and those of the P2 matrix, valid for N even, by

$$\Phi_{i_j} = (\pi/2n)[N + 1 - 2i][N + 2 - 2j] \tag{7.2.10}$$

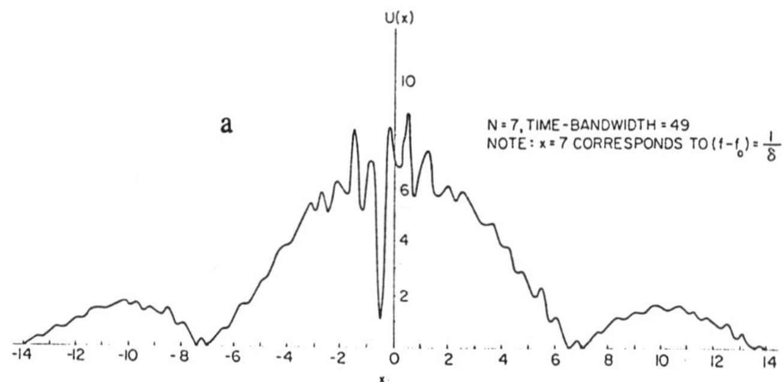

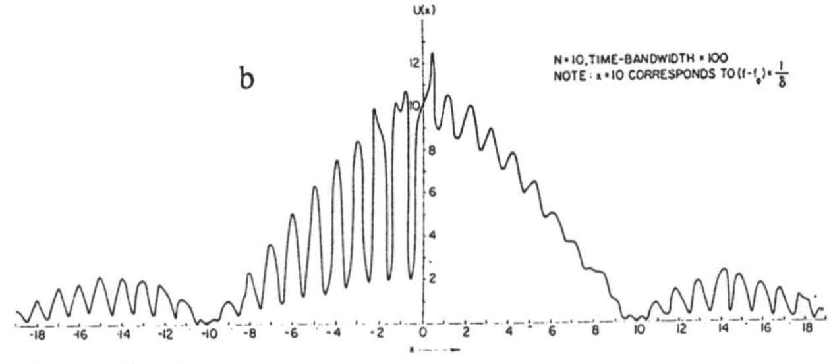

Figure 7.2.19 Frank-code spectra: (a) $N = 7$, (b) $N = 10$ [7.3].

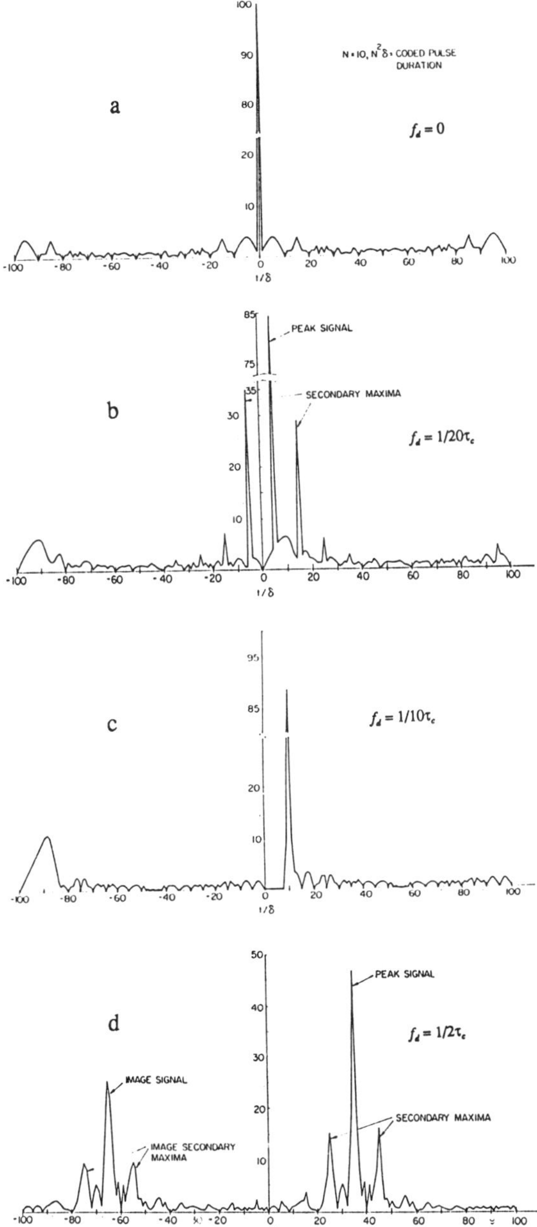

Figure 7.2.20 Frank-code matched-filter outputs vs. doppler shift, $N^2 = 100$ [7.3].

The P1 matrix yields a phase-coded waveform that is a rearrangement of a Frank code. Thus, for $N = 3$, the waveform phase code would be

$$0, -2\pi/3, -4\pi/3, 0, 0, 0, 0, 2\pi/3, 4\pi/3$$

For $N = 4$, the P2 phase sequence generated by Eq. (7.2.10) is

$$9\pi/8, 3\pi/8, -3\pi/8, -9\pi/8, 3\pi/8, \pi/8, -\pi/8, -3\pi/8,$$

$$-3\pi/8, -\pi/8, \pi/8, 3\pi/8, -9\pi/8, -3\pi/8, 3\pi/8, 9\pi/8$$

In the P2 code sequence each row of N phase shifts is symmetrical about 0°. The behavior of the matched filter outputs of the P1 and P2 codes is very similar to that of the Frank codes. The P-code spectra are more symmetrical than Frank-code spectra, with the amplitude indentations appearing on either side of the center frequency. This makes the P1 and P2 codes more tolerant to band limiting effects. Amplitude weighting of the receiver bandpass function to reduce range sidelobes can be employed with the P1 and P2 codes [7.12]. Sidelobe reduction weighting can also be used with the Frank codes. The matched filter outputs for the P1 and P2 codes, summarized in Figure 7.2.21, are identical to those of the Frank codes, being characterized by range/doppler coupling and the presence of large secondary responses and an image signal as a function of doppler shift.

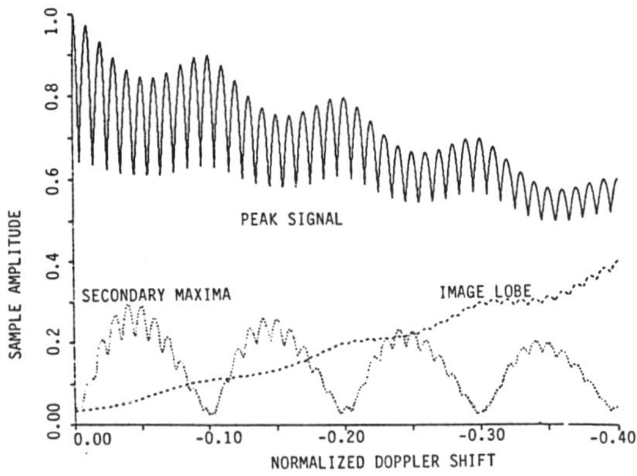

Figure 7.2.21 Doppler properties of Frank, P1, and P2 codes, $N^2 = 100$ [7.11].

The P3 and P4 codes are derived from the linear FM waveform [7.14]. These codes are more tolerant of doppler shift than the Frank and P1/P2 codes. The phase sequence of the P3 code is described by

$$[\Phi_i]_{P_3} = \pi(i-1)^2/(\tau/\tau_p) \tag{7.2.11}$$

Eq. (7.11) is a sampled version of a quadratic phase (i.e., linear FM) function. The phase sequence for the P4 code is

$$[\Phi_i]_{P_4} \;=\; \pi(i-1)^2/(\tau/\tau_p) - \pi(i-1) \qquad (7.2.12)$$

$$=\; [\Phi_i]_{P_3} - \tau(i-1)$$

Figure 7.2.22 provides a comparison of the zero-doppler and f_d=0.05/τ_p outputs for a Frank code and the P3 (or P4) code. Figure 7.2.23 plots the doppler behavior summary for the unweighted P3 and P4 codes.

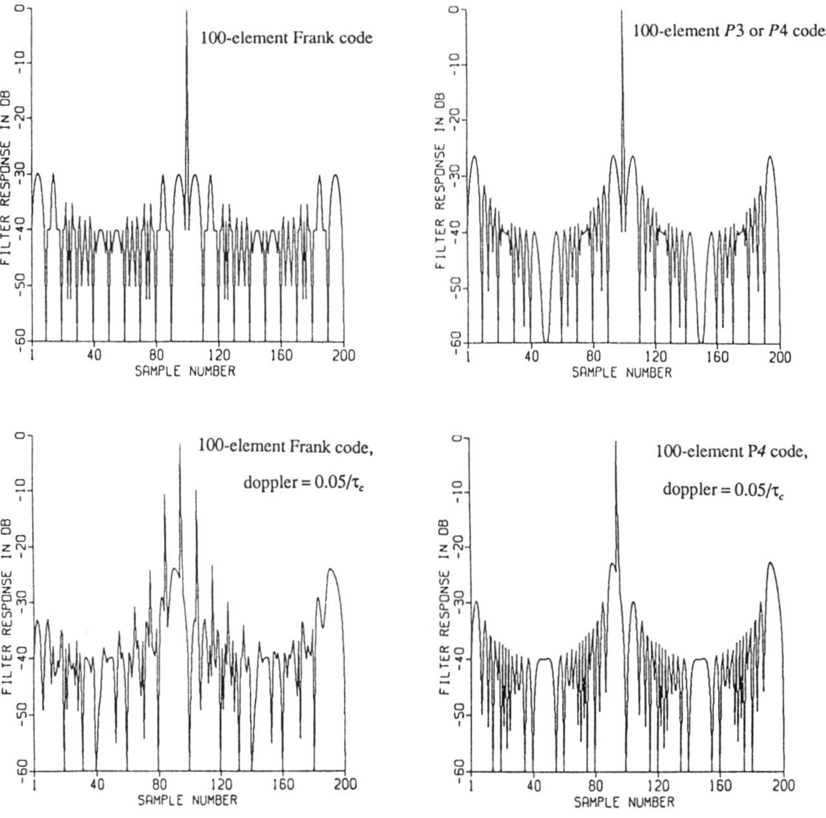

Figure 7.2.22 Comparison of Frank and P3/P4 codes
matched-filter outputs [7.13].

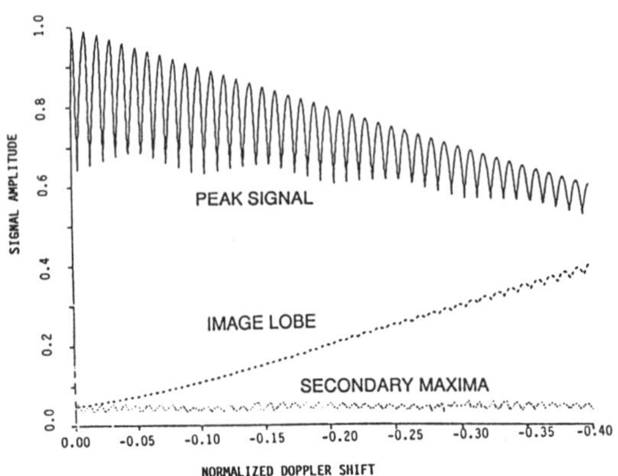

Figure 7.2.23 Doppler properties of 100-element P3 and P4 codes [7.13].

In summary, the polyphase waveforms represented by the Frank and P-codes are an important class of pulse compression signals that have much better time (range) sidelobe characteristics than unweighted linear FM, step-FM, or the m-sequences discussed in the following section. The choice of polyphase waveforms for radar use would depend on the expected doppler shift environment. One of the attractive features of a computer-controlled digital waveform generator and processor is that any of these waveforms, plus linear FM and m-sequences, can be realized in a single radar with appropriate software design. Additional details on Frank and P-codes, as well as on digital processor implementations for these waveforms, can be found in Lewis and Kretschmer [7.13].

Maximum-Length Phase-Code Pulse Compression

When radar signals are used as the primary means of distinguishing among rapidly moving targets having a significant range of radial velocities (or doppler shifts) the type of matched filter waveform required is one that has a maximum output for a specific target doppler shift and a zero, or very low output, for all other doppler shifts. This is the parameter estimation problem mentioned in Section 7.2.1. In general, this requires a waveform with random properties, such as that obtained with Gaussian noise. Processing a random noise waveform offers obvious implementation difficulties. However, certain binary sequences composed of equally subdivided segments having 0° (+) and 180° (-) phase shifts exhibit approximate randomness characteristics that make them suitable for radar waveform use.

One of the best approximations to random signals is obtained with binary (+,-) sequences that are known as *maximal length sequences (m-sequences)*. M-sequences are generated by n-stage *linear shift registers* (LSRs). They are designated as "maximal length" because they are the longest sequences that can be generated by an n-stage LSR, and have a sequence length $M = 2^n - 1$. Not all n-stage LSR's produce sequences of this length, and not all generated sequences of length $2^n - 1$ are m-sequences. Thus, it is important to state the m-sequence properties that distinguish them from other sequences of the same or shorter length. The first important property of an m-sequence (letting a one designate the + symbol, and a one the - symbol) is that the number of ones is always one more than the number of zeros. Thus, an m-sequence has $2^{n-1} + 1$ ones and 2^{n-1} zeros. The second property is that half the runs of symbols of the same kind are of length one, one-fourth are of length two, one-eighth are of length three, etc. The third significant property is that an m-sequence has a two-valued response (zero-doppler matched filter output) that is -1 for all $t \neq 0$, and N at $t = 0$.

The linear shift registers that generate m-sequences include digital logic to provide feedback paths to the shift register input. Each shift register can have an arbitrary initial loading of ones and zeros, except the all-zero case. When clock pulses are fed into the shift register input, a sequence of ones and zeros of length N is produced. The number and location of the feedback connections determines whether the sequence is maximal or not [7.14]. Figure 7.2.24 illustrates two six-stage LSR's that produce different m-sequences of length 63. The feedback connections can be represented by a polynomial function in x, such that if an x^n term appears, this indicates a feedback connection from the nth stage of the LSR. Thus, for Figure 7.2.24(a) the defining polynomial is $x^6 + x + 1$, and for Figure 7.2.24(b), the polynomial is $x^6 + x^5 + x^2 + x + 1$. One of the conditions for this polynomial to be the generating function for an m-sequence is that it have an odd number of terms and an even number of feedback connections. The initial loading of the shift register determines the starting point of each sequence, while the possible number of feedback connections determines the number of sequences that meet the properties cited above. Table 7.2.3 lists the possible number of m-sequences for $2 \leq n \leq 21$. The prime factors listed in Table 7.2.3 provide an indication of the comparative "cross-talk" properties of the sequences within each set. This can be an important factor if several radars in a region are operating with m-sequence waveforms, since it is desirable to minimize the interference among them. As a rule, the greater the number of prime factors in N, the larger the cross-talk between most pairs of signals in the set.

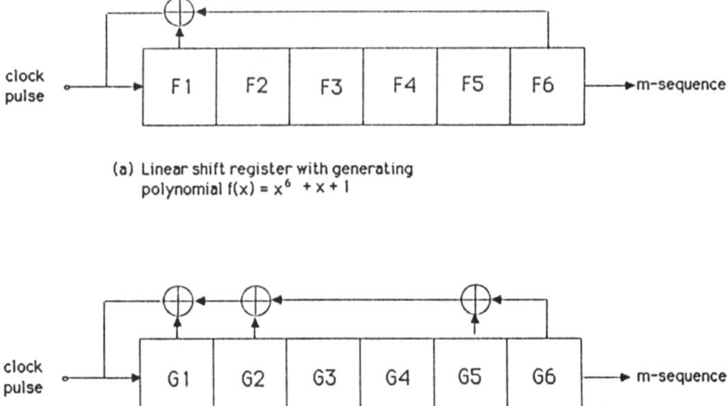

(a) Linear shift register with generating polynomial $f(x) = x^6 + x + 1$

(b) Linear shift register with generating polynomial $g(x) = x^6 + x^5 + x^2 + x + 1$

Figure 7.2.24 Examples of linear shift registers, $n = 6$, with different generating polynomials $f(x)$ and $g(x)$.

Table 7.2.3 Number of Maximal Length Sequences

No. of Stages n	Max. Seq. Length $2^n - 1$	No. Max. Length $M(n)$	Factors of $2^n - 1$
2	3	1	
3	7	2	
4	15	2	$3 \cdot 5$
5	31	6	
6	63	6	$3^2 \cdot 7$
7	127	18	
8	255	16	$3 \cdot 5 \cdot 17$
9	511	48	$7 \cdot 73$
10	1023	60	$3 \cdot 11 \cdot 31$
11	2047	176	$23 \cdot 89$
12	4095	144	$3^2 \cdot 5 \cdot 7 \cdot 13$
13	8191	630	
14	16383	756	$3 \cdot 43 \cdot 127$
15	32767	1800	$7 \cdot 31 \cdot 151$
16	65535	2048	$3 \cdot 5 \cdot 17 \cdot 257$
17	131071	7710	
18	202143	7776	$3^2 \cdot 7 \cdot 19 \cdot 73$
19	524287	27594	
20	1048575	2400	$3 \cdot 5^2 \cdot 11 \cdot 31 \cdot 41$
21	2097151	84672	$7^2 \cdot 127 \cdot 337$

(From Sarwate and Pursley [7.14])

Radar waveforms based on m-sequences can be periodic or aperiodic. If the waveform is periodic, this is equivalent to CW transmission and isolation must be provided between the transmitter and receiver. The periodic m-sequence is generated by continuously feeding clock pulses to the LSR input. After a sequence of length $N = 2^{n-1}$ has been generated the sequence repeats. When this sequence is received at the matched filter (or more probably a waveform correlator), the processed output is a series of compressed pulses as shown in Figure 7.2.25, having a peak-to-sidelobe ratio of N. The width of each compressed pulse is δ, and they are separated in time by $N\delta$, where $\delta = 1/($LSR clock rate$)$. The period of the m-sequence combined with the clock rate determines the maximum unambiguous range, although it is possible to extend the unambiguous range by using m-sequences with different clock rates and/or length and applying coincidence detection techniques.

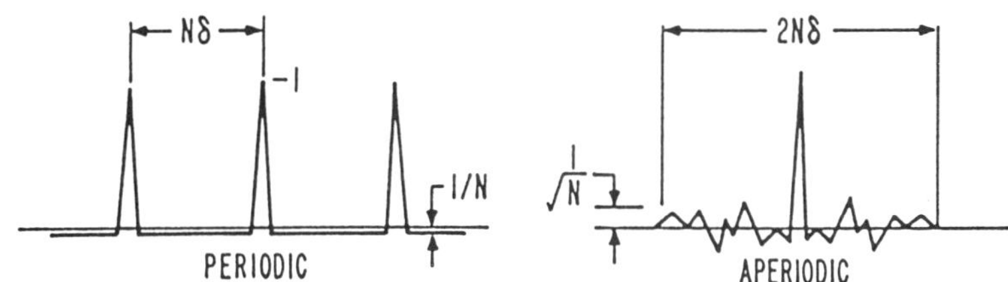

Figure 7.2.25 Zero-doppler matched-filter output
of periodic and aperiodic m-sequences.

An aperiodic m-sequence radar waveform is equivalent to a pulsed transmission of a single m-sequence period. The peak-to-sidelobe ratio at the matched filter output approaches $\sqrt{N}$ for the aperiodic sequences as N gets larger. This level can be acceptable for very long sequences. Figure 7.2.26 compares the sidelobe levels of periodic m-sequences of length 127 and 1023 to those of a linear FM matched filter and Hamming weighted outputs. Figure 7.2.27 illustrates the compressed pulse waveforms for two different sequences of length 15 and 31, showing that the sidelobe structure depends to some degree on the starting phase of the sequence as well as on its length.

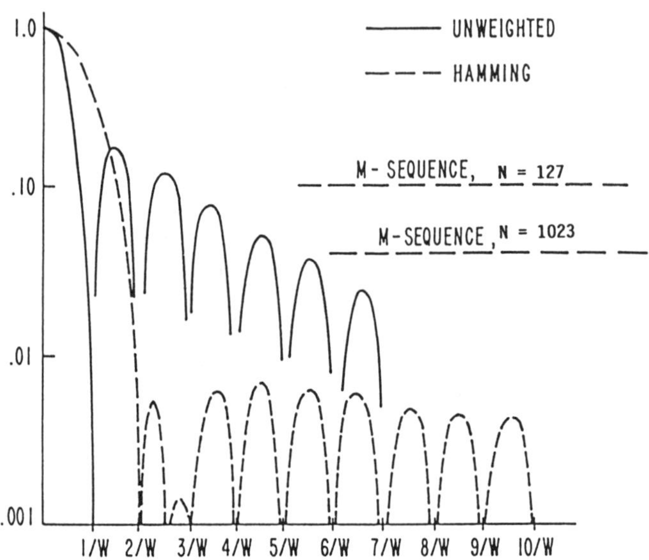

Figure 7.2.26 Comparison of compressed-pulse sidelobe levels:
Unweighted and Hamming weighted linear FM vs. periodic m-sequences
of length $N = 127$ and $N = 1023$.

If m-sequence waveforms are to be used in a target environment having a
large doppler spread it would be necessary, in theory, to implement a bank of
parallel matched filters to process these doppler-spread inputs, with the filter with
the largest output indicating a target at the doppler shift to which that filter is tuned.
One of the most important developments in the application of m-sequence (and
other binary phase code waveforms) to moving target detection was the invention
by R. M. Lerner of M.I.T. Lincoln Laboratory of the doppler matrix that replaces
a bank of complex matched filters with a simpler structure using standard phase
shift and resistor components [7.14]. This is shown in Figure 7.2.28. This type of
doppler matrix can also be readily implemented with digital microprocessors.

Very long m-sequences ($T\Delta f > 10^5$) have been used for deep-space probe
signals and in navigation satellite systems (GPS). In these cases the doppler effect
causes a time contraction (approaching target) or time dilation (receding target) of
the received signal as well as the doppler frequency translation $f_d = (2v_r/c)f_o$. As
a result, the LSR clock rates must be adjusted to compensate for the time contraction
or dilation associated with the doppler effect. If this is not done there will be
processing gain and resolution losses at the matched filter output.

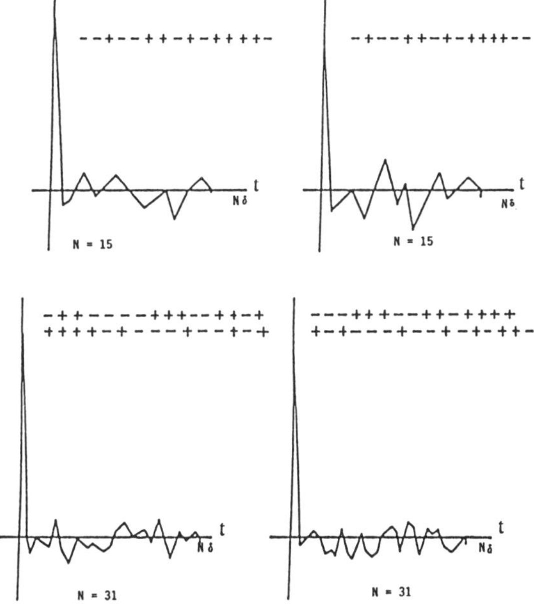

Figure 7.2.27 Matched-filter response functions of truncated maximum-length sequence phase code waveforms. Note effect of sequence order on sidelobes.

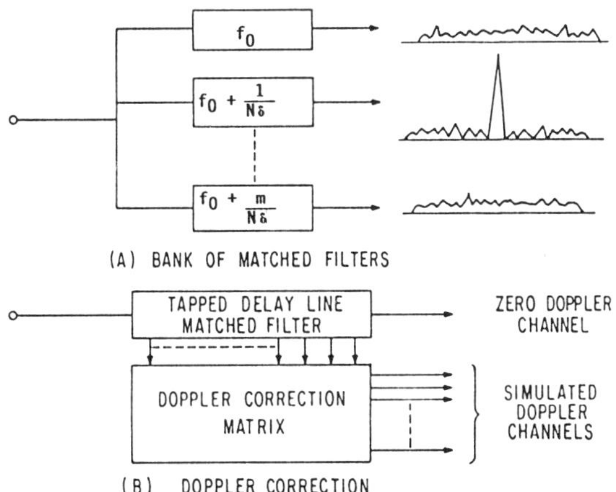

Figure 7.2.28 Range-doppler measurement.
(a) With a bank of matched filters centered at different doppler shifts.
(b) With a single matched filter and a doppler correction matrix.

Barker Codes

Barker codes [7.16] are binary phase shift waveforms similar to m-sequences, but with very short code lengths N. The time duration of a Barker code is $N\delta$, where δ is the subpulse interval. Barker codes have the property that the compressed pulse at the matched filter output is N units high in voltage and the time (range) sidelobes all have unit amplitude, as illustrated by Figure 7.2.29, resulting in a peak-to-sidelobe ratio of $20\log N$. The longest known binary Barker code has length $N = 13$ and thus has a peak-to-sidelobe ratio of 22.3 dB. Table 7.2.4 lists the binary Barker codes. These waveforms can be generated and processed by a tapped delay line or a digital processor. Since the Barker codes are very short, they have found limited application in long-range radars, which generally require much larger pulse compression ratios, but have been widely used in pulsed doppler airborne fine control radars. The F-15's AN/APG-63 radar incorporates Barker-coded waveforms. Barker codes have also served as building blocks for much longer codes that are products of two different Barker codes, or the product of a Barker code with another type of phase or frequency sequence.

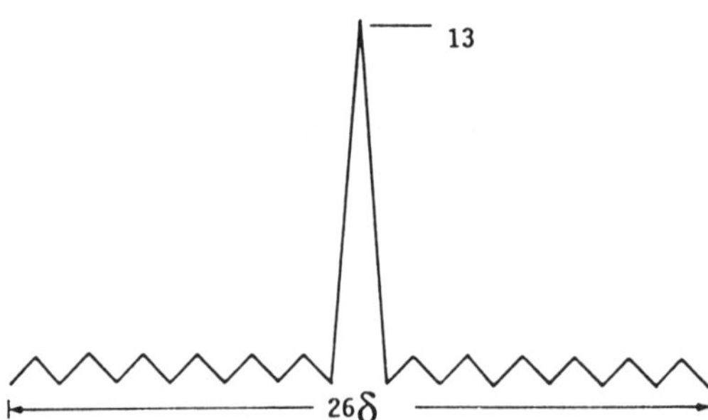

Figure 7.2.29 Matched-filter output of Barker code of length 13.

Table 7.2.4 Barker Code

Code Length	Code Sequence
2	+ - or + +
3	+ + -
4	+ + - + or + + + -
5	+ + + - +
7	+ + + - - + -
11	+ + + - - - + - - + -
13	+ + + + + - - + + - + - +

Complementary Codes

The principle of complementary-code pulse-compression waveforms is illustrated in Figure 7.2.30. Here, two separate matched filter outputs are seen to have equal amplitude time (range) sidelobes whose carrier frequencies are out of phase (indicated by the '+, -' signs), while the carriers of the main compressed pulse are in-phase. When these two waveforms are added, the sidelobes cancel and the mainlobes add coherently. Golay [7.17] has described complementary codes with binary phase sequences, and Welti [7.18, 7.19] has extended this concept to include polyphase codes. The matched filter outputs for these waveform pairs are sensitive to doppler shift. In addition, there are practical implementation problems in maintaining the theoretical sidelobe cancellation in an environment of multiple fluctuating target signals.

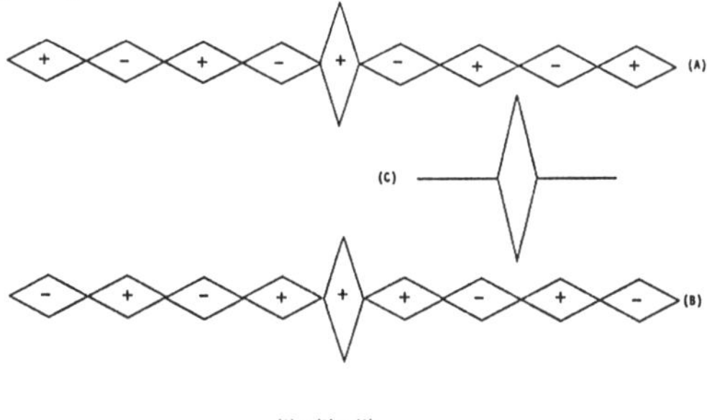

Figure 7.2.30 Range sidelobe cancellation with complementary sequences.

7.3 Moving Target Indication

Exploitation of the doppler effect in addition to basic pulse radar techniques permits achieving a level of moving-target-on-clutter visibility which far exceeds that of processing a single pulse-waveform echo signal. Early attempts at making use of the doppler information inherent in the radar echo signal from moving targets resulted in the capability labelled MTI (moving target indication). Typically, simpler pulse radar MTI techniques permit the extraction of a moving target echo from clutter which can be 20 to 30 dB greater in magnitude. MTI technology has since been developed with high levels of sophistication and complexity but pulsed doppler (PD) radar technology has matured as well.

The distinction between MTI and PD radar techniques is somewhat blurred in the literature. Literally, PD is an MTI technique. It may be helpful to consider MTI radars as pulse radars, configured to include some level of doppler processing, in order to isolate and detect a moving target, while PD radars are designed as coherent doppler radars which use a pulse waveform for transceive duplexing convenience. MTI and PD radars both incorporate range gating in the signal processing. Technology advances in oscillator stability and signal processing capability now permit PD radar to exceed the performance of any other type or radar in clutter, providing moving-target signals which may be some 100 dB smaller than the received clutter power.

A representative MTI processing technique is described below. PD radar techniques are discussed in Section 7.4.

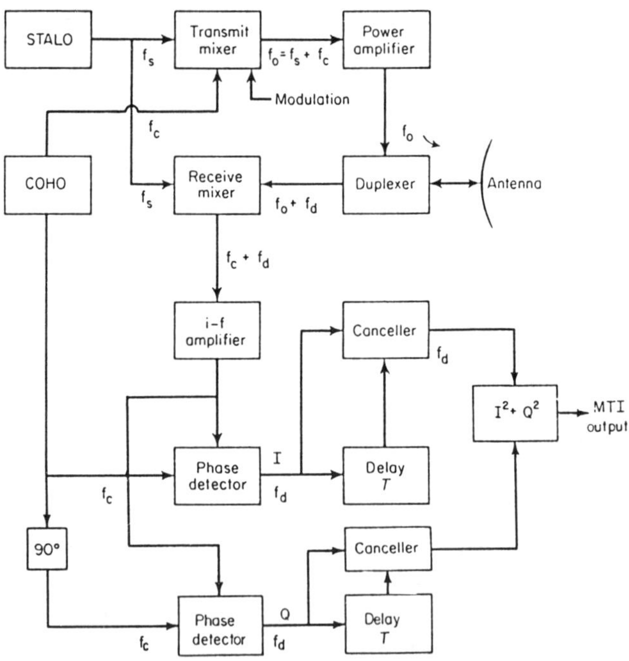

Figure 7.3.1 Coherent MTI radar system [7.20].
©Artech House, reprinted by permission.

7.3.1 A Typical MTI Radar

The block diagram of Figure 7.3.1 identifies the essential parts of a typical coherent MTI ground-based radar system. Two oscillators, the STALO (stable local oscillator) and the COHO (coherent oscillator) form the basis of a coherent transceive function and maintain sufficient stability on a pulse-to-pulse basis to assure that the stationary clutter appears at zero doppler frequency. The STALO is the primary radar frequency source and the COHO generates an IF offset and provides the I and Q (in-phase and quadrature) detector references. The I and Q detector outputs are inputs to the delay-line cancellers, which reject the zero frequency clutter and pass the moving-target doppler signals in a frequency baseband between zero and the PRF. Doppler frequencies greater than the PRF will be folded back into the baseband (*aliasing*). Intentional variations in the PRF can uncover targets at doppler frequencies that occur at multiples of the PRF. Blind speeds occur when the targets move 0, 1/2, 1, 3/2... wavelengths between consecutive transmitted pulses.

The use of I and Q processing precludes serious desensitization of the MTI process when the target pulse-train timing relative to the phase of the doppler signal results in a signal cancellation upon the subtraction in either of the delay-line cancellers. (It cannot happen in both I and Q channels simultaneously). This blind phase phenomenon is illustrated in Figure 7.3.2. Even with 10 pulses received over one-fourth of a doppler cycle ($f_d = f_r/40$), the output of a single canceller on the I channel can be some 10 dB below its full value.

7.3.2 Delay-Line Canceller

A common MTI doppler signal processor uses the *delay-line canceller*. Functionally, a canceller delays a copy of the input pulse train by one interpulse period and then subtracts the copy from the original. Stationary echos (clutter) which do not change from pulse to pulse are cancelled; moving targets which continually change phase at the doppler rate are not cancelled. Thus, the delay-line canceller performs as a filter with a frequency response as illustrated in Figure 7.3.3. The response is repetitive over multiples of f_r, the pulse repetition rate. Clutter spectra are shown which would fall into the cancellation response notches. Targets that move at velocities that result in doppler frequencies close to any multiple of the PRF would also be attenuated. In a similar manner, the response of two-delay cancellers in cascade is illustrated in Figure 7.3.3. Multiple cancellers broaden the rejection notches, require more equipment, and result in slight reduction in MTI system's response to moving targets.

The ideal MTI filter characteristic would be one that rejects the clutter spectrum without eliminating any moving targets. This ideal is not achievable in practice, but it is possible to synthesize delay-line filters with almost any desired frequency response. The basic technique employs a number of delay lines in cascade with feedback and feed-forward paths. One example of a synthesized delay-line filter response is illustrated in Figure 7.3.4; three-pole *Chebyshev low-pass filters* with 0.5 dB passband ripple are shown with three different clutter notch widths. Figure 7.3.5 describes the delay-line filter design necessary to achieve the middle delay-line filter characteristic of Figure 7.3.4.

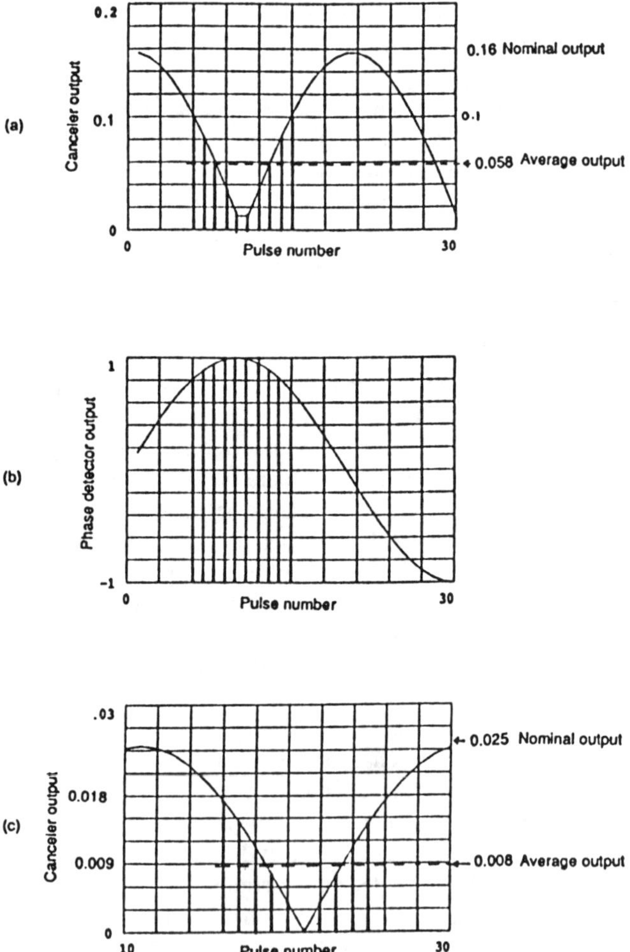

Figure 7.3.2 Blind phase phenomenon in MTI using only I channel. (a) Phase detector output over one doppler cycle. (b) Output of single-delay canceller on pulses 6—15. (c) Output of double-delay canceller on pulses 16—25.

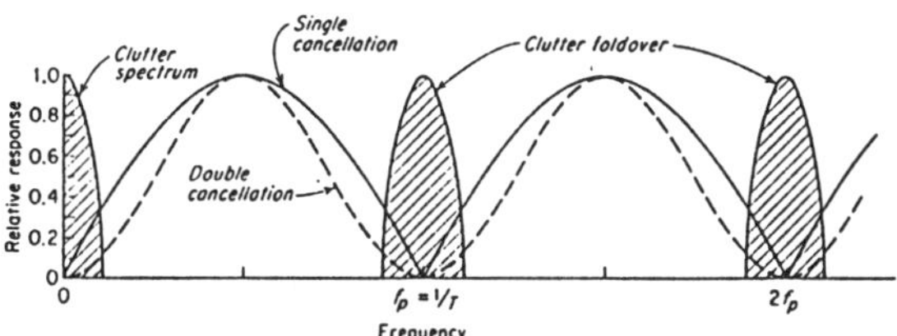

Figure 7.3.3 Relative frequency response of the single-delay-line canceller (solid curve) and the double-delay-line canceller (dashed curve). Shaded areas represents clutter spectrum [7.21].

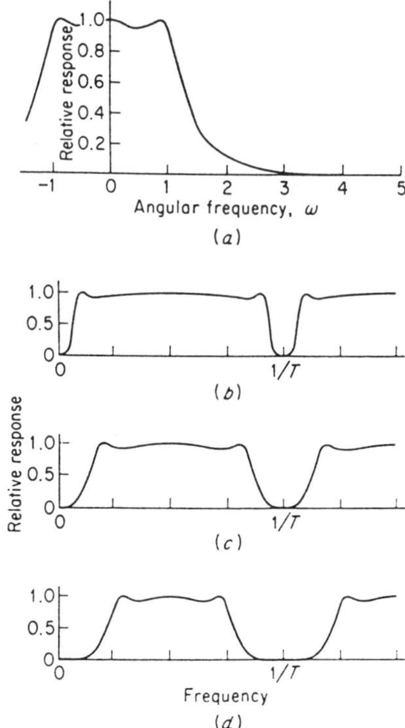

Figure 7.3.4 (a) Three-pole Chebyshev lowpass filter characteristic with 0.5 dB ripple in the passband. (b-d) Delay-line filter characteristics derived from (a) [7.21].

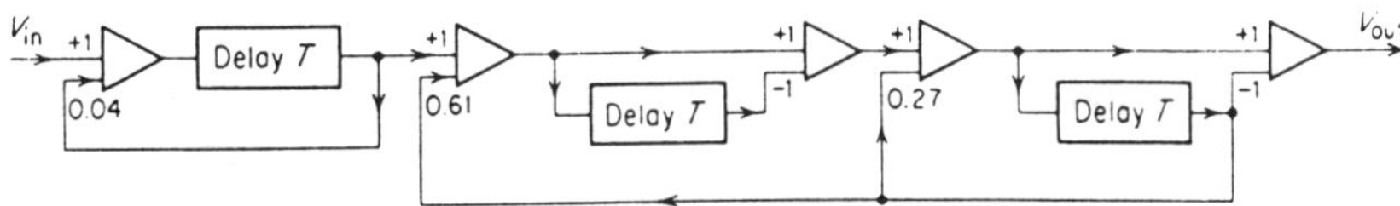

Figure 7.3.5 Form of the delay-line filter required to achieve the characteristic of Figure 7.3.4 [7.21].

The complexity associated with each of multiple delay lines and the requirement that all delay lines precisely match have limited most practical analog filters to not more than three delay paths. Digital techniques, however, permit realization of more complex filters.

7.3.3 MTI Filter Performance

The figure of merit for an MTI system is the *improvement factor* (I), a power ration which is defined as $I = r_o/r_i$, r_o is the filter output ratio of target signal to clutter, signal and r_i is the target-to-clutter signal ratio at the input to the receiver, averaged over all target speeds. This definition accounts for both the clutter attenuation and the average gain of the MTI system. It is therefore a valid measure of the MTI system response to clutter relative to the average MTI system response to targets.

The theoretical limitations in I for a scanning radar, using no-feedback coherent MTI cancellers, is illustrated in Figure 7.3.6. This assumes a linear system; that is, the echo signal envelope is identical to the two-way antenna voltage pattern. This assumption may be unrealistic for some practical MTI systems with relatively few hits per beamwidth. The scanning restriction indicated in Figure 7.3.6 does not apply to a radar that step-scans, such as a phased array.

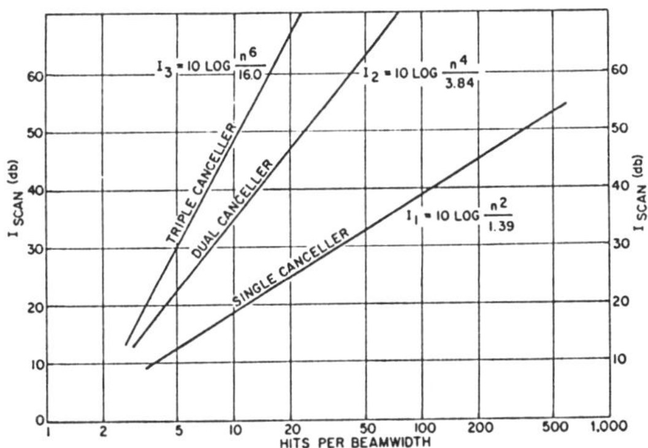

Figure 7.3.6 Theoretical MTI improvement factor due to scan modulation; gaussian antenna pattern; n = number of pulses within the one-way, half-power beamwidth [7.22].

The rms velocity spread of clutter also restricts the realizable improvement factor achievable in an MTI system. Figure 7.3.7 illustrates this restriction assuming a dual canceller for several clutter types and wavelength-PRF products; v_b is the first blind speed of the radar. It is assumed that the average velocity of rain and chaff clutter is compensated for such that the rain and chaff clutter are centered in the MTI filter rejection notch. The slopes of the lines in Figure 7.3.7 would be less for a single-delay canceller and more for a three-delay canceller. The use of limiters in the radar receiver may also restrict the full realization of the theoretical improvement factor in an MTI radar.

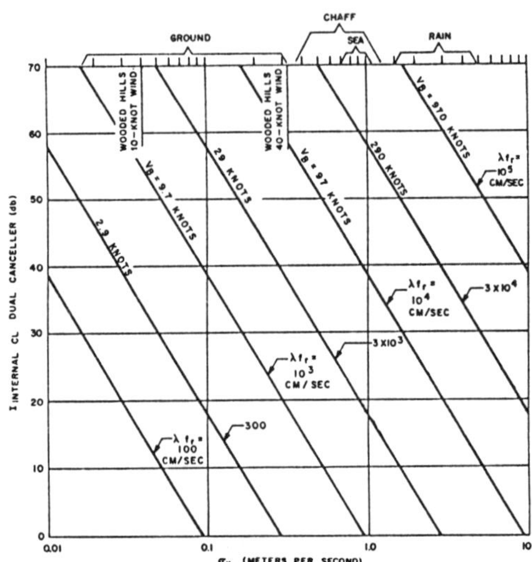

Figure 7.3.7 MTI improvement factor as a function of the rms velocity spread of clutter for a two-delay canceller (no feedback) [7.22].

Expressed in decibels, the parameter I is greater than radar subclutter visibility (SCV) by the clutter visibility factor, V_{oc}, which is approximately equal to Blake's visibility factor for noiselike background.

7.4 Pulse Doppler Techniques

A pulse doppler (PD) radar is designed to transmit and receive coherently in order to exploit the natural separation of target and clutter inherent via the doppler effect. Phase coherency between transmit and receive is maintained by the use of a common CW source. For transmission the source is offset, amplified, and then pulsed for time duplexing through a single antenna. The moving target and clutter echos are received, amplified, and processed with doppler filters.

A typical unambiguous clutter and target doppler spectrum received in an airborne high-PRF PD radar is illustrated in Figure 7.4.1. The transmitted spectrum is shown dashed, with a $\sin x/x$ envelope. The null-to-null width of the envelope in frequency is twice the inverse of the pulse width and the spectral lines are separated by the PRF. The shaded areas describe clutter which is shaped by the antenna beam and the dynamic geometry to the composite of clutter reflectors.

Figure 7.4.2 details the doppler spectrum of the central line which is usually that selected for processing; v_r and v_t are the radial speeds of the radar platform and target, respectively.

If the PD radar is not airborne but ground-based, the airborne clutter spectra shown would collapse into single lines and appear like the transmitted spectrum (i.e., no doppler displacements if clutter has no motion). Thus, the PD signal processing for the fixed radar would be simpler than for the airborne case.

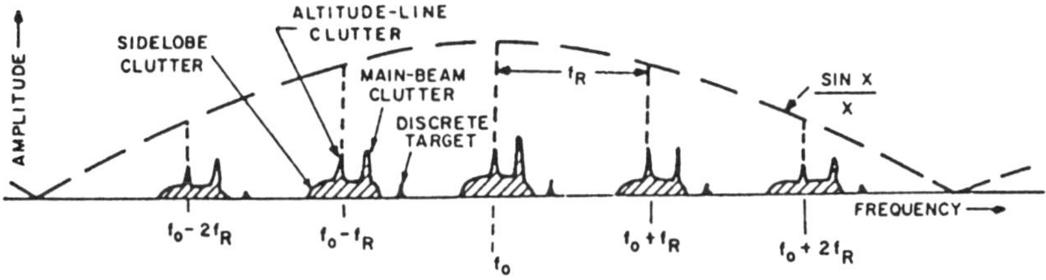

Figure 7.4.1 Clutter and target frequency spectrum from a horizontally moving platform [7.23].

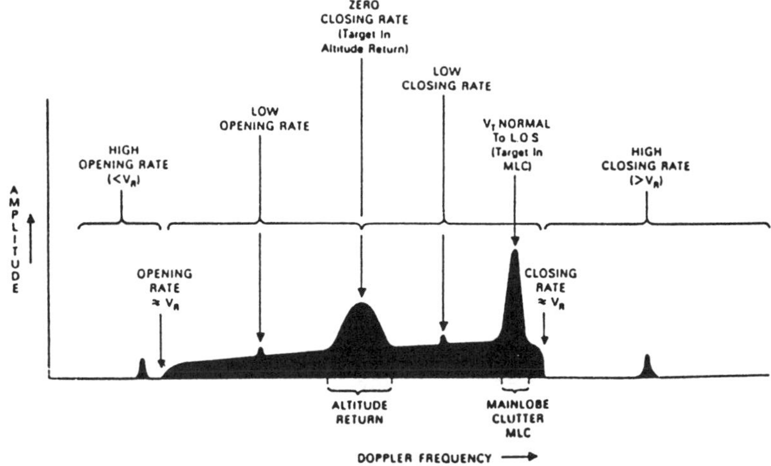

Figure 7.4.2 Doppler spectrum of high-PRF, airborne radar [7.24].

7.4.1 Clutter Rejection in Doppler Processing

Figure 7.4.3 shows a representative configuration of a PD radar for airborne use. Included are range-gating circuits, a single-sideband filter, main-beam clutter-rejection circuits, and a detection filter bank. Although only one receiver channel is shown, multiple range channels may be used to reduce *eclipsing loss* and to allow multiple-PRF range measurement in track-while-scan. To continuously track a single target, servomechanisms for velocity, range, and angle tracking can also be incorporated.

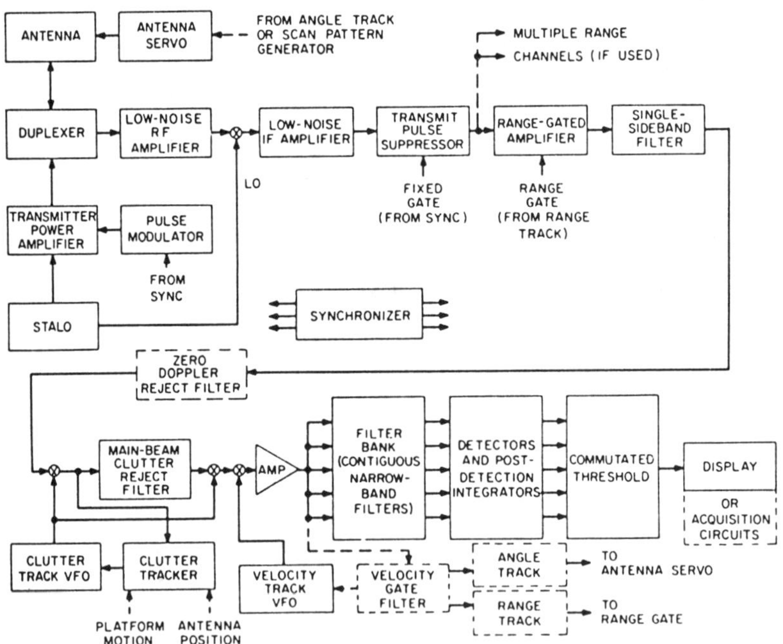

Figure 7.4.3 Typical pulse-doppler-radar configuration [7.23].

The transmit-pulse suppressor, in conjunction with the duplexer, reduces transmitter leakage into the receiver to a low level so that transmitter-noise sidebands do not degrade performance. The single-sideband filter has a bandwidth approximately equal to the PRF and has the effect of converting the input pulsed signal into a CW signal. Usually the carrier spectral line of the return signal is selected because it has the most power and does not shift if the PRF is varied. The signal-to-noise ratio (SNR) is not degraded by the single-sideband operation because the noise is also reduced in bandwidth. The advantage therefore gained is that the clutter and signal filtering are more easily performed on a single spectral line than on the entire spectrum. It should be noted that all time gating must precede the single-sideband filter.

The objective in designing the doppler processing circuits is to gate and amplify the moving-target signal while rejecting unwanted clutter and clutter noise-sidebands. Because some 200 dB of total echo amplification may be necessary in the receiver, it is necessary to strategically place clutter rejection filters to avoid saturating any part of the receiver. Note that clutter can often exceed the desired target signal by 80 to 100 dB.

If there were no clutter in a particular radar application, the moving-target echo could be extracted from noise alone by using simple filters, and an entire band of filters to cover the range ambiguity, whether analog or digital, would be inexpensive. With a clutter environment the narrowband target doppler filters may become one of the most expensive parts of the radar. Crystal lattice filters are common analog devices and multipoint FFT (Fast Fourier Transform) are commonly used in digital systems. Even with the best analog or digital filters, it is often necessary to reduce the clutter rejection burden on the doppler gate filters. This clutter rejection filter may be a fixed doppler bandpass filter or a tracking notch main-beam clutter rejection filter, as indicated in Figure 7.4.3.

There are two primary considerations in achieving clutter rejection: (1) the composite of all filtering to gate the target doppler signal must provide out-of-band suppression of the total clutter signal power to a level lower than that of the desired target signal and (2) the noise sidebands on the clutter signal which fall in the target gate must also be constrained to levels below the desired target signal. All-important in these two considerations is the doppler separation of target and clutter; this is determined by the target speed (V_t) and aspect angle, θ. A typical minimum value for $V_t \cos\theta$ in aircraft detection is 150 m/s, giving a 10 kHz doppler separation of clutter and target at X-band. The source is clutter sideband noise at a 10 kHz displacement (or above) is usually the stable microwave source. FM noise is always the dominant residual noise after good source design. Figure 7.4.4 illustrates a variety of sources available to a designer. If, for example, clutter were to be 100 dB above the target signal in a 1 kHz doppler gate, the source requirement might be 130 dB below the carrier in 1 Hz at the 10 kHz sideband. Therefore, it can be seen that some of the sources in Figure 7.4.4 would be unacceptable for application.

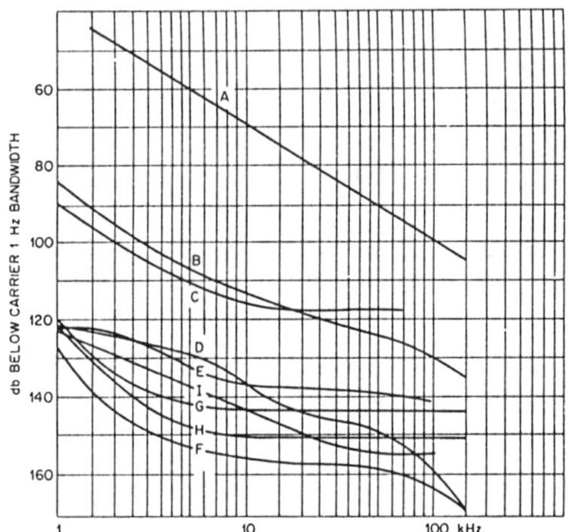

Figure 7.4.4 FM noise in microwave sources [7.25].

7.4.2 Modern PD Signal Processing

Modern search and fire control PD radars commonly employ signal processing techniques similar to those illustrated in Figure 7.4.5 for a track-while-scan (TWS) mode. In other modes, such as dedicated track, modified versions may be necessary. In any case, design is driven by the need to isolate the target from interferences (usually clutter) using range and doppler gating with matched filters.

Early in the processing, I and Q division precludes signal loss due to blind phases as previously discussed in MTI processing. Following the A/Ds, processing is digital. The summer (Σ) forms the mean amplitude during an interval equal to the pulse duration (matched filter criteria). Thus, time-delayed complex (I and Q) numbers are generated during the inter-pulse period which are the range-gated samples. These complex numbers are the inputs to the array processor (AP).

The AP processes in doppler with a parallel array of range gates; one channel for one range gate is shown from the input buffer. The buffer would also drive other range channels (not shown). The AP performs the indicated operations, one at a time but very rapidly, using identical instructions on each range-gated channel. The clutter canceller is a bandpass filter to all moving targets of interest. The weighting of the range-gated pulses during a coherency period lowers the sidelobes of the FFT. The FFT is an extremely efficient algorithm that generates a filter bank whose purpose is to improve signal-to-interference by coherent integration, resolve targets in doppler, and discriminate against clutter. The linear detector calculates the magnitude (M) of each filter output based on a simple linear approximation, rather than a more time-consuming square law calculation. The range and doppler correlator arrays the data and correlates dwell-to-dwell range and doppler data and resolves ambiguities. Targets are detected by employing a binary integration criterion which compares, within an antenna dwell, the number of coherent samples that exceed the threshold with the total number of samples per dwell.

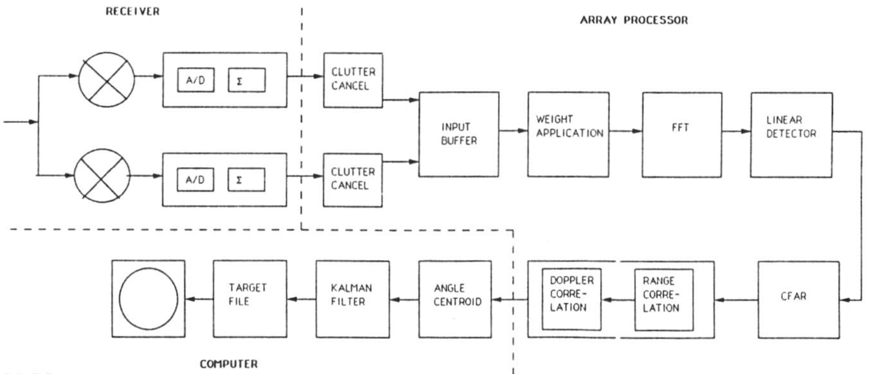

Figure 7.4.5 Track-while-scan mode (digital signal processor) [7.26].

At this point the AP passes the range-doppler data for the latest dwell to a computer which performs angle processing and establishes target track files using *Kalman filters*. The centroid of the amplitudes of the coherent samples, as the antenna scans past the target, is estimated by a simple centroiding algorithm based on the timing of the samples. The computer then interpolates and calibrates to find the azimuth direction that the centroid represents.

The range rate (doppler), together with range and azimuth data of the latest detection, is correlated with existing target track file predictions of the Kalman filter. If the latest data falls within a bound set by the variance of previous data, it is accepted. If not, a new target file is established.

7.4.3 PRF Selection

PD radars may be designed with high, medium, or low pulse repetition frequencies (i.e., HPRF, MPRF, or LPRF) for different applications. Selections of specific PRFs for particular application involves considerations of waveforms, ambiguity functions, matched filters, and signal processing relative to the range and doppler space of targets of interest and clutter. In high-PRF PD radar, by a common definition, no doppler ambiguity is permitted. The PRF must be high enough so that there can be no overlap of the target doppler spectra with adjacent spectral lines (i.e. $\mathrm{PRF}_{\min} = 2(2v_r + v_t)/\lambda$). If the minimum PRF is selected, then the maximum unambiguous range is given by $c/2\mathrm{PRF}_{\min}$, where c is the speed of light.

Because the receiver is gated off during pulse transmission, target signals received at this time are 'eclipsed.' With a high PRF a target may be eclipsed many times in the range interval of interest, but because of relative target speed, eclipsing cannot be total. An eclipsing loss, equal to the reciprocal of $(1 - D_u)$, where D_u is the duty cycle on the average, must be accounted for. In the absence of clutter spectra (i.e., a ground-based radar looking up or an airborne radar looking into space) the selection of the PRF may be based on expected target doppler frequencies alone. However, clutter spectra are usually present in an airborne application, if not from the mainlobe, then from the antenna sidelobes, and the spectral occupancy of clutter can extend out to the doppler frequency of the moving radar platform on each side of a PRF line.

Figure 7.4.6 illustrates typical clutter spectra for low, medium, and high PRF waveforms in a moving radar system. At LPRFs the mainbeam clutter occupies most of the available doppler spectrum, resulting in little usefulness except for perhaps slow moving ground targets. At MPRF there is a more useful doppler clear region with competing sidelobe clutter. At HPRFs there is a significant clutter-free region where high-speed closing, forward-hemisphere targets can be detected. Tail chase and opening targets compete with the sidelobe clutter region which is larger at HPRFs than at MPRFs. MPRF modes can be ambiguous in both range and doppler, but can complement HPRF modes by providing better performance when targets are in a tail aspect. Solving the range ambiguity for a true range measurement is possible at MPRFs but may not be practical at HPRFs. Thus, it is common in airborne PD radar search to interleave HPRF and MPRF waveforms in order to best cover the expected range of target parameters while accepting some minimum number of blind zones in the range-doppler space of interest.

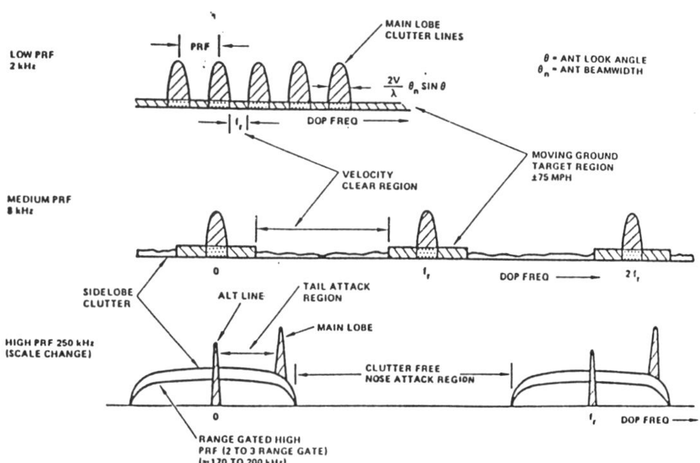

Figure 7.4.6 Typical clutter spectra: low, medium, high PRF PD radar [7.26].

7.4.4 Loss Factors in PD Radar

As with MTI systems, there are several loss factors which will be encountered in PD radar processing.

Filter Matching Loss, L_{mf}. This is the loss relative to matching to the envelope of the input pulse train, $a_0(t/t_o)$. This loss is controlled by the filter bandwidth and the weighting applied to the pulses to obtain the desired filter sidelobes. In many cases, the coherent processing is carried out over a time $t_f < t_o$, and a loss from this must be included either as an increase in L_{mf} or as a noncoherent integration loss $L_i(n')$, where $n' = t_o/t_f$. In the modified Blake chart, the signal energy was calculated for t_f, and the detectability factor $D(n')$ includes the required integration loss.

Range Gate Matching Loss, L_m. When a range gate is used at the input to the doppler filters, the combined response of this gate and the preceding IF filter H_1 should match the transmitted pulse. If a wideband IF is used with a rectangular gate, the range gate loss will be

$$L_m = \tau_g/\tau, \tau_g > \tau$$

$$= \tau/\tau_g, \tau > \tau_g \tag{7.4.1}$$

If a short sampling strobe is used in place of the range gate, L_m is determined by the preceding IF filter.

Range Straddling Loss, L_{er}. The range gates or strobes will be spaced by approximately τ, but a signal having a peak that arrives other than in the center of the gate or strobe will not be passed with full amplitude. The amount of straddling loss depends on the pulse shape and gate duration, as well as on the spacing. For rectangular pulses and gates, preceded by a wideband IF, the loss will be as shown in Figure 7.4.7. Overlapping gates may be used to reduce this loss, at the expense of greater processor throughput.

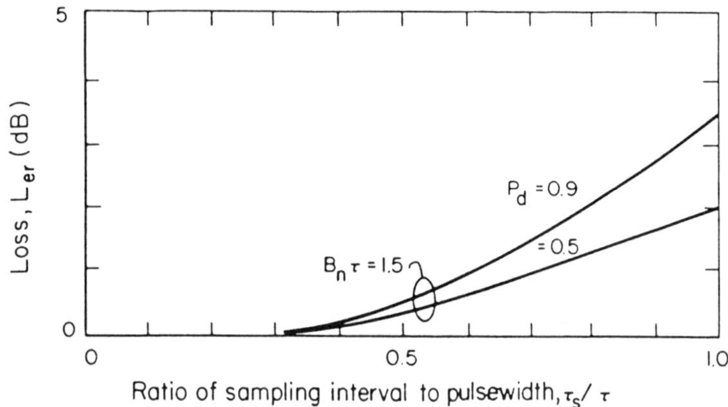

Figure 7.4.7 Range straddling loss for rectangular pulses [7.20].

Filter Straddling Loss, L_{ef}. Signals may also be centered at frequencies between the doppler filters, introducing a filter straddling loss (Figure 7.4.8). When heavily weighted filters are used, they will overlap within the -3 dB points, making this loss insignificant.

Angle Straddling Loss, L_{ea}. When the antenna scans continuously, and burst processing is used, the center of the burst may not coincide with the angle of the target. As a result, the peak of the burst envelope may straddle two processing intervals, introducing an angle straddling loss. This loss is 2.0 dB for continuous processing intervals $t_f = t_o$, when using a 40 dB filter weighting. When the observation time is divided into two or more processing intervals, there is negligible straddling loss. Overlapping processing intervals may be used, but this is inconsistent with burst-to-burst PRF or RF agility, so the usual choice is to make $t_f < t_o/2$ to minimize this straddling loss.

Eclipsing Loss, L_{ec}. Medium- and high-PRF systems will lose some target pulses during search as a result of eclipsing by the transmitted pulse (and sometimes also by short-range clutter). On the average, the fraction of energy lost will be Δ_r /R_u. Accordingly, for search or target acquisition with probabilities near 0.5, the loss will be

$$L_{ec} = \frac{R_u}{(R_u - \Delta_r)} \tag{7.4.2}$$

where $\Delta_r = \tau c/2$. For higher probabilities, the loss will increase, and may be found by integrating P_d over all target ranges and increasing signal power until the desired value is achieved, with the burst-to-burst diversity procedure to be employed in the system.

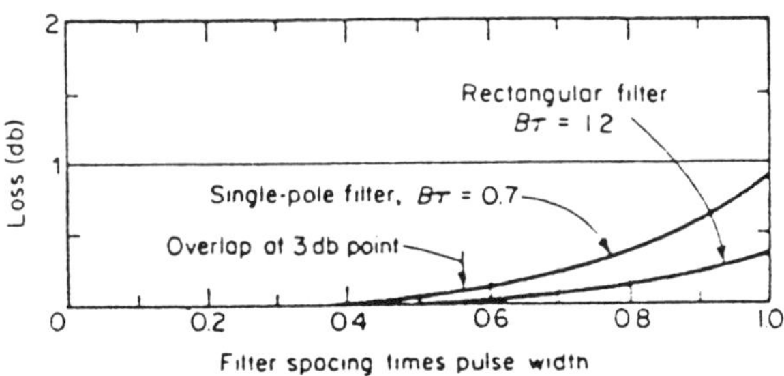

Figure 7.4.8 Filter straddling loss for unweighted burst [7.20].

Velocity Response Loss, L_{ev}. Low- and medium-PRF systems will lose some targets in the clutter rejection notch. The calculation is similar to that for eclipsing. For P_d near 0.5,

$$L_{ev} = \frac{v_b}{(v_b - \Delta_v)} \tag{7.4.3}$$

where Δ_v is the width of the rejection notch.

Other losses, including those caused by collapsing of coordinates, noncoherent integration following doppler filtering, and binary integration, will be similar to those in MTI and other systems.

Transient Gating Loss, L_{eg}. All PD systems must transmit enough pulses, at the beginning of each burst and before processing begins, to ensure that the echoes from clutter at maximum range have been received. During these transmissions, the processor must be gated off to prevent adding into the weighted output a sample which has no clutter echo. When an analog filter is used to reduce clutter prior to A/D conversion, time t_t must also be allowed for the initial transient at throughput of this filter to decay. The transient gating loss is expressed in terms of the total transient gating time t_g and the total burst duration t_f:

$$L_{eg} = 1 + t_g/t_f \tag{7.4.4}$$

where $t_g = t_{dc} + t_t$. The gating time t_g is usually about 1 ms, and this means that bursts shorter than about 5 ms will have significant gating loss.

7.5 Impulse Radar

7.5.1 Introduction

Very short pulse radar systems (e.g., nanosecond duration transmissions) were developed in the early 1970s primarily for detecting targets at short ranges; the primary advantage of short pulse radar systems is very fine range resolution (e.g., one foot or less) at very low cost [7.27, 7.28].

The transmitted pulses here are produced differently than in conventional radar systems, as shown in Figure 7.5.1. A very short duration, or video pulse approximating an impulse, is used to shock excite a transmitting antenna cavity. The resulting transmitted signal is, essentially, the impulse response of the antenna and generally resembles several cycles of RF energy as shown in Figure 7.5.2, where the apparent frequency depends on the dimensions of the antenna cavity and the properties of the excitation source. Hence, the description, "impulse" radar. Others have termed such systems as a "baseband" radar or reflectometer based on the modulation source. Still others refer to the transmission as "carrier-free" or, generically, as "ultra-wideband," when the fractional bandwidth of the signal is greater than 30%.

Receiving antennas and threshold detectors for these types of transmissions also have some special requirements. The antennas should be broadband (e.g., the reciprocal of the pulse width) with an ability to focus energy while the receivers must be capable of detecting small amounts of energy over a short time duration and rejecting outside clutter or interference via nanosecond range gating.

Impulse radar systems using avalanche transistor transmitters and a tunnel diode as a receiver threshold device have a range of several hundred feet on walking human targets (e.g., intrusion detection) to perhaps 5,000 feet on large targets such as ships (e.g., ship docking application). By the use of focused antennas, laser activated switch transmitters, coherent processing receivers, and target cross section, the range can possibly be extended to 25 miles. As the system range increases beyond what can be accomplished using the low cost avalanche transistor transmitter and tunnel diode front-end, so does the competitive advantage of this type of system over other radar techniques diminish.

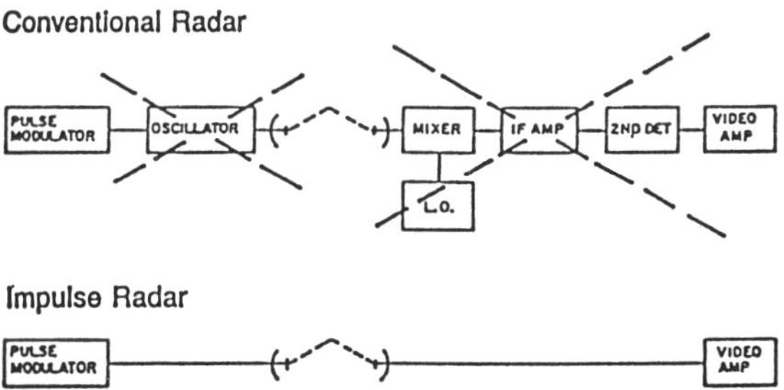

Figure 7.5.1 Description of an impulse radar system.

Impulse radars have been used extensively for ground probing (e.g., for finding buried pipes), intrusion detection, and ship docking. Other applications such as braking-alert for automobile safety and low flying missile detection are currently under investigation [7.29—7.32]

7.5.2 The Radar Equation

The radar equation as developed in Chapter 1 concerns single frequency parameters; for example, wavelength, antenna gain, and target cross section. Clearly, these parameters are frequency dependent, having both an amplitude spectrum and phase function associated with the Fourier transform of their system functions. Fortunately, one can still use the radar equation as a guide in determining power and range limitations. Generally, the radar equation serves as an upper bound on power and range; it is an optimistic answer. If the amplitude spectrum of the target (at different angles of incidence), the antenna properties, and the receiver threshold detection mechanism are well defined, an exact solution is possible. Generally, a computer program with target signature algorithms is necessary to determine threshold levels where quantitative solutions are required.

7.5.3 Baseband Transmitters

A simple way to generate a short pulse is by the use of the avalanche transistor circuit shown in Figure 7.5.3. A trigger pulse causes the transistor to "avalanche," effectively shorting the collector, base, and emitter together [7.33]. The charge stored in the transmission line is then discharged into the load resistance, R_0. Typically, avalanche transistors produce pulse outputs from 20 to 200 volts with rise times ranging from 100 psec to 1 nsec: Generally, the higher the output voltage, the slower the rise time. Not that a DC power supply that supplies an output equal to somewhat greater than twice the output voltage is required for pulse generation. Avalanche transistors may also be connected in *Marx generator* configuration where capacitors in a network are charged in parallel but discharged in series. In this manner, a voltage output of 1000 volts or greater at approximately 1 nsec rise time has been realized [7.34].

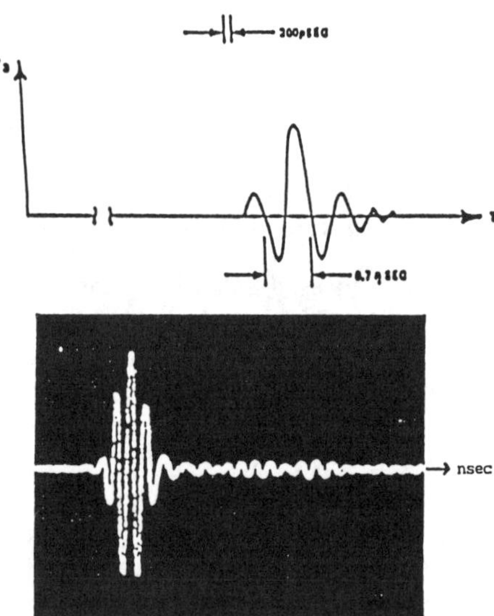

Figure 7.5.2 (a) Transmitted waveform of a typical L-band impulse radar system. (b) Actual transmit-receive response of L-band radar system.

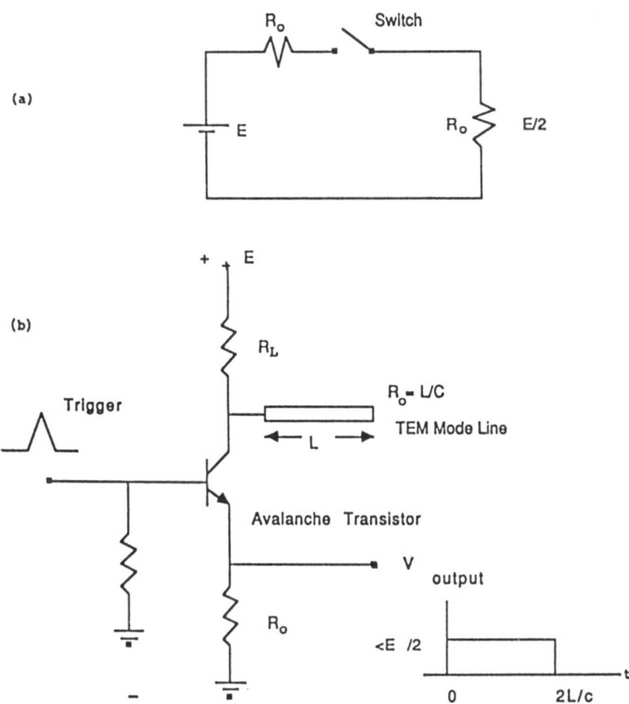

Figure 7.5.3 Baseband modulation sources.

For lower voltage applications the rise time of the output pulse can be "sharpened" by the appropriate use of step recovery diodes. For example, a tandem connection of five step recovery diodes can be employed to produce a 200 volt, 200 psec rise time pulse from an avalanche source [7.34].

In place of the low-cost avalanche transistor, a solid state switch excited by a pulsed laser source can also be used to discharge the source capacitor. Here, the source voltage can be thousands of volts, and it has been reported that rise times of 100 ps or less have been achieved using this approach. The cost of a laser activated switch and the associated equipment required is much higher than the avalanche transistor approach; for example, more than 100 times greater [7.35].

7.5.4 Antennas

To radiate pulses without distortion, the system function of an antenna should have a flat amplitude spectrum and linear phase function. For ultra wideband radiating structures this is not possible. Antennas generally have a high pass (e.g., $j\omega$) amplitude spectrum. Certain antennas, for example the log periodic and the spiral, have a flat amplitude spectrum, but a highly non-linear phase function. The result is that these antennas suffer from severe short-pulse dispersion and are inappropriate for application to ultra-wideband transmissions.

The effect of short-pulse signals in antenna systems becomes observable wherever the propagation delay of the signal on the structure is comparable to the rise time or the duration of the pulse. When this situation occurs, only certain portions of the antenna are involved in the radiation process at any one time.

For example, consider the radiation field of a dipole of total length $2h$, radius a, and driven by a source of internal impedance Z_g. If $h >> a$, so that circumferential variations can be neglected, then it can be shown that in the broadside direction the impulse response of such a structure can be approximated by a train of impulses of varying weight and of spacing h/c seconds, where c is the speed of light [7.36]. In any time window less than h/c seconds in duration, the impulse response can be represented by a single impulse. This means that any time-limited source waveform of duration less than h/c seconds can be transmitted and replicated during this time window without distortion. It can be shown via the Rayleigh-Carson theory that the receive response of an antenna is the integral of the same element used in the transient mode [7.37].

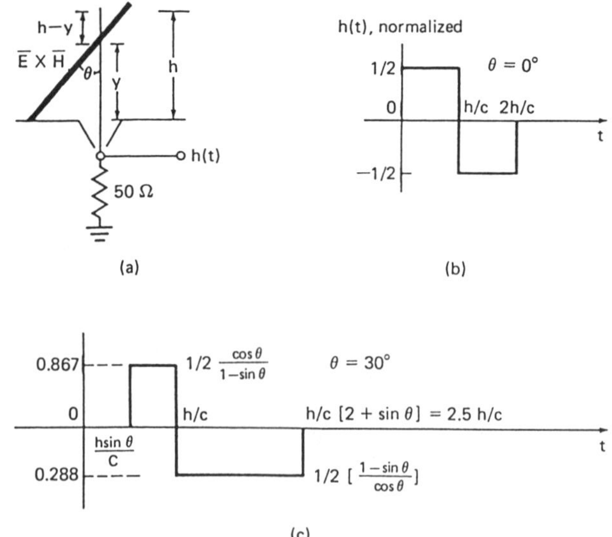

Figure 7.5.4 Impulse response of matched dipole: (a) configuration, (b) $\theta = 0°$, and (c) $\theta = 30°$ [7.38].

By proper source matching or end loading of a radiating element, one can force the impulse response to be time limited, as shown in Figure 7.5.4 , for a dipole in the receive mode. Forcing an antenna to have a time-limited (TL) impulse response means that any dispersion of the incident signal cannot exceed the duration of the impulse response: the resulting response of the antenna to the incident signal wave shape depends on the function behavior of the TL impulse response, but dispersion depends on duration alone [7.38].

To narrow the beamwidth of the antenna, one can build an array of non-dispersive elements and perform time domain steering to assure coalescence in the far field. The "sidelobe" time domain residue level of such an antenna is governed by the number of elements in the array. For example, with equal time delay feeding each of N elements in an array, the radiated signals all coalescence at boresight in the far field. Off boresight, the residue level can be as low as 20 log $1/N$ [7.39].

Another approach for beam focusing involves the use of a low dispersive feed to illuminate a dish or reflecting surface as the antenna aperture. If the time dependence of the aperture distribution is independent of space, then at boresight one can expect the far field response to be, simply, the derivative of the aperture illumination.

7.5.5 Receivers

Two types of receivers are employed to detect very short pulses. The first employs a tunnel diode threshold detector and binary integrator. The second, is a more conventional coherent processor with in-phase and quadrature channels for detection. The latter provides about 20 dB more sensitivity at the cost of complexity. The choice of receiver depends on application needs.

The tunnel diode (TD) is a two-terminal device with an I-V curve, having a negative resistance region, as shown in Figure 7.5.5. By operating the TD in a bistable mode and close to the negative resistance region at point A, a small incident signal can cause the diode to switch from stable state A to stable state C. The stretched voltage swing Δv provides the input to conventional TTL logic circuitry for detection. For optimum sensitivity, point A is maintained as close as possible to the switching point as established by a constant-false-alarm closed loop. Noise spikes ultimately determine that operating point. For example, if the hits obtained from noise alone are fed to a register that assign $a + 1$ each time a noise spike exceeds the threshold and a zero when it does not, then outputs from, for example, 32 stages of the register can be continually summed. By appropriately feeding the register, a sliding average of detector hits in any group of 32 consecutive periods are summed. The threshold is arbitrarily lowered until on the average two or three hits due to noise alone are contained in the register: the transmitter is disabled during this time. A DC voltage derived from the sum port is used via a feedback loop to maintain the threshold voltage at point A. When the transmitter is enabled, the threshold for establishing the presence of a target is set in a separate register. Then, for example, a target is said to be present when 27 or more of the 32 stages of the second register are filled [7.40]. The TD is range gated to reduce system collapsing loss. The optimum duration of the range gate for detection is equal to the rms spread of the pulse transmission. Without a preamplfier, the sensitivity based on TD noise alone is about 1 mV on a half cycle of incident microwave energy.

A coherent processor employing a local oscillator or burst generator can also be used for reception. Here, the received RF short-pulse signal is mixed with a coherent oscillator operating at the same "apparent" center frequency is mixed using a back diode. A back diode is a doped TD having an almost perfect diode I-V curve, as is also shown in Figure 7.5.5 [7.41]. A 90° phase-shifted LO signal is fed to a second back diode mixer creating a "quadrature" channel. The resulting baseband signal in each channel is range gated, fed to low pass filter-amplifer, boxcar circuit, and bank of doppler filters. The bandwidths of the doppler filters are determined by the expected dynamics of the target. The voltage output from the doppler filter in the in-phase and quadrature channel is squared, summed, and fed to a threshold circuit for processing [7.21, pp. 437-438].

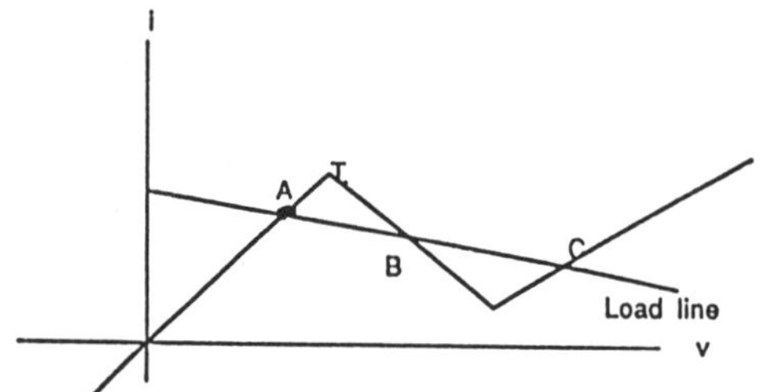

(a) Low frequency two-terminal properties of a tunnel diode.

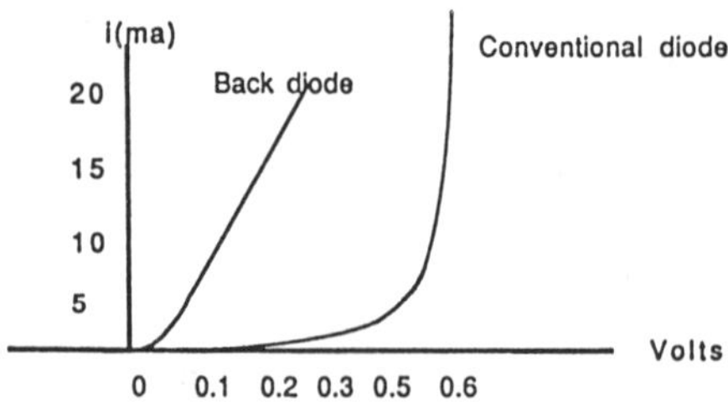

(b) Typical silicon back diode vs conventional diode.

Figure 7.5.5 I-V properties of tunnel and back diodes [7.24].

7.6 References

[7.1]   D. O. North, "An analysis of the factors which determine signal/noise discrimination in pulsed carrier systems," *RCA Laboratories Tech. Rpt. PTR-6C*, 25 June 1943; reprinted in *Proc. IEEE* **51**, pp. 1015-1027, 1963.

[7.2]   P. M. Woodward, *Probability and Information Theory, with Applications to Radar*, McGraw-Hill, 1953.

[7.3]   C. E. Cook and M. Bernfeld, *Radar Signals - An Introduction to Theory and Application*, Academic Press, New York, 1967.

[7.4]   C. E. Cook and J. Paolillo, "A pulse compression predistortion function for efficient sidelobe reduction in a high power radar," *Proc. IEEE*, **52**, pp. 377-389, 1964.

[7.5]   M. A. Jack, P. M. Grant, and J. H. Collins, "The theory, design, and application of surface acoustic wave Fourier-transform processors," *Proc. IEEE*, **68**, pp. 450-468, 1980.

[7.6]   *Handbook of Acoustical Signal Processing*, Andersen Laboratories, Bloomfield, CT, 1988.

[7.7]   T. T. Taylor, "Design of Line-Source Antennas for Narrow Beamwidths and Low Sidelobes," *IRE Trans.*, **AP-3**, pp. 16-28, 1955.

[7.8]   E. L. Key, E. N. Fowle, and R. D. Haggarty, "A method of designing signals of large time-bandwidth product," *IRE Intern. Conv. Record* Pt. 4, pp. 146-155, 1961.

[7.9]   C. E. Cook, "A class of nonlinear fm pulse-compression signals," *Proc. IEEE*, **52**, pp. 1369-1371, 1964.

[7.10] R. L. Frank, "Polyphase codes with good nonperiodic correlation properties," *IEEE Trans.*, **IT-9**, pp. 43-45, 1963.

[7.11] F. F. Kretschmer and B. L. Lewis, "Doppler properties of polyphase coded pulse compression waveforms," *IEEE Trans.*, **AES-19**, pp. 521-531, 1983.

[7.12] B. L. Lewis and F. F. Kretschmer, "A new class of polyphase pulse compression codes and techniques," *IEEE Trans.*, **AES-17**, pp. 364-372, 1981.

[7.13] B. L. Lewis and F. F. Kretschmer, Jr., "Linear frequency modulation derived polyphase pulse compression codes, *IEEE Trans.*, **AES-18**, pp. 637-641, 1982.

[7.14] D. V. Sarwate and M. B. Pursley, "Crosscorrelation properties of pseudorandom and related sequences," *Proc. IEEE*, **68**, pp. 593-619, 1980.

[7.15] R. M. Lerner, "A matched filter detection system for complicated doppler shifted signals," *IRE Trans.*, **IT-6**, pp. 373-380, 1960.

[7.16] R. H. Barker, "Group synchronization of binary digital systems," in *Communication Theory*, W. Jackson, ed., pp. 273-287, Academic Press, New York, 1958.

[7.17] M. J. E. Golay, "Complementary series," *IRE Trans.*, **IT-7**, pp. 82-87, 1961.

[7.18] G. R. Welti, "Quaternary codes for pulsed radar," *IRE Trans.*, **IT-6**, pp. 400-408, 1960.

[7.19]   R. Wilson and J. Richter, "Generation and performance of quadriphase Welti codes for radar and synchronization of coherent and differentially coherent PSK," *IEEE Trans.*, **COM-27**, pp. 1296-1301, 1979.

[7.20]   D. K. Barton, *Modern Radar Systems Analysis,* Artech House, 1988.

[7.21]   M. I. Skolnik, *Introduction to Radar Systems,* McGraw-Hill, 1980.

[7.22]   W. W. Shrader, "MTI Radar," Chapter 17 in *Radar Handbook,* M. I. Skolnik, ed., McGraw-Hill, 1970

[7.23]   D. H. Mooney and W. A. Skillman, "Pulse-doppler radar," Chapter 19 in *Radar Handbook*, (M. I. Skolnik, ed.), McGraw-Hill, 1970.

[7.24]   G. Stimson, *Introduction to Airborne Radar,* Hughes Aircraft Company, El Segundo, CA, 1983.

[7.25]   W. K. Saunders, "CW and FM radar," Chapter 16 in *Radar Handbook*, M. I. Skolnik, ed., McGraw-Hill, 1970.

[7.26]   W. Hardenburg, Westinghouse, 1989.

[7.27]   R. M. Lerner, "Ground radar system," U.S. Patent #3,831,173.

[7.28]   A. M. Nicolson and G. F. Ross, "A new radar concept of short range application," *IEEE International Radar Symposium Record*, Washington, D.C., pp. 146-151, April 1975.

[7.29]   L. C. Chan, D. L. Moffatt, and L. Peters, "A characterization of subsurface radar targets," *Proceedings IEEE,* Vol. 67, No. 7, pp. 991-1001, 1979.

[7.30]   G. F. Ross, *et al.,* "ABR for area protection of facilities," *DNA-TR-88-127,* Limited Distribution, May 1988.

[7.31]   C. L. Bennett and G. F. Ross, "Time domain electromagnetics and its applications," *Proceedings IEEE,* Vol. 66, No. 3, pp. 229-318, 1978.

[7.32]   E. Miller, *et al., Time Domain Measurements in Electromagnetics,* Van Nostrand Reinhold Company, New York, 1986.

[7.33]   J. Millman and H. Taub, *Pulse, Digital, and Switching Waveforms,* McGraw-Hill, New York, pp. 198-203, pp. 473-474, pp. 508-512, 1965.

[7.34]   J. Andrews, "Fast pulse generator study," Chapter 4, pp. 101-104, *Time Domain Measurements in Electromagnetics, op. cit.*

[7.35]   P. Paulis, et al., "The generation of microwave pulses by optoelectronically switched resonators," *IEEE Journal of Quantum Electronics*, Vol. QE-22, No. 1, pp. 108-111, January 1986.

[7.36]   R. W. P. King and H. J. Schmidt, "The transient response of linear antennas and loops," *IRE Transactions A&P*, AP-10m, pp. 222-228, 1962.

[7.37]   H. J. Schmidt "A survey of time domain techniques in antenna theory," plenary talk, 1967 *IEEE NEREM* Paper, Boston, MA, 1967

[7.38]   G. F. Ross, "A time domain criterion for the design of wideband radiating elements," *IEEE Transactions A&P*, Vol. 16, No. 3, p. 355, 1968.

[7.39]   G. F. Ross and L. Susman, "Array antenna signal processing systems," U.S. Patent #3,714,655.

[7.40]   A. M. Nicolson and R. Brophy, "Detector having a constant false alarm rate and method for providing same," U.S. Patent #3,755,696.

[7.41]   J. Millman and H. Taub, *op. cit.,* pp. 455-457.

Chapter 8

RADAR PROPAGATION

This chapter summarizes the effects on radar performance of the atmosphere and the surface underlying the signal path between the radar and the target. There are four different phenomena by which the propagation path influences the radar:

(a) **Atmospheric Attenuation**: Molecules of atmospheric gases, precipitation particles, and electrons in the ionosphere absorb energy from the radar wave as it passes through the atmosphere. This same coupling of EM waves to the atmosphere allows kinetic energy in the atmospheric particles to be coupled back into the radar receiver as thermal noise.

(b) **Surface Reflection**: Radar waves striking the earth's surface are reflected to add to the direct radiation field, causing either reinforcement or cancellation of the field that would have been present in free space.

(c) **Diffraction**: Waves passing close to the surface cause a diffraction field, which modifies the direct radiation field above the surface and may also cause energy to be propagated into the shadow zone behind an obstacle.

(d) **Refraction**: As the waves pass through atmospheric regions of varying refractive index, they are bent from a straight line into curved rays, making the target appear to be at a location different from its true position.

Thus, propagation phenomena have three different effects on radar performance: the strength of the signal, compared to that observed in free space, may be reduced or enhanced; the receiver noise level may be increased; and the apparent location of the target, as measured by the angle from which the signal arrives at the radar, the time delay of the signal, and its doppler shift, may be changed. These effects are in addition to any backscattering of energy which produces clutter in the radar receiver.

8.1 Atmospheric Attenuation

8.1.1 Clear Atmosphere

Curves and tabulated data on the atmospheric attenuation of radar waves at sea level, for different frequency bands, were given in Section 2.2.1. At a given frequency, the attenuation coefficient k_α, which expresses the magnitude of the attenuation in dB/km, varies with the density of the atmospheric gases (including water vapor), and the actual path attenuation can be plotted as a function of elevation angle and slant range to the target, as in Figures 8.1.1 and 8.1.2. All the curves have an initial slope equal to the sea-level attenuation coefficient. As the rays pass upward into thinner portions of the atmosphere, the slopes decrease, leveling off at a value which depends only on elevation angle for ranges beyond the stratosphere.

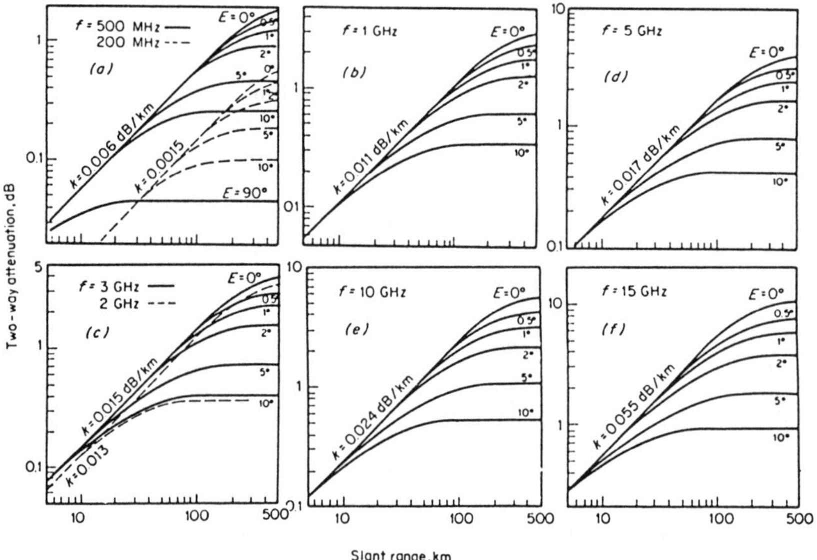

Figure 8.1.1 Atmospheric attenuation vs. range for different
elevation angles and frequencies 0.2 to 15 GHz.

To calculate the atmospheric attenuation for a path that starts at a given altitude above sea level and extends upwards, the procedure is to read two values from Figure 8.1.1 or Figure 8.1.2: one for a path from sea level to the target, and a second from sea level to the radar altitude, both at the same elevation angle. The difference is the radar-to-target attenuation.

For example, assume a K_u-Band radar operating from an aircraft or mountain at 3 km altitude, viewing a target at 2° elevation and 50 km range. A path leaving sea level at 2° elevation will pass through 3 km altitude (over a flat earth) at 86 km range. For the actual spherical earth, the range-height-angle charts presented later in this chapter give a range of 77 km, but the difference will not generally be significant. The attenuation for the 136 km path from sea level to the target is 3.2 dB, and for the 86 km path from sea level to the radar is 2.6 dB, leaving 0.6 dB as the radar-to-target attenuation. The attenuation coefficient at K_u-Band is much less at 3 km altitude than at sea level, partly because of the reduced air density but also because the water-vapor content of the atmosphere is greatly reduced at the high altitudes.

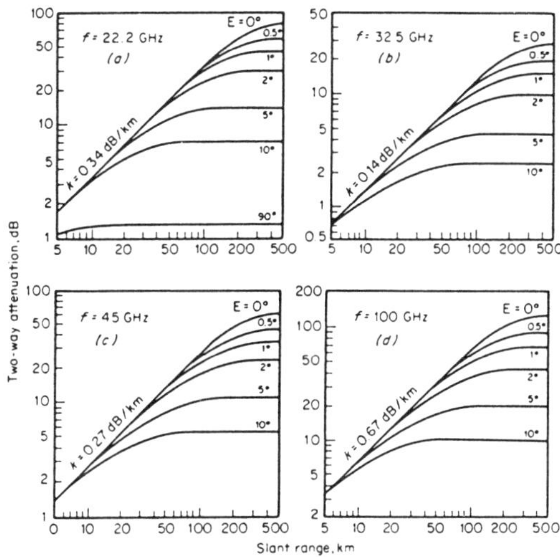

Figure 8.1.2 Atmospheric attenuation vs. range for different elevation angles and frequencies 20 to 100 GHz.

For an air-to-surface path, the curves may be read for the corresponding surface-to-air path, since the attenuation does not depend on which end of the path has the original source of the energy. Attenuations at frequencies not shown in Figures 8.1.1 and 8.1.2 can be estimated by locating the attenuation coefficient in Figure 2.2.2 and scaling from the curve for the closest radar band. Actual attenuations in upper microwave and mmw bands are sensitive to the water-vapor content of the air, which varies considerably with climate.

8.1.2 Attenuation in Precipitation

It was shown in Section 2.2.1 that frequencies above X-band are subject to increasingly severe weather effects, limiting their use in the atmosphere to short ranges. Two other precipitation effects can be considered.

Figure 8.1.3 shows the attenuation coefficient for clouds and fog, in which the particle sizes are much smaller than for rain. The density of condensed particles is given in g/m^3, with approximate figures for optical visibility in m. It can be seen that mmw radar (e.g., W-Band, 95 GHz) can penetrate for considerable distances through clouds where optical visibility is limited to a few hundred meters. For example, with 0.032 g/m^3 clouds optical visibility is limited to about 600 m, while W-Band has only 0.3 dB/km attenuation (added to the clear-air value of 0.8 dB/km).

Another effect of precipitation is the radar signal loss due to formation of water films on radomes. Table 8.1 shows values of this loss as a function of film thickness for S- and K$_u$-Band systems using radomes or exposed reflectors [8.1]. The exposed reflector loss is minimal, because the water film lies against a conducting surface where the electric field drops to zero. However, somewhere between the RF source and the reflector there must be a feed, which will usually contain or be contained within a window or radome surface. Preventing the formation of water films on these surfaces is important if reliable operation in precipitation is required.

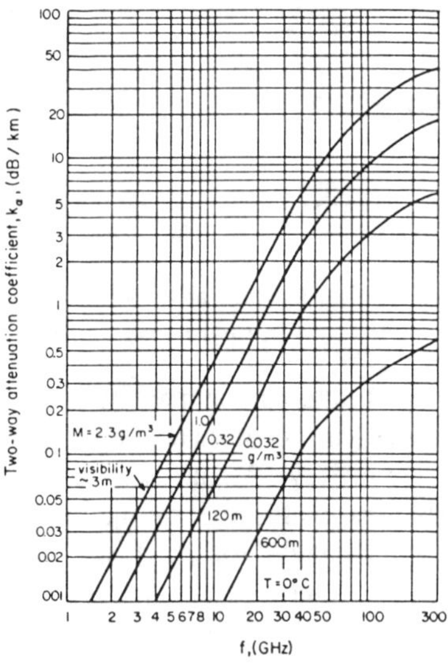

Figure 8.1.3 Attenuation in clouds and fog.

8.1.3 Ionospheric Attenuation

Attenuation in the ionosphere is a problem for radars operating in the HF and VHF bands, but seldom exceeds 1 dB at frequencies above 100 MHz. A more serious problem for UHF radars is the *Faraday rotation effect,* in which the plane of any linear polarization is rotated as the signal passes through the ionized layer. The effect is nonreciprocal, meaning that instead of the effect canceling, the two-way rotation is twice the one-way value, leading to potential mismatch in the polarization of the receiving antenna relative to that of the transmitter. Some systems use circular polarization to overcome this problem.

Table 8.1 Radar Signal Losses in Water Films

Film Thickness (mm)	Radome Loss (dB)		Reflector Loss (dB)	
	3.6 GHz	16 GHz	3.6 GHz	16 GHz
0.05		3.2		
0.13	1.1	5.3	≤0.01	0.03
0.25	2.6	8.7	≤0.01	0.2
0.38	4.2	10.9	≤0.01	0.9
0.5	5.6	12.3	≤0.01	2.7

8.1.4 Atmospheric Noise Temperature

The radar range equation was discussed in Chapter 3, and equations were given for calculation of the system noise, in terms of an equivalent input temperature T_s. One component of this temperature is the sky temperature T_a', shown in Figure 3.1.5.

When precipitation or clouds are present, their absorption of energy from the radar wave (signal attenuation) will increase the sky temperature. The resulting value can be found as:

$$T_a' = T_p(1 - 1/\sqrt{L_{\alpha t}}) \tag{8.1.1}$$

where $T_p \approx 290$ K is the physical temperature of the atmosphere, given by the weighted average over the attenuating regions, and $L_{\alpha t}$ is the two-way attenuation through the entire atmosphere, including the precipitation or clouds. However, unless the radar receiver uses a very low-noise RF amplifier, the sky temperature will not play an important part in determining the equivalent input temperature.

8.2 Surface Reflection Effects

8.2.1 Reflection from a Flat Surface

The geometrical relationships between the direct and reflected rays from a flat surface are shown in Figure 8.2.1. For a radar antenna height h_r, target range R, and altitude h_t, the following equations apply:

$$\text{Target elevation angle,} \quad \theta_t = \sin^{-1}\left(\frac{h_t - h_r}{R}\right) \approx \frac{h_t - h_r}{R} \tag{8.2.1}$$

$$\text{Grazing angle,} \quad \psi = \sin^{-1}\left(\frac{h_t + h_r}{R}\right) \approx \frac{h_t - h_r}{R} \tag{8.2.2}$$

$$\text{Pathlength difference,} \quad \delta_0 = R_1 + R_2 - R = R\left(\frac{\cos\theta_t}{\cos\psi} - 1\right) \approx \left(\frac{2h_r h_t}{R}\right) \tag{8.2.3}$$

$$\text{Range to reflection point,} \quad x_0 = h_r \cot\psi \approx \frac{R h_r}{h_r + h_t} \tag{8.2.4}$$

Because the pathlength difference is usually very small, differences in signal attenuation are negligible, and the reflected ray arrives at the target with an amplitude, relative to the direct ray, given by the product of the reflection coefficient of the surface, ρ, and the ratio of antenna voltage gain for the reflected signal, $f(-\psi)$, to that for the direct signal, $f(\theta_t)$.

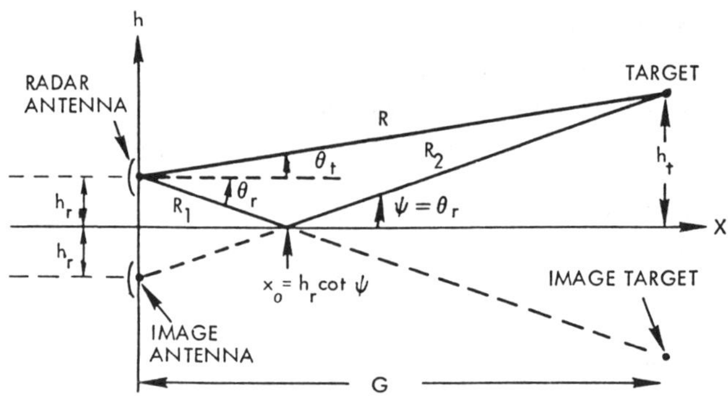

Figure 8.2.1 Geometry of specular reflection from a flat surface.

Figure 8.2.2 shows the magnitude of the reflection coefficient for a smooth surface, as a function of the grazing angle and the surface material, for horizontal and vertical polarization of the radar wave. This *Fresnel reflection coefficient,* ρ_0, is sensitive also to radar frequency in the case of salt water. It can be seen that horizontal polarization leads to strong reflection at grazing low angles (extending up to 20° or 30° for water surfaces). For vertical polarization, the coefficient drops steeply toward zero at the *Brewster angle* ψ_B, after which it rises again but with reversal of phase.

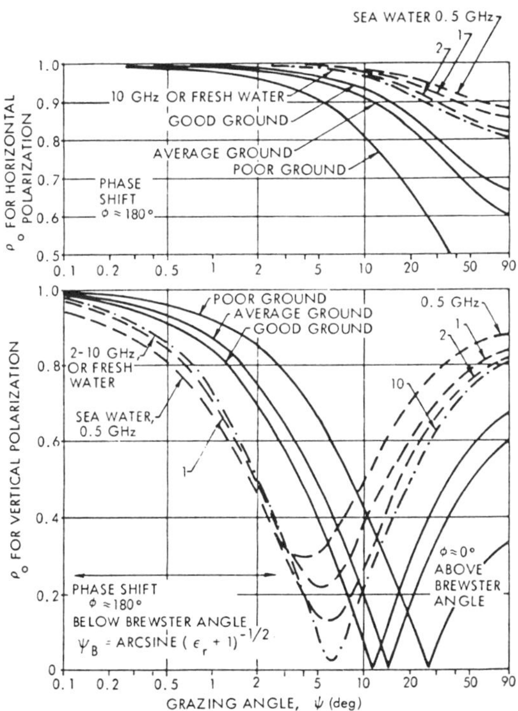

Figure 8.2.2 Fresnel reflection coefficients ρ_0
for horizontal and vertical polarizations.

8.2.2 Pattern-Propagation Factor

In the radar equation (see Chapter 3), a pattern-propagation factor F is used to account for the effects of surface reflections and other departures from maximum antenna gain and free-space conditions. When the vector addition of the direct and reflected signals is considered, this factor becomes

$$F \;=\; |f(\theta_t) + f(-\psi)\rho D \exp(-j\alpha)| \qquad (8.2.5)$$

where $f(\theta)$ is the voltage pattern of the antenna, relative to the on-axis gain, at elevation angle θ, $f(\psi)$ is the voltage pattern at the angle of the reflected signal, D is the divergence factor for the spherical earth, and α is the phase angle of the reflected wave:

$$\alpha \;=\; (2\pi\delta_0/\lambda) + \phi \qquad (8.2.6)$$

where ϕ is the phase angle of the reflection coefficient. This formulation for F applies only when $\delta_0 > \lambda/6$. For smaller path differences, the effects of diffraction are dominant.

Combining (8.2.5) and (8.2.6), expressing the result for F^2, and assuming a broad antenna pattern, we obtain an equation for the propagation factor without pattern effects:

$$F^2 \;=\; 1 + \rho^2 + 2\rho\cos(\phi + 4\pi h_r\theta_t/\lambda) \qquad (8.2.7)$$

At low angles over reflective surfaces ($\rho \approx 1$, $\phi \approx \pi$), this factor becomes:

$$F \;=\; 2|\sin(\pi\theta_t/\theta_n)| \qquad (8.2.8)$$

The radar detection coverage then extends to twice the free-space range at the peaks of the lobes, and is reduced to zero at the nulls, which occur at elevation intervals given by

$$\theta_n \;=\; \lambda/2h_r \qquad (8.2.9)$$

A typical plot of radar coverage, showing the effects of reflection lobing, was given in Figure 2.2.3.

8.2.3 Spherical Earth Reflection

The modifications to Figure 8.2.1 and Eqs. (8.2.1) - (8.2.4), required to describe paths over the spherical earth, are discussed in [8.2, Appendix A], which also gives a simple calculator program for calculation of the reflection point and the heights of the two terminals above the plane tangent to the earth at that point. For short-range, over-land paths, the variations in local surface heights are such that earth's curvature is relatively unimportant. Radars operating from high ground sites or airborne platforms may require calculations which include not only the spherical earth geometry, but evaluation of the divergence factor D in (8.2.5). This is discussed in [8.3, pp. 2-39 to 2-44].

Reflection from a Rough Surface

Rough surfaces on land or sea can be described by the parameter σ_h, the rms deviation from the smooth, spherical earth. The regular (specular) reflection coefficient of a rough surface is the product of the Fresnel coefficient, ρ_0, and a *specular scattering coefficient*, ρ_s, given by:

$$\rho_s^2 \;=\; \exp\left[-\left(\frac{4\pi\sigma_h}{\lambda}\right)^2\right] \tag{8.2.10}$$

This coefficient is plotted in Figure 8.2.3, as a function of grazing angle ψ, in mr, for different degrees of surface roughness. When the specular reflection is reduced by the rough surface, it is replaced by a *diffuse reflection* component, causing the resultant signal to vary rapidly about its free-space value but adding little to the average signal power.

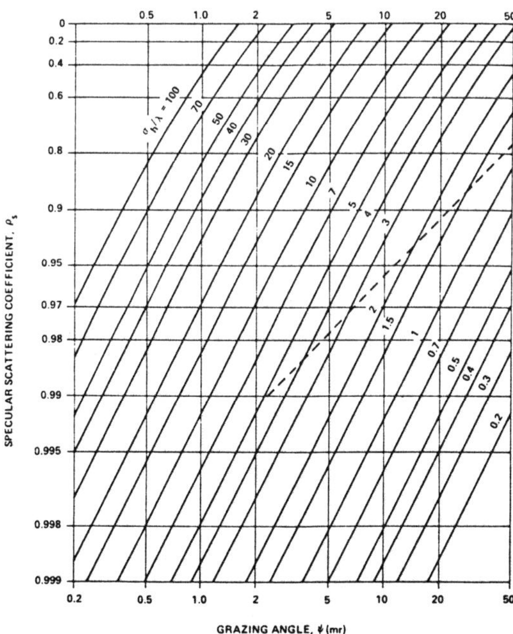

Figure 8.2.3 Specular scattering coefficient vs. grazing angle
for different values of surface roughness.

8.2.4 Reflection from Vegetation

Land surfaces are often covered by a layer of vegetation, which absorbs some of the radar energy. The reduced amplitude of the reflected signal can be described by a further multiplicative factor ρ_v, the *vegetation factor.* There are no adequate models for this factor, but values measured in microwave bands range from 0.3 for grassy surfaces to 0.03 for dense vegetation.

8.2.5 Multipath Errors

One consequence of surface reflections, in addition to modification of the magnitude of the resultant signal received from the target, is the introduction of changes in its apparent angle of arrival. The radar antenna sees not only the target, as a source of signal, but also the "image" of the target on the underlying surface. In the case of specular reflection, the image may be an inverted replica of the real target, separated in elevation by the angle $\theta_t + \psi$ (from Figure 8.2.1). Over rough surfaces this specular image may be replaced or supplemented by a diffused image, appearing as a vertical band of reflections extending from the horizon to a point well below the target (Figure 8.2.4).

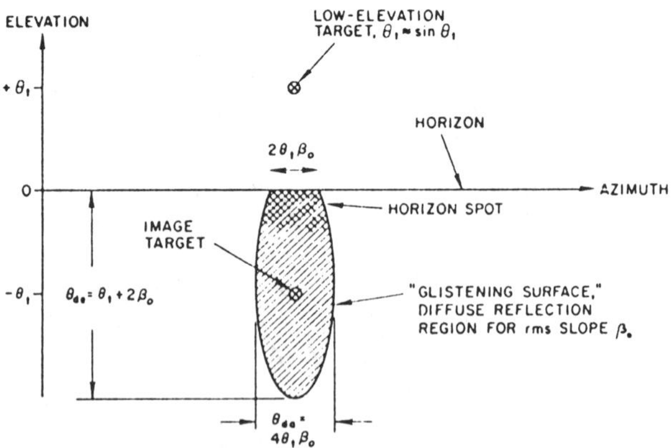

Figure 8.2.4 Radar view of a target over a slightly rough surface.

In tracking or measuring the target position, the radar estimates the direction to the target as the angle of a line normal to the phasefront received at the antenna. When the target appears over a reflecting surface, the reflections cause ripples in this phasefront, and the antenna attempts to find the best plane-wave fit to the distorted wavefront (Figure 8.2.5). As the phases of the reflected signal components change, the tilt of this equivalent plane wave changes, causing **multipath error** to appear (primarily in the elevation measurement). Errors will also appear in measured range, and for irregular surfaces in azimuth as well.

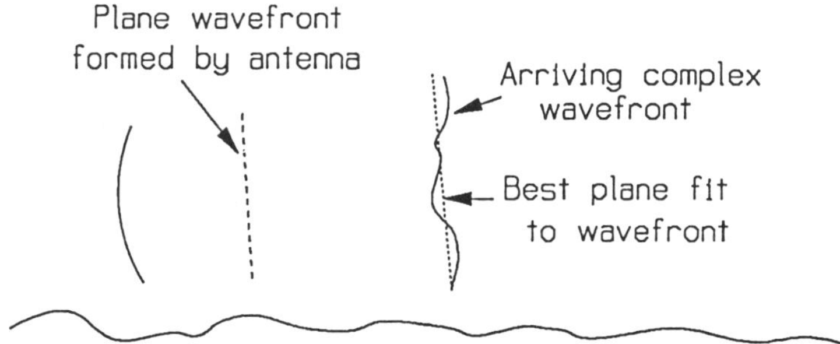

Figure 8.2.5 Measurement of angle of arrival of a complex wavefront.

The magnitude of the measurement errors caused by multipath reflections from the surface will be discussed in Chapter 11.

8.3 Diffraction

8.3.1 Diffraction over the Smooth Spherical Earth

Signals from low-altitude targets, if they appear below the first lobe of the reflection pattern, will be reduced by the **smooth-sphere diffraction** effect. Typical plots of the one-way propagation factor are shown in Figure 8.3.1 for high and low microwave frequencies. The received field strength drops steadily as the target approaches and goes below the horizon but, unlike the optical case, the signal does not disappear abruptly. Radars operating in the lower frequency bands start to lose the signal at elevation angles well above the horizon. For higher radar frequencies the signal above the horizon is less affected, and the drop-off near the horizon becomes steeper. While there is some propagation beyond the horizon, the two-way loss is such that in the normal radar frequency bands there is no useful operation in that region. In the lower part of the VHF spectrum, experiments have shown that **ground-wave** transmission can provide some coverage beyond the normal horizon.

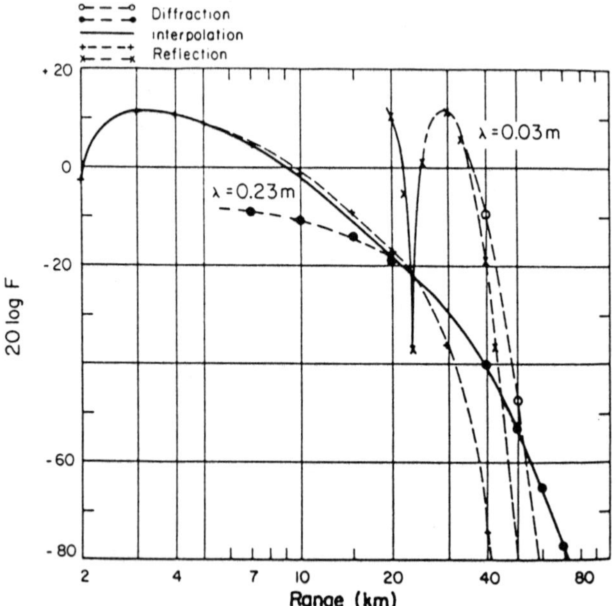

Figure 8.3.1 Typical curves of propagation factor vs. range
for low-altitude target.

The procedures for calculating the propagation factor below the first reflection lobe have been discussed in [8.3, pp. 2-43 to 2-45], [8.4], and [8.2, Appendix A]. The calculator program in this last reference can be used for the frequency bands (above 100 MHz) used by most radars.

8.3.2 Diffraction beyond a Knife-Edge Obstacle

When the path from the radar to the target passes over or through an obstacle other than the bulge of the spherical earth, propagation will be governed by *knife-edge diffraction*. The effect was illustrated in Figure 2.2.5, for an obstacle (mask) consisting of a tree line, a hill, or a structure rising above the spherical earth. The term *knife edge* should not be interpreted, in radar bands, as requiring a sharp obstacle. When the obstacle is at a range R_1 near the radar ($R_1 \ll R$, where R is the range to the target), the maximum radius of curvature of an obstacle giving knife-edge diffraction can be found as [8.5, 1979]

$$r \; < \; 0.0024\sqrt{R_1^3/\lambda} \tag{8.3.1}$$

Most natural surface features, as well as man-made structures and tree lines, will meet this criterion. Criteria for full illumination of the obstacle by the radar wave are discussed in [8.2, pp. 301-302].

The magnitude of the diffracted field from a knife edge can be estimated by determining the diffraction parameter v, using the geometry of Figure 8.3.2, where

$$v = 2\sqrt{\frac{R_1 + R_2 - R}{\lambda}} = \sqrt{\left(\frac{2R}{\lambda}\right)} \tan\alpha_1 \tan\alpha_2 \qquad (8.3.2)$$

The direct path can pass above or below the top of the obstacle. The diffracted field strength, relative to that of the direct ray, is plotted in Figure 8.3.3. If there is a clear path to the target, the diffracted field adds vectorially to this, causing minor ripples in the resultant field. For $v = 0$, the diffracted field is out of phase with the direct field, causing a resultant - 6 dB relative to free space (-12 dB two-way). If the path is blocked by the obstacle, only the diffracted field reaches the target, and this field is described by $A(v)$ as plotted. The two-way radar signal then has a propagation factor $F^4 = -40 \log A(v)$.

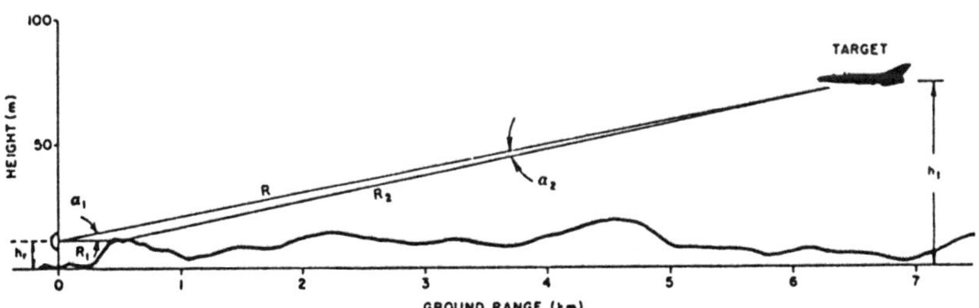

Figure 8.3.2 Target track over a nearby obstacle.

As previously shown in Figure 2.2.5, the presence of a knife-edge component can actually increase the signal strength relative to that observed over the spherical earth. In communications this is known as ***obstacle gain***, and successful microwave links have been established by directing the antennas at the top of a ridge or mountain which blocks the direct path between terminals. In radar, the two-way loss prevents effective coverage except on targets (and clutter) very near the masking angle, but in some cases the deterioration of coverage is much less than that predicted by smooth-sphere diffraction theory.

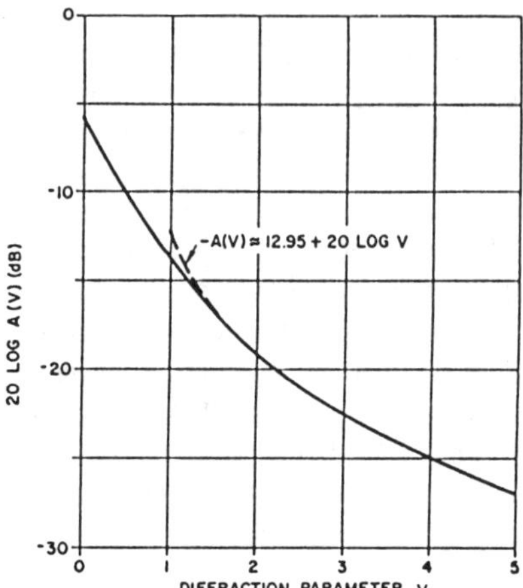

Figure 8.3.3 Amplitude of knife-edge diffraction component.

8.4 Atmospheric Refraction

Radar waves passing through the atmosphere are bent downward by the changing refractive index of the troposphere, and then again by the ionosphere. Three effects are produced in this process:

(a) Regular refraction, resulting from the gradual reduction in refractive index with altitude, causing range and elevation angle errors;

(b) Tropospheric fluctuations, resulting from random variations in refractive index, causing slowly varying errors in all measured coordinates; and

(c) Ducting, resulting from steep gradients in refractive index (usually near the surface), creating low-loss propagation paths to some regions not normally reached by the rays, and leaving holes in the coverage of other regions.

8.4.1 Exponential Reference Atmosphere

The refractive index of the troposphere for most radar frequencies can be expressed as

$$N \;=\; (n-1)\times 10^{6} \;=\; \left(\frac{77.6}{T}\right)\left(P + \frac{4810p}{T}\right) \tag{8.4.1}$$

where T is temperature in kelvins, P is pressure in millibars, p is the partial pressure of the water-vapor component, and n is the refractive index.

At sea level, N is typically between 300 and 350, and the mean and extreme profiles with altitude are as shown in Figure 8.4.1. The exponential reference atmosphere, which is plotted as a dashed line on the figure, is described by

$$N(h) \;=\; 313\exp(-0.1439h) \;\approx\; 313\exp(-h/7) \tag{8.4.2}$$

where h is in km above sea level. At low altitudes, the nearly constant gradient of refractive index produces ray bending which results in an ***effective earth's radius*** $ka = (4/3)a$, where $a = 6.5 \times 10^6$ m, the actual radius of the earth. Radar coverage charts based on the 4/3 radius factor are generally accurate enough to describe coverage for surface-based radars, and charts based on the exponential reference atmosphere are also available [8.6].

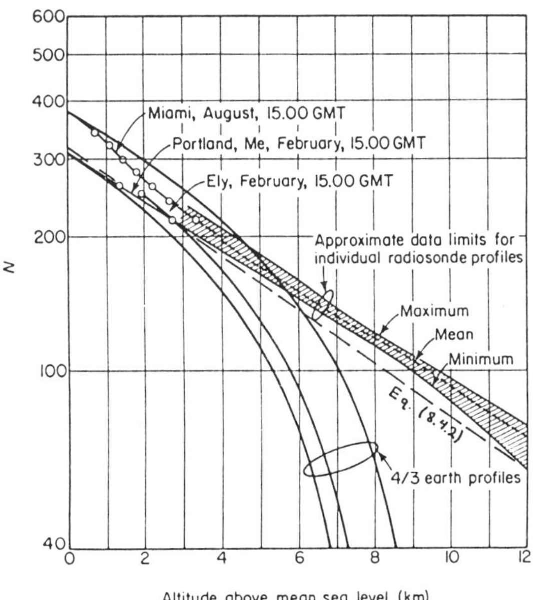

Figure 8.4.1 Refractivity vs. altitude, compared
to the 4/3 earth radius profiles [8.7].

8.4.2 Elevation and Range Bias Errors

The errors in measured target elevation and range data, caused by refraction in the exponential reference atmosphere, are shown in Figures 8.4.2 and 8.4.3. The surface refractivity is taken as 313 N-units, and the errors may be scaled directly for other values of refractivity, whether these result from different meteorological conditions or from elevation of the radar site.

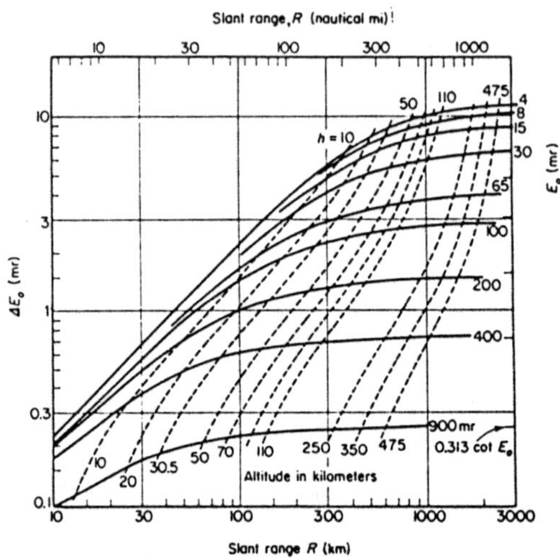

Figure 8.4.2 Elevation bias error vs. range for exponential reference atmosphere, $N_s = 313$.

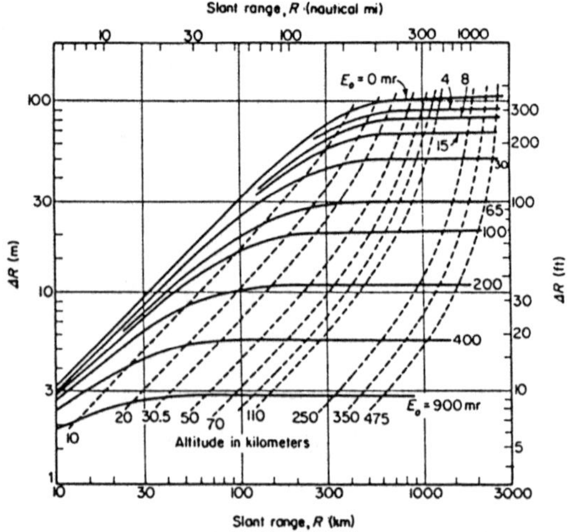

Figure 8.4.3 Range bias error vs. range for exponential reference atmosphere, $N_s = 313$.

These bias errors may be corrected, to a large extent, by subtracting the predicted error shown in these figures, using a table look-up procedure or approximating equations. The residual errors, after scaling for actual surface refractivity, will be about 5% of the original error.

8.4.3 Tropospheric Fluctuations

Irregular variations in refractivity occur in the troposphere, with scale sizes from fractions of a meter to hundreds of kilometers. The phasefront reaching the radar receiving antenna will have a varying tilt and local ripples which depend on these fluctuations. Figure 8.4.4 shows the expected rms error in azimuth measurements caused by these fluctuations, as a function of aperture width (or baseline length of an interferometer system).

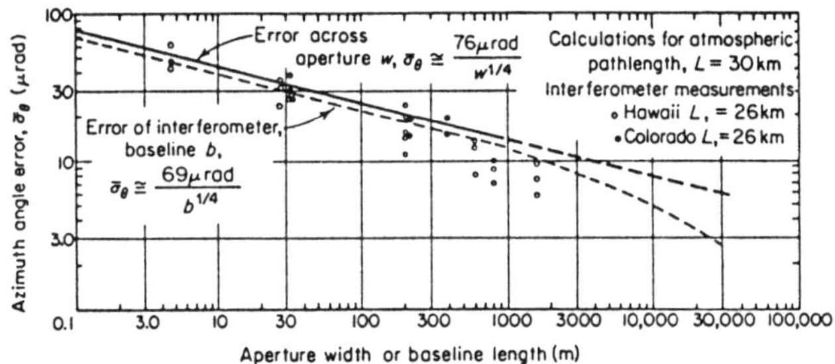

Figure 8.4.4 Azimuth error from tropospheric fluctuations, as a function of aperture width or baseline length.

8.4.4 Ducting

The reference atmosphere has a refractivity gradient $dN/dh = -45$ N-units per km of altitude, near sea level. When this gradient increases beyond -157 N-units per km, rays leaving the antenna will be trapped in a ***duct***, within which they may be propagated for great distances with low loss, as in a waveguide. Data on the occurrence of ducts, and their heights, has been collected and made available in a computer program [8.8]. This program also provides calculations of the propagation factor vs. range, and ray-tracing plots which show the regions of enhanced and reduced coverage.

8.4.5 Range-Height-Angle Charts

The standard means of describing the detection coverage of a search radar, or the acquisition envelope of a tracking radar, is to plot the contour of some constant detection probability in range-altitude space, using an earth surface contour which has been curved so that ray paths appear as straight lines (see Figure 2.2.3). For most cases, these charts can be drawn using the effective earth's radius factor of 4/3. The program VCCALC [8.6] permits these charts to be drawn for most practical combinations of maximum target range and altitude. Calculations and personal computer programs for radars operating at altitudes well above sea level are given in [8.10] and [8.11].

8.5 References

[8.1]   B. C. Blevis, "Losses due to rain on radomes and antenna reflecting surfaces," *IEEE Trans.* **AP-13**, No. 1, January 1965, pp. 175-176.

[8.2]   D. K. Barton, *Modern Radar System Analysis*, Artech House, 1988.

[8.3]   L. V. Blake, "Prediction of Radar Range," Chapter 2 in *Radar Handbook* (M. I. Skolnik, ed.), McGraw-Hill, 1970.

[8.4]   M. L. Meeks, *Radar Propagation at Low Altitudes*, Artech House, 1982.

[8.5]   D. K. Barton, "Low-altitude tracking over rough surfaces I: Theoretical predictions," *IEEE Eascon Record*, 1979, pp. 224-234.

[8.6]   J. E. Fielding and G. D. Reynolds, *VCCALC: Vertical Coverage Calculation Software and Users Manual*, Artech House, 1988.

[8.7]   B. R. Bean and G. D. Thayer, "CRPL Exponential Reference Atmosphere," National Bureau of Standards Monograph No. 4, U.S. Government Printing Office, October 29, 1959.

[8.8]   H. V. Hitney, A. E. Barrios, and G. E. Lindem, "Engineer's Refractive Effects Prediction System (EREPS)," Technical Document 1342, Naval Ocean Systems Center, July 1988.

[8.9]   L. V. Blake, "Machine Plotting of Radio/Radar Vertical-Plane Coverage Diagrams," NRL Report 7098, Naval Research Laboratory, June 1970.

[8.10] W. A. Skillman, *Radar Calculations Using the TI-59 Programmable Calculator*, Artech House, 1983.

[8.11] W. A. Skillman, *Radar Calculations Using Personal Computers*, Artech House, 1984.

Chapter 9

SEARCH AND TARGET ACQUISITION

A *search radar* is a radar used primarily for the detection of targets in a particular volume of interest. In the process of detecting the target, its location is indicated to some degree of accuracy. After detection, target position and velocity may be determined by *track while scan,* an automatic target tracking process in which the radar antenna and receiver provide periodic video data as inputs to computer channels which follow individual targets.

A *tracking radar* is a radar whose primary function is the automatic tracking of targets. Prior to tracking, the radar must acquire the target through search of a limited region in which the target is known or suspected to lie. Hence, the maximum range of both search and tracking radars will be determined by their ability to detect targets while scanning a volume of space. That volume often approaches a hemisphere for search radar, and may be as small as one beamwidth for a tracker which has been designated onto a particular target.

This chapter discusses the search and target detection process for radars of different types: *two-dimensional (2D)* and *three-dimensional (3D)* air-search radars, surface search radars, ground-mapping radars, and tracking radars during their acquisition mode. The relationships between search performance and the radar parameters are presented, to provide the starting point for test and evaluation of these radars.

9.1 Two-Dimensional Air-Search Radar

9.1.1 Coverage Pattern

The 2D radar scans in azimuth only, using a fan beam which extends from the horizon to the highest elevation at which targets are to be detected. When the targets are to be kept in view (e.g., for track-while-scan operations) as they approach the radar, the beam pattern may be shaped to follow the *cosecant-squared* shape (Figure 9.1.1). In this pattern, the mainlobe provides maximum range R_m from the horizon to an elevation angle $\theta_1 \approx \arcsin(h_m/R_m)$, where h_m is the maximum altitude of intended targets. To prevent high-altitude targets from flying above the beam into a *cone of silence* above θ_1, the upper side of the beam is extended with reduced gain to an angle θ_2, often between 30° and 60°. The antenna gain between θ_1 and θ_2 is proportional to $\csc^2\theta$ of the elevation angle.

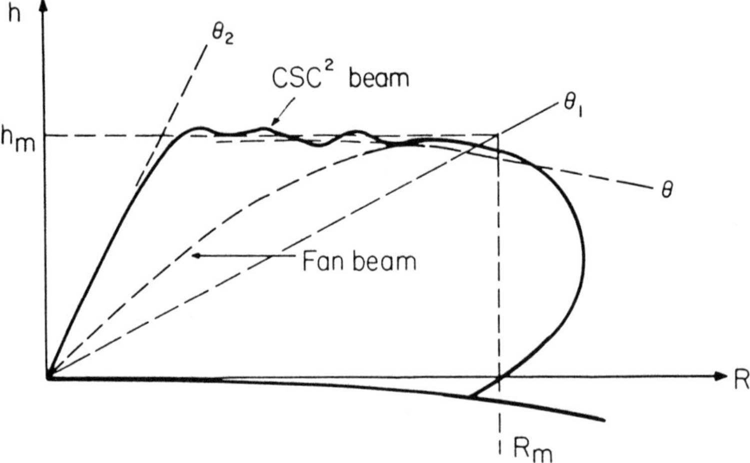

Figure 9.1.1 Coverage pattern of radar with $\csc^2$ antenna.

In the radar with $\csc^2\theta$ coverage, a target which is detectable at θ_1, at range R and altitude $h = R_1\sin\theta_1$, will maintain a constant signal strength and detection probability until it reaches $R_2 = h\sin\theta_2$. This follows from the radar equation, in which S varies as G^2/R^4:

$$S = KG^2/R^4 = K(\csc\theta^2)^2/R^4 = K/(R\sin\theta)^4 = K/h^4 \tag{9.1.1}$$

Airborne radars, operating from high altitudes, may use the same type of antenna with the $\csc^2\theta$ extension on the lower edge of the mainlobe, in which case θ can be taken as the depression angle from horizontal.

9.1.2 Aperture Size for 2D Radar

The dimensions of the radar antenna for 2D radar are limited by the wavelength and the beamwidths required for coverage of the volume. The portion of the antenna which produces the main elevation lobe has a maximum height given by

$$h \le K_\theta\lambda/\theta_1 \tag{9.1.2}$$

where $K_\theta \approx 1.2$ radians $= 70°$ is the beamwidth constant (see Section 6.1) and the small-angle assumption $\sin\theta \approx \theta$ has been used. Additional antenna structure, extending the physical height, may be needed to produce the $\csc^2\theta$ pattern, but this does not contribute to antenna gain or effective aperture size. The fact that the extra aperture must be illuminated to provide high-angle coverage actually reduces the gain and effective aperture of that portion of the reflector which produces the mainlobe. For example, if the coverage is to extend to altitude $h_m = 20$ km at range $R_m = 150$ km, we have $\theta_1 = 0.13$ radian $= 7.6°$, and the main portion of the antenna will be limited to a height $h = 9\lambda$. At S-band, this corresponds to about 0.9 m. The distorted portion of the aperture might increase the over-all height to 1.4 m. If one-third of the transmitter power is applied to illumination of the high-angle region, the effective height of the antenna will be reduced to 0.6 m.

The width of the antenna determines its azimuth beamwidth, which must be narrow enough to provide the required azimuth resolution, but wide enough to obtain the beam dwell time needed for MTI or doppler processing. In the absence of a mechanical constraint (e.g., transportability or aircraft mounting), the azimuth beamwidth will normally be the minimum value which can provide a given dwell time $t_o = n/f_r$, where n is the number of pulses (hits per scan) needed for doppler processing and f_r is the PRF. The available dwell time is related to the search frame time t_s, the azimuth sector A_m scanned during this time, and the azimuth beamwidth θ_a.

$$t_o = t_s \theta_a / A_m \tag{9.1.3}$$

From this, the maximum width of the antenna will be

$$w \leq K_\theta \lambda / \theta_a = K_\theta \lambda t_s / A_m t_o \tag{9.1.4}$$

Example of 2D Aperture Calculation

For example, if $f_r = 1000$ Hz to obtain an unambiguous range of 150 km, and $n = 24$ is required for doppler processing, while scanning $A_m = 2\pi$ radians in $t_s = 6$ s, the dwell time must be at least

$$t_o = 24/1000 = 0.024 \text{ s}$$

The azimuth beamwidth in this case must be at least

$$\theta_a = 0.024 \times 2\pi/6 = 0.025 \text{ radian} = 1.4°$$

The maximum width of the antenna is then

$$w = (1.2/0.025)\lambda = 48\lambda$$

At S-band, this corresponds to about 5 m. The physical area of the aperture in this band, for the $\csc^2$ coverage discussed above, is then limited to about 7 m^2, and the effective aperture area will be only about 2 m^2 for the usual illumination efficiencies. Typical values for θ_a will be between 1° and 2° for 2D radars in which the antenna width is not constrained by physical factors such as transportability.

Effective Beamwidth and Aperture

The effective elevation beamwidth, used to compute antenna gain or effective aperture area, can be found as

$$\theta_m = \theta_1 (2 - \theta_1 \cot \theta_2) \tag{9.1.5}$$

At worst, this will be twice the width θ_1 of the mainlobe, but in a typical case (e.g., $\theta_1 = 10°$, $\theta_2 = 30°$) it will be about 1.7 times as great, corresponding to a *cosecant-squared loss* of 2.3 dB, relative to the antenna which would provide coverage only to θ_1.

The effective aperture which can be used in a given 2D radar application can then be expressed in terms of the angle sectors to be covered, the frame time, and the wavelength

$$A_r = (\lambda^2 / A_m \theta_m)\,(t_s / t_o)/L_n \tag{9.1.6}$$

where $L_n \approx 1.2$ is the beamwidth constant used in Eq. (1.16) and θ_m is the effective elevation coverage sector (equal to θ_1 in the absence of $\csc^2$ coverage). This equation establishes the fact that the allowable effective aperture area in 2D radar application varies inversely with the solid angle of the coverage and the time-on-target, and directly as the frame time and the square of wavelength.

9.1.3 Required Average Power for 2D Radar

The search radar equation (2.4.7) may now be solved for average power required in a 2D radar when the maximum allowable antenna dimensions are used. Thus

$$P_{av} = \frac{4\pi R_m^4 A_m^2 \theta_m^2 k T_s D L_s L_n t_o}{\lambda^2 t_s^2 \sigma} \tag{9.1.7}$$

Because the allowable aperture varies as λ^2, the required power varies inversely as λ^2 rather than being independent of wavelength as in the original search radar equation. For this reason, as well as for better clutter rejection, 2D radars will normally use the longest wavelength which is consistent with the mechanical constraints on antenna size.

Using the values from the example of Section 2.4.3, but with $\csc^2$ coverage to 30°, Eq. (9.1.7) gives a required $P_{av} = 10\,\text{kW}$ at S-band. This is consistent with the results in Table 2.4.1, when the product $P_{av} A_r$ is increased by 1.8 to provide the $\csc^2$ coverage with $A_r = 2.5\ \text{m}^2$. The high ratio of power to aperture area is indicative of the fact that S-band is too high a frequency for economical radar system design to meet this coverage requirement.

The 2D search radar equation as given in (9.1.7) can be used in evaluating the maximum expected performance in a benign environment of radars at different frequencies, with respect to operational requirements, e.g., coverage, frame time, target cross section, propagation conditions, and detection performance (related directly to the detectability factor D). When specific radar parameters are known, range calculations using the Blake chart provide a more direct result.

9.1.4 Multipath Lobing in 2D Radar

Figure 2.2.3 shows the actual detection coverage obtained with a 2D antenna pattern having a broad elevation beam. The maximum detection range is almost doubled, but deep null regions extend into the desired coverage. The pattern-propagation factor F is calculated from Eqs. (8.2.5) - (8.2.8), as a function of elevation angle and surface reflection coefficient. For early-warning applications, where maximum range is desired, the 2D search radar will normally use horizontal polarization to obtain the greatest possible reflection coefficient. For air traffic control and similar applications requiring consistent coverage within the free-space detection envelope, vertical polarization will be used to minimize the reflection coefficient (Figure 8.2.2). In addition, antennas for these radars will be designed with a sharp reduction in gain for negative elevation angles, to avoid illumination of the surface.

9.1.5 Clutter Considerations in 2D Radar

In 2D radar, each target must compete with the underlying surface clutter and with any volume clutter (e.g., rain or chaff) lying within the broad elevation beam. The average cross sections of surface clutter, viewed by surface-based radars with typical beamwidths near 1.5° and pulsewidths of 1 μs, were calculated in Section 2.2.3 as 9 m^2 for land and 1.26 m^2 (at X-band) for sea. Land clutter peaks up to 10^4 m^2 were also predicted, requiring improvement factors exceeding 50 dB or special methods for rejection of clutter peaks. Use of MTI or other doppler processing is essential in 2D air search radars. Many modern systems use the moving target detector (MTD) type of pulsed doppler to achieve the required improvement in this clutter environment. This method requires approximately $n = 20$ hits per scan as a minimum for proper operation, while MTI systems can operate with n as low as 10 for mechanically scanned antennas, or 3 for electronically step-scanned antennas.

Weather clutter can be expected up to some maximum altitude h_c ($\approx$ 5 km), usually below the maximum altitude h_m of the radar coverage. For this case, the volume within the 2D radar resolution cell, given by Eq. (2.4.9) for clutter which completely fills the beam, becomes

$$V_c = (R\theta_a/L_p)h_c \tag{9.1.8}$$

In our 2D example, with $\theta_a = 0.015$ radian $= 0.86°$ and $\tau_n c/2 = 150$ m, this volume becomes $V_c = 1.3 \times 10^9$ m^2 at $R = 150$ km. In rain falling at a rate of 3 mm/h, the reflectivity at S-band is $\eta_v = 3 \times 10^{-9}$ m^2/m^3, giving a rain cross section of 3.8 m^2, about 6 dB above a 1 m^2 aircraft target. From 20 to 25 dB improvement in S/C ratio is required for detection in this case. In civil air traffic control radar, a major portion of this improvement is often obtained by using an antenna with *circular polarization*, which reduces the rain echo by about 17 dB while reducing target signal by only 3 dB.

If strong multipath lobing is present, volume clutter will be enhanced by a factor $F_c^4 \approx 6$, or $+8$ dB, requiring even greater improvement factors. At the same time, the circular polarization of the wave reaching the clutter may be degraded by surface reflection, reducing the circular polarization improvement below the theoretical 17 dB.

Chaff clouds can extend to altitudes above the weather ceiling (e.g., $h_c = 15$ km), and because of their random orientation cannot be canceled with circular polarization. In addition, the velocity spread of the chaff spectrum is often 30 to 40 m/s, preventing effective doppler rejection in radars with blind speeds below about 150 m/s. Special waveforms (medium or high PRF doppler bursts) or longer wavelengths ($\lambda > 0.3$ m) are required for successful operation in chaff.

Use of MTI processing is not very effective against rain or chaff, because of the wind velocities at which the scatterers are moving. Moving target detector (MTD) processing may be effective if the total velocity spread of the clutter spectrum is a small fraction (e.g., 1/4) of the blind speed v_b of the radar waveform. The blind speed is given by

$$v_b = f_r \lambda / 2 \tag{9.1.9}$$

and the performance of MTI or MTD systems depends on the ratio of v_b to clutter velocity spread (see Chapter 7). This favors use either of the highest available PRF or the longest available wavelength, for clutter rejection.

9.1.6 Wavelength, Waveform, and Dwell Time for 2D Search Radar

The wavelength λ, which does not enter into the original search radar equation, is seen to be a critical factor in 2D radar, first because of the limitation on antenna height. Required average power, for a given detection range, will vary as λ^{-2}, giving a great advantage to the lower frequency bands in cases where the antenna size is not limited by physical factors. Secondly, if doppler techniques are to be used for rejection of volume clutter, the longer wavelengths can provide the needed high blind speed, while maintaining an adequate unambiguous range.

The dwell time required in 2D radar is determined by the clutter environment. Since rain and chaff must be considered in any military environment, the dwell time must allow for $n \approx 20$ hits per dwell, dictating a relatively slow scan of the coverage volume. Radars with fewer hits per scan cannot be expected to operate effectively in rain or chaff environments.

9.2 Three-Dimensional Air-Search Radar

The advantages of 3D radar were outlined in Section 2.4.9, and included the relaxation of the antenna height limitation, the benefits of elevation resolution in rejecting clutter and jamming, and the provision of height data on all detected targets.

9.2.1 Methods of Providing 3D Coverage

The 3D radar coverage is provided in one of three ways:

(1) A single pencil beam is scanned in a raster (Figure 2.4.1a) to cover each elevation position in sequence during the time required to scan through the azimuth beamwidth;

(2) A stack of elevation beams is generated (Figure 2.4.1d) to simultaneously cover all elevation positions during this time; or

(3) A more limited stack of beams is time-sequenced to cover two or more sectors which constitute the total elevation coverage.

Scanning Pencil Beam

The scanning pencil beam method requires the least receiving and processing hardware, since only one channel is used. The beam must scan rapidly in elevation, and electronic scanning antennas of the frequency- or phase-scanning type (Section 6.6) will normally be required. Because of the rapid scan, the dwell times in each beam will be quite limited. In order to provide time for even one pulse in each beam, the PRI is usually varied to match the constant-altitude coverage pattern similar to that of the $\csc^2$ antenna (Figure 9.1.1). The total time required to scan the elevation sector within a single azimuth beamwidth can be calculated as that which would provide full-range coverage (using $t_r = 2R_m/c$) over an elevation sector

$$\theta_m' = \sin\theta_1(1 + \ln\sin\theta_2 - \ln\sin\theta_1) \tag{9.2.1}$$

For example, with $\theta_1 = 7.6°$, $\theta_2 = 30°$, the time would be that required to scan at maximum range to $\theta_m' = 18°$. Using an elevation beamwidth $\theta_e = 1°$, with an unambiguous range of 150 km ($t_r = 1$ ms), $n = 1$ hit per beam would require 18 ms per azimuth beamwidth. The lowest beam could receive a 3-pulse burst for MTI if this time were extended to 20 ms per azimuth beamwidth, giving an azimuth scan rate of $50\,\theta_a$/s. An antenna rotating in azimuth, with $\theta_a = 1.2°$ would complete one rotation in 6 s.

Partial Stacked Beams

Some frequency-scanned 3D radars have been designed to overcome the dwell-time limitation of the scanning pencil beam by transmitting m subpulses, at different frequencies, during each pulse. The resulting m beams cover a sector which is a significant fraction of the total elevation coverage, permitting many pulses to be transmitted before sequencing to the next elevation sector. The number of receiver and processor channels is equal to m, and each of these may use MTI or other doppler processing, increasing the complexity of the system. In return for this complexity, however, the azimuth scan rate may be increased while still retaining MTI capability and the other ECCM benefits of multiple-pulse processing.

For example, a partial stack of seven 1° beams could be used to obtain full-range coverage to $\theta_1 = 7°$. Sequencing into three higher sectors would provide the 30° total coverage (when considering the off-broadside broadening of the elevation beamwidth). The time required would be somewhat greater than that predicted by Eq. (9.2.1), because the PRI for each sector would be determined by the range to the lowest beam in that sector. However, in many cases a single pulse would be sufficient to provide detection in the upper sectors, allowing more than seven pulses per dwell for the lowest sector.

Stacked Beams

The stacked-beam system requires the greatest number of receiver and processor channels, but does not require electronic scanning. The antenna uses either a reflector with multiple feed horns, or a planar array with a multiple-beam feed matrix. Dwell-time considerations are identical to those of the 2D radar, providing the maximum value of n for all beams in the stack.

9.2.2 Power-Aperture Product for 3D Search Radars

The 3D radar normally adjusts its resources to provide detection coverage with a constant maximum altitude, similar to the $\csc^2$ coverage of the 2D radar (Figure 9.1.1). In order to calculate the required power-aperture product, an effective elevation sector θ_m for full-range coverage must be computed. If the full gain of the antenna is used for receiving at each elevation, the transmitted energy per beam would follow a $\csc^4$ function, leading to an effective elevation sector

$$\theta_m = (4/3)\left[1 - \frac{(\sin\theta_1/\sin\theta_2)^3}{4} \right] \tag{9.2.2}$$

The maximum value for this angle is $\theta_m = (4/3)\theta_1$, as compared to $2\theta_1$ for the $\csc^2$ antenna. In most cases, receiving beams in the upper elevations are broadened, reducing their gains and requiring greater diversion of transmitted power into the upper coverage.

Scanning Pencil Beam

To save time in scanning the elevation sector, this type of 3D radar is often designed to broaden the elevation beamwidth at the higher angles. This reduces the effective receiving aperture, and requires that the energy per beam be increased to a value intermediate between the csc^4 and csc^2 relationship. The ratio θ_m/θ_1 is then between 4/3 and 2.

Partial Stacked Beams

In a frequency-scanned antenna, broadening of the elevation beamwidth is not an option. Therefore, Eq. (9.2.2) will apply. When phase scanning is used, the upper beams may be shaped as desired, trading transmitter power for time.

Stacked Beams

The transmitting pattern for a stacked-beam system must illuminate the entire elevation sector, but can be controlled to adjust the upper coverage contour. In addition, the receiving beams can be broadened at high elevations to reduce the number of channels required. Thus, the effective elevation sector can vary between $(4/3)\theta_1$, for constant receiving beamwidth and csc^4 transmitting gain function, and $2\theta_1$, for receiving and transmitting gains both following $csc^2\theta$. It should be noted that the increased dwell time and number of pulses per beam leads to an integration gain. This compensates for the decreased transmitting gain for the broad elevation beam. The slight increase in system loss, if some noncoherent integration is used rather than MTD coherent integration, is a small disadvantage compared to the processing advantages of the n-pulse burst.

9.2.3 Wavelength and Waveform for 3D Search Radar

The 3D antenna area can be increased in proportion to the number of elevation beams used. Thus, there is no firm limit on the aperture size that can be used at any wavelength. The choice of wavelength will be determined primarily by performance requirements in clutter, and by mechanical limitations for transportable radars. Most 3D search radars operate at S-band, to achieve high resolution and accuracy in height measurement, but in this band the clutter rejection capability is severely limited. The largest transportable 3D radars use L-band, reducing the rain clutter input and making possible, at least in stacked-beam systems, reasonable performance against all types of clutter.

Stacked-beam radars can use MTD processing with bursts of 8 to 16 pulses. The other systems are generally limited to such techniques as 3-pulse MTI bursts in the lower beam positions, and often to single-pulse waveforms in the upper beams.

The evaluation of 3D search radar poses difficult problems, since the performance may vary greatly over the different beams and receiving channels. In the most extreme case, separate calculations are needed for each beam in each type of environment.

9.3 Surface Search Radar

Surface search radar is a special form of 2D radar in which the elevation coverage is that necessary to illuminate the earth's surface over the specified range coverage. Typical applications include:

(1) Marine navigation and collision avoidance,

(2) Sea search and rescue,

(3) Battlefield surveillance, and

(4) Fire control against surface targets on land and sea.

In most of these applications, the targets of interest are unresolvable in doppler from the surrounding surface features, and target detection is dependent on the amplitude of the target or on the radar achieving sufficient resolution to discriminate in favor of desired target shapes. There is no unique requirement for minimum dwell time, and the size of the antenna is limited primarily by mechanical considerations.

9.3.1 Shipborne Surface Search Radar

The height of the antenna above the surface is small, relative to the minimum search range, and hence the beamwidth can be made as narrow as permitted by antenna size and the accuracy of the antenna stabilization system. If the antenna platform is not stabilized, the beamwidth must be at least $20° = 0.35$ radian to accommodate pitch and roll of the ship. The solid angle to be searched is then

$$\psi_s \approx 0.35 A_m \approx 2 \text{ steradians (for } 360° \text{ scan)}$$

At maximum detection range, the target almost always lies below the first lobe of the multipath reflection pattern, and the pattern-propagation factor F^4, included in Eq. (2.4.7) as a component of L_s, is

$$F^4 \approx (4\pi h'_t h'_r / R\lambda)^4 \tag{9.3.1}$$

The search radar equation for average power required becomes

$$P_{av} = \frac{2\pi R_m^8 \lambda^3 K_\theta \theta_m^2 kT_s D L_s L_n}{(4\pi)^3 h_t'^4 h_r'^4 \sigma t_s w} = \frac{0.76 R_m^8 \lambda^3 kT_s D L_s L_n}{(4\pi)^3 h_t'^4 h_r'^4 \sigma t_s w} \tag{9.3.2}$$

where h_t' and h_r' are target and radar heights above a plane tangent to the sea at the specular reflection point, w is the width of the antenna, and L_s now includes search losses other than F^4. It is assumed that a line-of-sight path exists between the antenna and the target, when using the 4/3 earth radius (or other value for non-standard refractive conditions, [9.2]). The location of the specular reflection point, and the corresponding target and antenna heights can be calculated, for any effective earth's radius, using a computer program [9.1, Appendix A].

The fact that R^8 now appears in the numerator indicates the rapid increase in required power to overcome the null caused by multipath reflections. These reflections also lead to the fourth power of heights in the denominator, and the λ^3 in the numerator. Surface search radar over the sea is best carried out at the highest frequencies permitted by atmospheric and rain attenuation, using antennas mounted as high as possible on the ship. In practice, X-band is used most often, with average powers in the order of 100 W. Change in power from that level has a relatively small effect on detection range, because of the dominant R^8 relationship.

9.3.2 Ground-Based Surface Search Radar

The ground-based surface search radar is governed by factors similar to those for shipborne radar, but the broad elevation beamwidth is needed only for operation from land vehicles over rough terrain. If the antenna platform can be fixed and leveled, a taller aperture can be used to reduce the search coverage angle and increase the height h of the receiving aperture. The search radar equation for required power becomes

$$P_{av} = \frac{R_m^8 \lambda^5 k T_s D L_s}{(4\pi)^2 h_t'^4 h_r'^4 \sigma t_s w h^2} \tag{9.3.3}$$

The advantage of short wavelengths is increased because the elevation coverage and beamwidth may be reduced, for a given aperture height h. The azimuth resolution, for given aperture width w, is also improved at shorter wavelengths. Hence, if the user is willing to accept the potential attenuation in bad weather, the use of K_a-band or mmw radar for ground-based surveillance can be an attractive option.

9.3.3 Airborne Surface Search Radar

When the radar platform is airborne, the line-of-sight limit to range is increased, and free-space propagation can be obtained at shorter ranges. The performance of surface search radar is then dominated by power-aperture product, surface clutter, and (at low depression angles over land) by local terrain masking (from hills or trees). The standard search radar equation (2.4.7) can be used to calculate the required power-aperture product.

Surface targets may be moving slowly or fixed, making resolution in doppler difficult or impossible. High resolution in the spatial coordinates is essential for detection of small targets. When considering the limited size of antennas which can be mounted on most airborne platforms, this forces the designer into the high microwave or millimeter-wave bands. Pulse compression, for high resolution in range, is also necessary in most cases. Large antenna dimensions can be achieved by mounting an array along the side of the vehicle, in which case electronic scanning must be used.

An example of techniques which provide high performance in airborne surface search is the AN/APS-116 radar [9.3], designed for detection of periscopes against the sea background. The submarine at periscope depth moves at velocities too low to permit resolution of the periscope from sea clutter by doppler processing. Instead, a combination of high resolution in azimuth, very high resolution in range, and scan-to-scan integration is used in the AN/APS-116 to detect the $1.0\,\mathrm{m}^2$ periscope. At X-band, clutter from a medium sea can be characterized by a reflectivity parameter $\gamma = 0.01$. With an azimuth beamwidth $\theta_a = 2.4° = 0.042$ r, a range resolution $(\tau_n c/2) = 0.5$ m, and a radar altitude of 1000 m, the clutter cross section will be

$$\sigma_c = \frac{h_r \theta_a \tau_n c}{L_p \ 2}\gamma = \frac{1000 \times 0.042}{1.33}\times 0.5 \times 0.01 = 0.16\,\mathrm{m}^2$$

Starting with a ratio $S/C = 1/0.16 = +8$ dB, and integrating over 5 s at 5 scans/s, the target becomes detectable with high probability. Integration during a single scan is of limited value because the sea clutter remains correlated for tens of milliseconds. To permit scan-to-scan integration of moving targets in a system with very high range resolution, the signal processing makes use of thresholding and stretching of the signal on each scan, followed by binary integration over the several scans.

9.4 Ground-Mapping Radar

9.4.1 Real-Aperture Approach

Surface search over land is often directed toward mapping of the land, to locate fixed features or slowly moving vehicles. The "target" becomes a resolution element on the surface whose radar cross section is

$$\sigma = \Delta_r \Delta_x \gamma \sin\psi \tag{9.4.1}$$

where Δ_r is the range resolution depth, Δ_x is the cross-range resolution width, γ is the surface reflectivity parameter for elements which are to be detectable above thermal noise, and $\psi = \sin^{-1} h_r/R$ is the grazing angle to the surface (neglecting the effects of earth's curvature). Then, since $\Delta_x = R\theta_a$, we can write (neglecting the beamshape factor L_p which is small)

$$\sigma = \Delta_r h_r \theta_a \gamma \tag{9.4.2}$$

The equation for required average power, in order to detect surface elements having minimum reflectivity γ at maximum range R_m, becomes

$$P_{\mathrm{av}} = \frac{4\pi R_m^4 A_m \theta_m^2 k T_s D(1) L_s L_n}{\lambda^2 t_s \Delta_r \gamma h_r} \tag{9.4.3}$$

where θ_m is the elevation coverage angle (modified as necessary for $\csc^2$ coverage) and A_m is the azimuth sector to be covered. Note that $\lambda/\theta_m = h/K_\theta$ represents the height of the aperture, and that the required power is inversely proportional to the square of this value. The width of the antenna does not enter into the equation, although a limitation on width will affect the wavelength which can be used for a given cross-range resolution.

A typical coverage in the elevation plane for this type of radar is shown in Figure 9.4.1. Because the upper edge of the beam is not horizontal, (9.1.5) cannot be used to calculate an effective elevation coverage angle. Instead, this angle is given by

$$\theta_m = \theta_e + \sin^2\theta_1(\cot\theta_1 - \cot\theta_2) \tag{9.4.4}$$

where θ_e is the elevation beamwidth, θ_1 is the depression angle to the lower edge of the mainlobe, and θ_2 is the depression angle to the lower edge of the $\csc^2$ region.

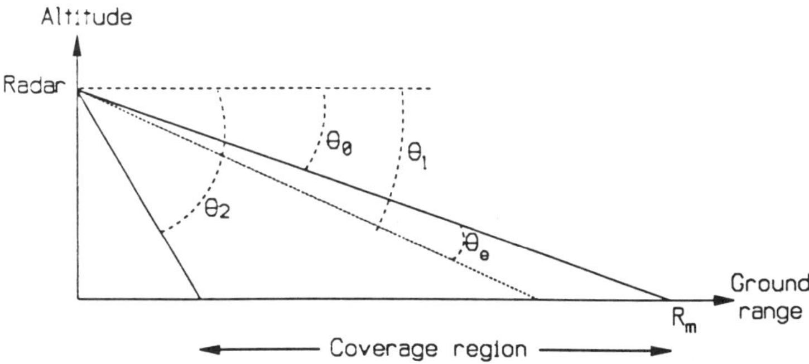

Figure 9.4.1 Typical ground-mapping radar coverage.

A typical requirement for ground mapping might be to obtain 10 m × 100 m cells out to 50 km range, with $\gamma = 0.01$ producing a detectable signal +10 dB above noise. The azimuth resolution must be $100/50000 = 0.002$ r, requiring the use of K_a-band radar. An elevation beamwidth $\theta_e = 2.3° = 0.04$ r is sufficient, with $\csc^2$ shaping on the under side of the mainlobe as shown in Figure 9.4.1. The effective elevation beamwidth can be calculated from (9.1.5) only if one edge of the beam is horizontal. For depressed beams, the effective beamwidth is $\theta_m = \theta_e +$

$\theta_1(1-\theta_1\cot\theta_2)$. The effective beamwidth for $\theta_1 = 0.1$, $\theta_2 = 0.2$ is then $5° = 0.09$ r. Ground coverage is from $h_r/\theta_2 = 15$ km to $R_m = 50$ km. The parameters listed in Table 9.1 might be assumed.

Table 9.1 Parameters of Real-Aperture Ground-Mapping Radar

Maximum range, $R_m = 50$ km	Surface reflectivity, $\gamma = 0.01$
Aircraft height, $h_r = 3000$ m	Azimuth sector, $A_m = \pi/2$
Elevation sector, $\theta_m = 0.09$ r	Wavelength, $\lambda = 0.0086$ m
Range resolution, $\Delta_r = 10$ m	Input temperature, $T_s = 1000$ K
Detectability factor, $D(1) = 10$	Search loss factor, $L_s = 15$ dB (includes $L_\alpha = 5$ dB for clear air)
Beamshape factor, $L_n = 1.33$	Frame time, $t_s = 2.5$ s

With these parameters, the average power requirement is
$$P_{av} = 1220 \text{ W}$$
The selected waveform might be
$$P_t = 30 \text{ kW}, \quad f_r = 3000 \text{ Hz}, \quad \tau = 13 \text{ μs}, \quad B = 20 \text{ MHz}$$
and the antenna dimensions would be
$$w = K_\theta\lambda/\theta_a = 1.15 \times 0.0086/0.002 = 5 \text{ m}$$
$$h = K_\theta\lambda/\theta_e = 1.15 \times 0.0086/0.035 = 0.28 \text{ m}$$

At K_a-band, if operation in clouds were required, the search loss factor would be increased by perhaps 5 dB, requiring a peak power level of 100 kW if maximum range were to be maintained. In rain, range would be substantially reduced.

The antenna would be mounted along the fuselage of the aircraft, and electronically scanned to cover the 90° azimuth sector. Almost 1000 electronically controlled phase-shifting elements would be required, each coupled to about 45 radiating elements in the vertical plane. The $\csc^2$ beam shaping would be provided by proper adjustment of phase in the 45 radiating elements. Back-to-back antennas might be needed to permit operation while flying in either direction parallel to the border of the mapped region. Thus, even for rather limited mapping resolution, the real-aperture system is large and relatively expensive.

9.5 Synthetic-Aperture Approach

The synthetic-aperture radar (SAR) system achieves very high resolution in azimuth (and hence in the cross-range dimension Δ_x) by gathering and processing samples of a signal transmitted and received from many points along a flight path of length $L = t_o v_a$ (Figure 9.4.2). During this time, ground points within the real beamwidth θ_a are observed. The signal samples are stored over the observation time t_o, and the processing permits formation of high-resolution beams of width

$$\Delta_a = K_\theta' \lambda / L \sin \alpha \qquad (9.5.1)$$

Here, $K_\theta' \approx 0.5$ is the beamwidth constant for SAR operation, whose value is about half that for real-aperture operation. The reason for this is that each signal sample from a point on the synthetic aperture has a phase resulting from two-way propagation to the target area, doubling the sensitivity of the SAR system to angle differences between target scattering points. In some literature on SAR, a processing factor $m = 1/K_\theta'$ is used to describe the effect of weight for low sidelobes.

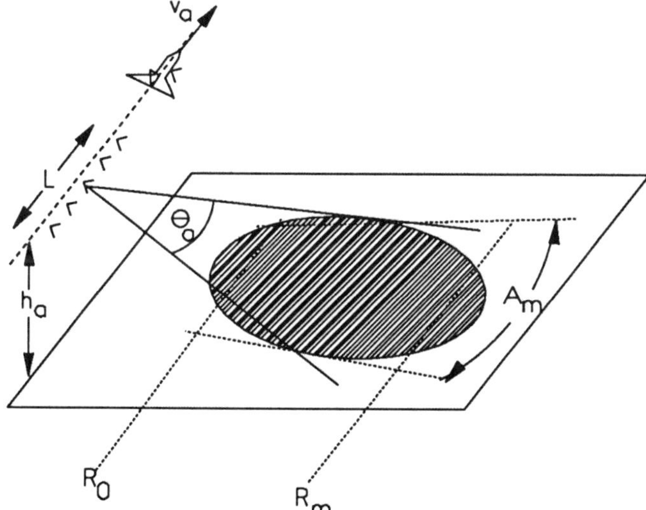

Figure 9.4.2 Geometry of SAR operation.

At a range R, the cross-range resolution Δ_x for the SAR system is

$$\Delta_x = R \Delta_a = R K_\theta' \lambda / L \sin \alpha \qquad (9.5.2)$$

The best resolution results from coherent processing of all data received while the real beam illuminates the target area, in which case $L = R\theta_a$ and

$$\Delta_x = K_\theta' \lambda / \theta_a \sin \alpha = w K_\theta' / K_\theta \sin \alpha \qquad (9.5.3)$$

where w is the width of the real aperture used to obtain the data. Thus, the best resolution, when operating in a direction normal to the platform velocity vector ($\alpha = 90°$), is equal to half the width of the physical antenna on the SAR platform, and this resolution is independent of range to the target area.

9.5.1 SAR Range Equation

The radar equation for SAR can be derived from the search radar equation (2.4.7) by substituting

$$\sigma = \Delta_r \Delta_x \sigma^0 = \Delta_r \Delta_x \gamma h_r / R_m$$

$$t_s = t_o = R_m \theta_a / v_a$$

$$A_m = \theta_a = K_\theta' \lambda / \Delta_x \sin \alpha$$

The result gives the average power required to form a map showing surface regions of reflectivity γ or greater as

$$P_{av} = \frac{4\pi R_m^4 \theta_m^2 v_a K_\theta' k T_s D(1) L_s L_n^2}{\lambda \Delta_r \Delta_x^2 \gamma h_r \sin \alpha} \tag{9.5.4}$$

The resolution is best when samples over the entire observation time t_o are combined coherently in the processor, but the resulting map or image has a severe *speckle,* corresponding to the Rayleigh amplitude distribution which results inevitably from coherent addition of many scattering sources within each resolution cell. Improved visibility of target features is obtained by forming several overlapping maps of the same target area, and adding them noncoherently (after processing) to smooth the speckle. The resolution cell size (in cross-range) of such maps is then increased in proportion to the number of maps processed during t_o. However, in some cases the stability of the system would not have permitted effective coherent processing over the entire time, in any case, and the noncoherent addition yields improved visibility and simplified processing at no cost in resolution.

9.5.2 Example of Synthetic Aperture Radar

Consider the previous ground-mapping problem, but with a requirement for higher resolution, e.g., 3 m × 3 m out to maximum range $R_m = 50$ km. This performance would clearly be impossible with a real-aperture system. An aircraft platform might fly at $v_a = 200$ m/s, with the radar beam looking normal to the velocity vector to form a strip map. In 250 s the map would cover 50 km in length along the flight path.

Weighting of the data for sidelobe reduction might lead to $K_\theta' = 0.8$, and if noncoherent addition of $n = 4$ maps is desired the physical aperture size would be

$$w = (\Delta_x K_\theta \sin \alpha)/n K_\theta' = 3 \times 1.15 \times 1/3 \times 0.8 = 1 \text{ m}$$

Since there is no requirement for very short wavelength in SAR, the wavelength would be chosen for convenience in operation, compatibility of the platform with the size of the vertical aperture needed for elevation coverage, and required stability in the radar oscillators and platform inertial monitoring. A typical choice would be X-band ($\lambda = 0.03$ m), which would give a physical beamwidth

$$\theta_a = K_\theta \lambda/w = 1.15 \times 0.03 / 1 = 0.035 \text{ radian} = 2°$$

The aperture height, to generate an elevation beamwidth of 0.04 r = 2.3° would be about 0.9 m.

The average power requirement can be calculated from (9.4.8), using $nK_\theta' = 3.2$ in place of K_θ' yielding

$$P_{av} = 30 \text{ W}$$

This low power is typical of SAR operation, and is made possible by the very long coherent integration times used in such systems. In practice, if the 0.9 m aperture height were inconvenient, or if allowance were needed for aircraft roll angle, the antenna height could be reduced to about 0.3 m, and power increased. The increase in power would not be a factor of three, because the wider beam would make $\csc^2$ coverage unnecessary.

9.6 Tracking Radar Acquisition

A final topic in the search radar subject area is the acquisition of targets by a narrow-beam tracking radar. When a phased array antenna is used in the tracker, an acquisition scan can be implemented to cover a broad volume in space, extending even to search of the full quadrant covered by the scanning capability of the array. The limitation in this case is imposed by the time budget, along with the power-aperture product, of the radar system. In most cases, whether electronic or mechanical scan is used, the tracker acquisition scan will cover a narrow sector within which a target has been designated by an external surveillance sensor.

9.6.1 Target Designation Data

Designation of the target can come from several types of external sensors, with corresponding differences in the requirements for acquisition scanning.

Optical Designation

A co-located optical instrument will provide designation in both azimuth and elevation, eliminating the need for acquisition scan by the antenna. The acquisition range of the radar is then found directly from the radar equation, setting the observation time t_o equal to the time in which a given probability of acquisition P_a is to be obtained. The detectability factor D_x is calculated for this probability, and beamshape losses are included in the equation to account for possible errors in angle designation. The radar must have the capability to detect targets in the beam at any range and with any doppler shift, since these coordinates cannot be determined by the usual optical sensor.

2D Search Radar Designation

A co-located 2D search radar can provide range and azimuth designation data to the tracker, but target elevation (and usually radial velocity) will be unknown. The tracker acquisition scan will cover the elevation sector in which targets might be located, and some azimuth scanning may also be required to cover the $\pm 3\sigma$ error of the search radar data. Because range data are available, the tracker may restrict its acquisition sensing to a few range gates surrounding the designated range, permitting it to use a low threshold without encountering excessive numbers of false alarms. The search radar equation is applied directly, with the solid angle ψ_s calculated from the widths of the elevation and azimuth sectors.

When the search radar is remotely located, its data are converted to rectangular coordinates, the origin shifted to the tracker site, and reconverted to polar coordinates. The absence of elevation data makes it necessary to ignore target altitude, and ground range $G = R\cos E$ from the search radar is estimated as slant range R, causing an error in the rectangular coordinates which increases with target height. Hence, the azimuth scan sector for acquisition may be larger than for co-located search radar.

3D Search Radar Designation

The designation data from 3D radar include target elevation or height. This reduces the sector which the tracker acquisition scan must cover, in both elevation and azimuth coordinates. As the accuracy of the designation data improves, the scan sector can potentially be reduced to a single beam position, within which a narrow range interval is sensed by the detection circuits. The resulting acquisition range is then limited by the ability of the signal processor to integrate efficiently over periods in the order of a second or more.

Jammer Strobe Designation

When the search system is jammed, it can designate the jamming vehicle as a target in angle (azimuth, for 2D radar; azimuth and elevation, for 3D radar). A co-located tracker can then scan at that angle (or fix its beam up the strobe, for designation by a 3D search radar). If the tracker frequency is clear of jamming, target detection should be possible at the maximum tracker detection range. In most cases, the tracker frequency will also be jammed, and the target jam strobe will be detected by the tracker immediately after designation.

Jammer designation from a network of search radars, using triangulation, can provide range and azimuth data to the tracker, permitting decisions as to the degree of threat and whether the threat is within engagement range. With 2D radars, there remains the problem of *ghosts,* formed by multiple intersections of strobes from two or more jammers. This problem is reduced in 3D radars because the strobes in azimuth and elevation may not cross at locations other than those of true targets.

9.7 References

[9.1] D. K. Barton, *Modern Radar System Analysis*, Artech House, 1988.

[9.2] H. V. Hitney, A. E. Barrios and G. E. Lindem, *Engineer's Refractive Effects Prediction System (EREPS)*, Naval Ocean Systems Center Technical Document 1342, July 1988.

[9.3] J. M. Smith and R. H. Logan, "AN/FPS-116 Periscope-Detecting Radar," IEEE Trans. **AES-16**, No. 1, January 1980, pp. 66-73.

Chapter 10

TRACKING RADAR SYSTEMS

10.1 Measurement Fundamentals

10.1.1 Detection, Resolution, and Interpolation

Any radar detection of a target provides some information on target coordinates, although the data may be very coarse. The PPI display of a conventional 2D search radar gives range and azimuth data in the form of an intensified blip whose width in azimuth is approximately the beamwidth of the antenna, and whose range extent is that of the processed pulse as it leaves the receiver, possibly stretched by limited display bandwidth or excessive CRT spot size. The simplest type of automatic radar measurement consists of reporting the azimuth and range resolution cell in which the target is detected.

More refined estimates of target location are made by interpolation within a cell or between two cells. The human operator places a cursor scale at the apparent center of the PPI blip to obtain interpolation to a fraction of the beamwidth. An automatic target extractor may integrate signal pulses in contiguous, fixed gates, and make an estimate based on the relative amplitudes; or it may perform a continuous integration and produce an estimate when the output begins to decline from its peak value. These interpolation processes can be applied to search, track-while-scan, or tracking radars.

The tracking radar is distinguished by the fact that its antenna beam position or receiver range gate follows the point in radar space in which the target signal is expected to appear, so that interpolation may be performed more efficiently and accurately. Modern tracking radars may time-share the tracking function among many targets, and may interlace tracking and search modes with a single antenna, so that total dedication to a single target may no longer characterize the tracking radar.

10.1.2 Basic Measurement Processes

Radar operates in four-dimensional space, resolving the target in the following coordinates:

- Two angles (often azimuth and elevation);
- Range, measured by the time delay of the received signal; and
- Radial velocity, measured by the doppler shift of the signal.

Measurement of target position is performed by comparing the signal amplitude in two channels, as illustrated in Figure 10.1. The voltage response of a channel, in a radar coordinate which is denoted by z in the figure, is described by $f(z)$. The two channels are offset from a measurement axis z_0 by $-z_k$ and $+z_k$, such that the difference between them becomes an indicator of the off-axis position of the target position.

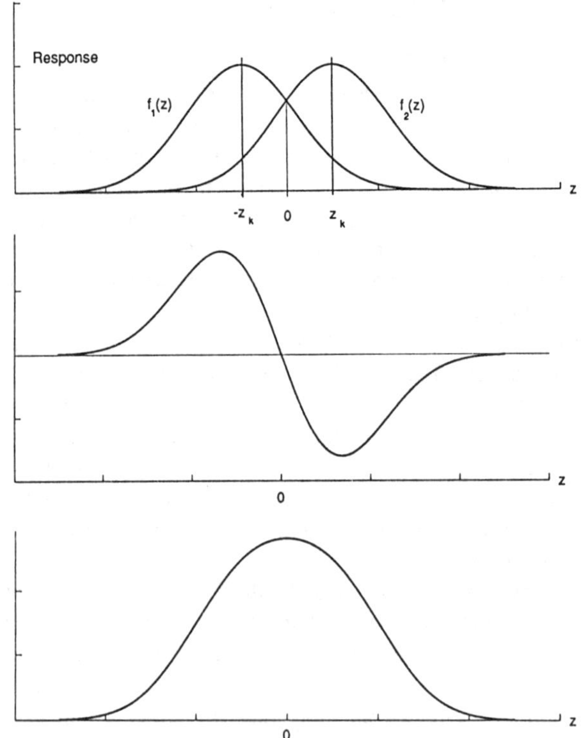

Figure 10.1 Basic measurement process. (a) Offset channels. (b) Difference response. (c) Sum response.

In order that this difference signal indicate the target position, independent of signal strength, the two channels are added to form the sum channel, and the difference output is divided by this sum to *normalize* the measurement.

The sensitivity of the measurement system is determined by the slope of the normalized difference curve, at the point where it crosses the axis. This slope is further normalized by the half-power (or -3 dB) width of the sum response, z_3, to obtain a dimensionless measurement slope

$$k_z = -\left(\frac{d(\Delta/\Sigma)}{d(z/z_3)}\right)_{z=0} \tag{10.1.1}$$

10.1.3 Precision of Measurement

If the difference channel of the measurement system contains a random interference component of power I_Δ, competing with a signal whose power in the sum channel is S, the resulting measurement error on a single sample of that interference may be described by a standard deviation

$$\sigma_{zI} = \frac{z_3}{k_z\sqrt{2(S/I_\Delta)}} \tag{10.1.2}$$

The factor of two within the radical results from the fact that the interference power is assumed to be divided equally into components in phase and in quadrature with the sum signal, only one of which passes the error channel detector.

When the number of independent samples of interference averaged during a measurement is n_e, the variance of the error is reduced by this number, giving

$$\sigma_z = \frac{z_3}{k_z\sqrt{2(S/I_\Delta)n_e}} \tag{10.1.3}$$

The number of samples n_e depends on the averaging time t_o of the measurement system, and the correlation time t_c of the interference.

10.2 Angle Tracking and Measurement

10.2.1 Sequential and Simultaneous Lobing

The two channels used to perform the measurement may be generated either sequentially or simultaneously. In the case of angle measurement, the channels are beams produced by an antenna, and in Figure 10.2 a single receiver channel is switched *sequentially* between two adjacent feed horns (or antenna ports) to sample these beams. The receiver output for each beam is stored, and the average stored values are formed into sum and difference outputs. The sum output provides for display of the target (and range measurement), and also is used to normalize the difference output in a dividing circuit. This normalized difference output may be used, in a tracking radar, as the error signal of a servo loop to control the antenna position.

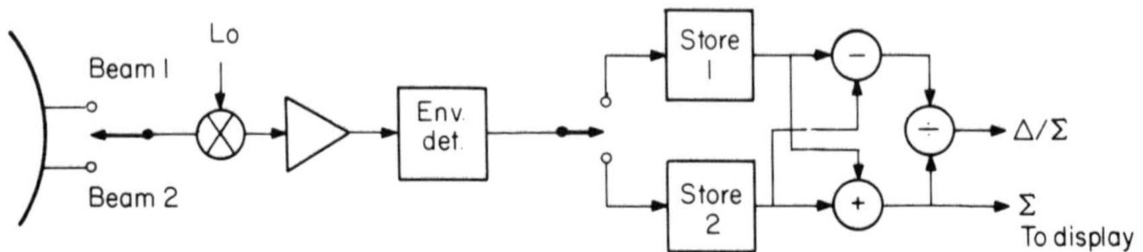

Figure 10.2 Sequential measurement of angle
using divider normalization.

Figure 10.3 shows the block diagram of a *simultaneous* lobe comparison (monopulse) angle measurement system with AGC normalization. The two offset beams are combined in an **RF hybrid** to produce the sum and difference channels simultaneously on each received pulse. Two identical receiver channels are used, the sum channel leading to an envelope detector from which display and ranging data are obtained, along with the AGC control voltage. The same AGC controls the difference channel in an open-loop fashion, providing normalization. The difference channel output is converted to bipolar video (baseband) by phase-sensitive detection, using the sum-channel pulse as a reference. This output, smoothed over many pulses, may be used as the error signal of an angle servo loop in a tracking radar.

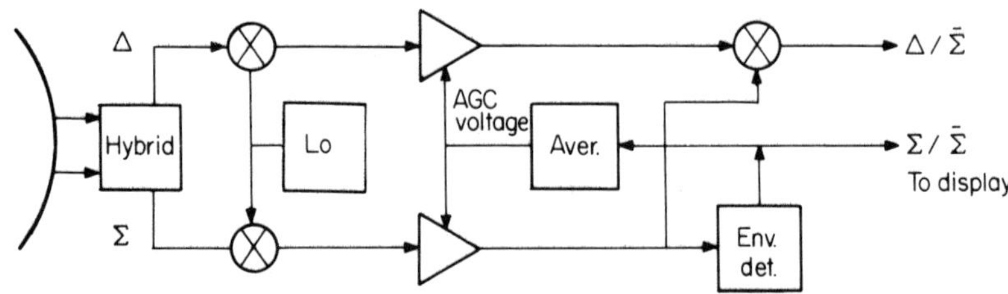

Figure 10.3 Simultaneous lobe comparison (monopulse) system
using AGC normalization.

The Conical-Scan Tracker

Figure 10.4 shows the elements of a typical conical-scan tracker. The beam is offset from the antenna (tracking) axis and is rotated rapidly around that axis, producing sinusoidal modulation of the received signal envelope (Figure 10.5). The modulation is recovered after the envelope detector, and demodulated to DC error signals by phase detection, using a reference voltage synchronized to the conical scan. The received energy is shared between two coordinates, half contributing to azimuth sensing and half to elevation sensing.

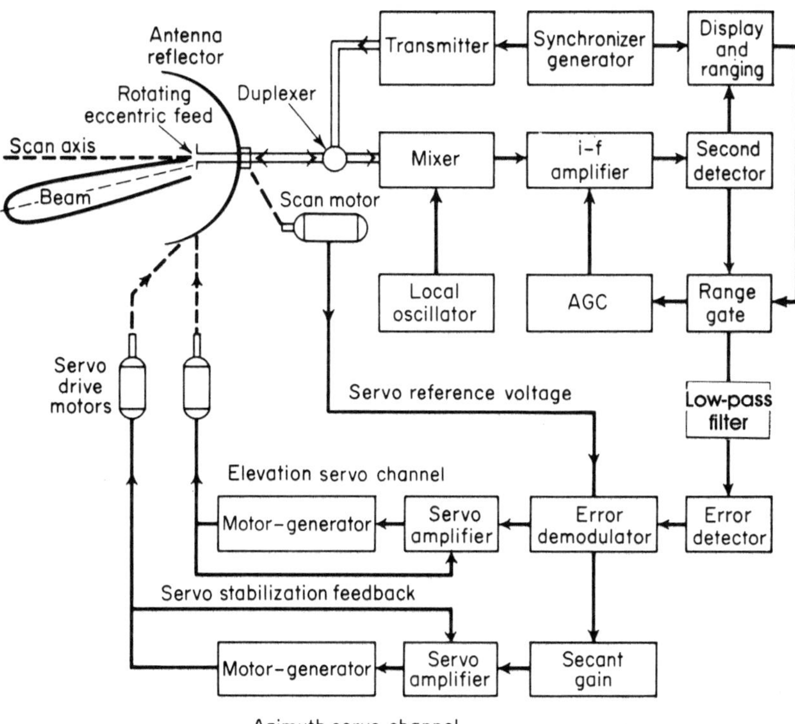

Figure 10.4 Block diagram of conical-scan tracker.

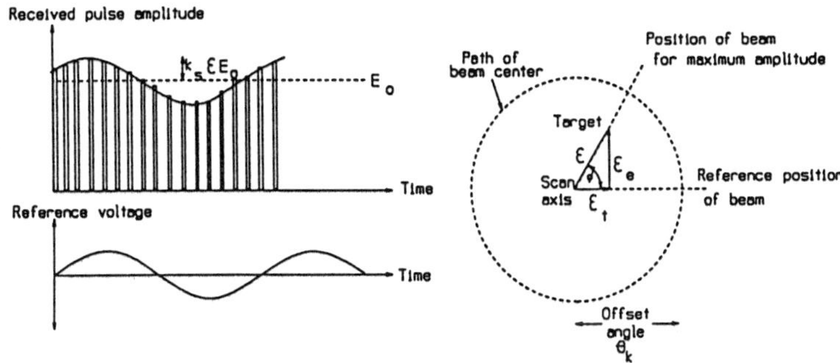

Figure 10.5 Error detection in conical-scan radar.

Conical-Scan Response to Thermal Noise

The tracking error caused by thermal noise in a conical-scan radar is given by:

$$\sigma_\theta = \frac{\theta_3}{k_s\sqrt{E/N_0}} = \frac{\theta_3}{k_s\sqrt{(S/N)n}} = \frac{\theta_3\sqrt{L_k}}{k_s\sqrt{(S/N)_m n}} \tag{10.2.1}$$

where θ_3 is the half-power beamwidth, k_s is the error slope, E/N_0 is the signal-to-noise energy ratio in the receiver, S/N is the signal-to-noise power ratio of a single pulse, n is the number of noise samples averaged by the servo, $(S/N)_m$ is the snr at the peak of the beam, and L_k is the *crossover loss*. The number of noise samples is the number of pulses received during the averaging time of the servo, $n = f_r t_o = f_r/2\beta_n$, where β_n is the (one-sided) servo bandwidth. Values of error slope for two-way conical scanning antennas are shown in Figure 10.5. For the two-way case, the optimum value of $k_s/\sqrt{L_k}$ is near 1.4, obtained at a squint angle $\theta_k \approx 0.3\theta_3$.

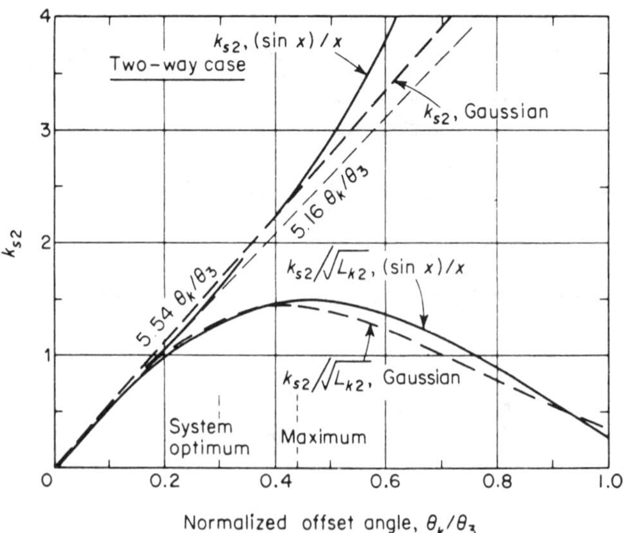

Figure 10.6 Conical-scan error slope (two-way case).

Conical-Scan Response to Target Scintillation

A limitation of the conical-scan tracker is its sensitivity to amplitude scintillation of the target. If the target scintillation spectrum contains components at the scan frequency, an interfering component will be produced. In a typical case, the rms error is about $\theta_3/50$, if the interference is due to normal target scintillation. This error can be much larger, and lead to loss of track, if induced by ECM modulations.

10.2.2 Sector-Scanning Systems

Precision approach radars (PARs) of ground-controlled approach and landing systems often use the scanning procedure illustrated in Figure 10.7. Two separate beams are generated, one scanning in azimuth and one in elevation, and usually time sharing a single transmitter and receiver. Each beam is narrow in the scanned coordinate, and broad enough in the other coordinate to cover a good part of the scanning field of the other system. In each coordinate, an intensity-modulated CRT produces a narrow blip on a range-angle display which is calibrated with range marks and the established approach path. The operator interpolates visually to the center of each blip, and advises the pilot of his deviation from the intended approach.

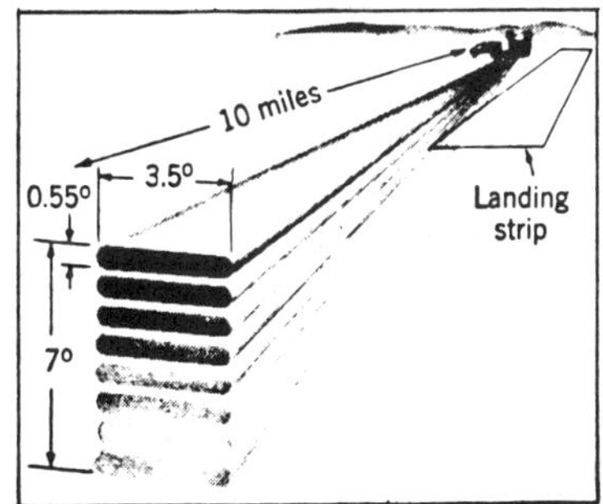

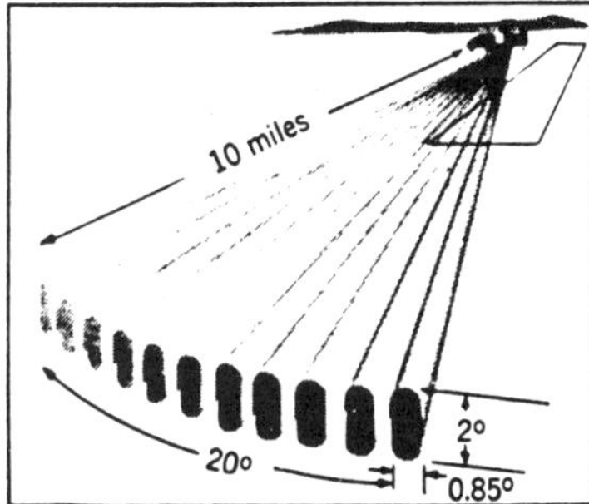

Figure 10.7 Antenna scan coverage of typical precision approach radar.

Signal Processing for Sector-Scanning Radar

A procedure similar to the visual centroid estimation is used with automatic tracking gates (angle gates) in some foreign equipment for missile command guidance, and in one coordinate the procedure is commonly applied for track-while-scan in 2D search radars and nodding height finders. The output pulse train from a linear-scanning radar appears as in Figure 10.8(a), modulated by the two-way pattern of the antenna. The number of pulses used in the measurement is $n = f_r t_o$, where t_o is the time required to move the beam through the one-way beamwidth θ_3. There are two ways of estimating the centroid of the received pulse train. In Figure 10.8(b) the pulse train is integrated, and the times of upward and downward threshold crossings are recorded. The centroid (delayed by the time constant of the integrator) is the midpoint of the two crossings. In Figure 10.8(c) a *split angle gate* is used, integrating the pulses with positive weight during the first half, and negative weight during the second half, as shown in Figure 10.8(d). When the gate is centered on the pulse train, the gate output goes to zero.

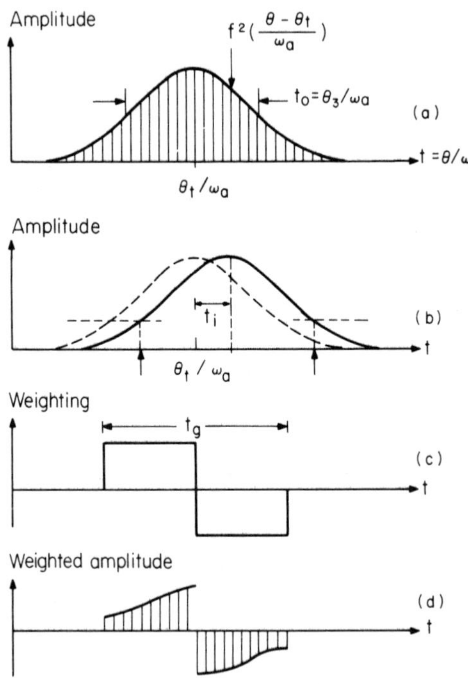

Figure 10.8 Waveforms in sector-scanning radar. (a) Envelope of pulse train. (b) Integrated envelope. (c) Split angle gate. (d) Pulse train at gate output.

Response of Sector Scan to Thermal Noise

The tracking error caused by thermal noise in a linear-scanning radar can be expressed as

$$\sigma_\theta = \frac{\theta_3}{k_p\sqrt{2E/N_0}} = \frac{\theta_3\sqrt{L_p}}{k_p\sqrt{2(S/N)_m n}} = \frac{0.5\theta_3}{\sqrt{(S/N)_m n}} \tag{10.2.2}$$

where θ_3 is the half-power beamwidth, k_p is the error slope, E/N_0 is the signal-to-noise ratio in the receiver, $(S/N)_m$ is the snr at the peak of the beam, L_p is the beamshape loss, and n is the number of pulses received between the half-power points of the antenna pattern.

Response of Sector Scan to Target Scintillation

As with all sequentially sampled systems, target scintillation causes an error in measurement. Figure 10.9(a) shows the two-way antenna pattern, which would give the envelope of the pulse train from a steady target. If the target amplitude varies as in Figure 10.9(b), the received envelope will appear as in Figure 10.9(c), with its centroid displaced. The resulting error, normalized to beamwidth, is shown in Figure 10.10 as a function of n_e-1, the number of independent amplitude-difference samples received during the observation time.

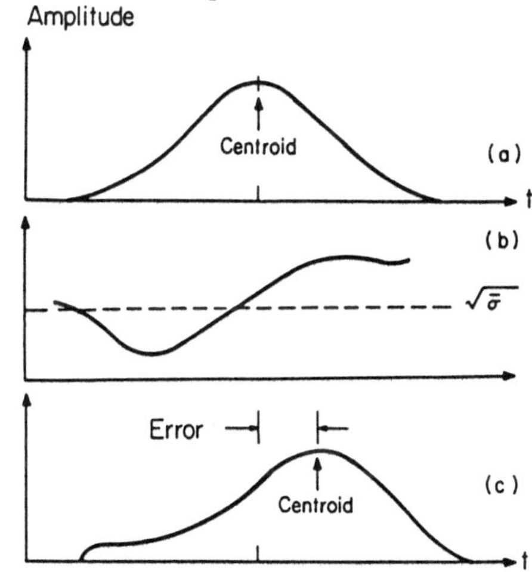

Figure 10.9 Waveforms in scanning across scintillating target.
(a) Two-way antenna pattern. (b) Typical target amplitude vs. time.
(c) Resulting envelope of received signal during scan.

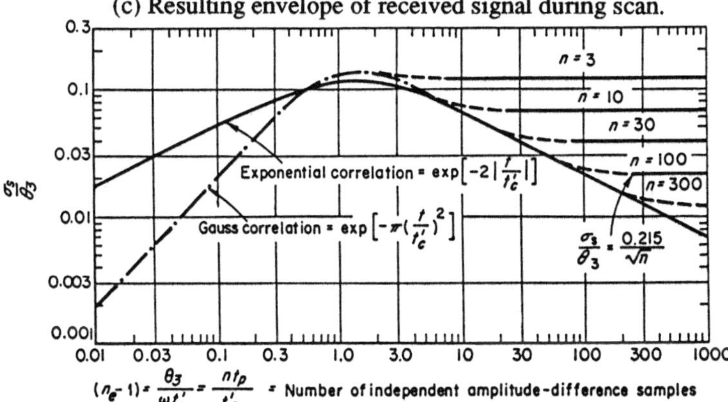

Figure 10.10 Scintillation error in sector scan radar [10.3, p. 180].

As an example of scintillation effects, assume a target of length $L_x = 15$ m, at a range $R = 30$ km and flying at 300 m/s normal to an X-band radar beam of width $\theta_3 = 1°$ which scans a sector $\Delta_a = 10°$ in $t_s = 0.5$ s. The observation time will be

$$t_o = t_s \theta_a / \Delta_a = 0.5 \times 1/10 = 0.05 \text{ s}$$

The target correlation time will be determined by the angular rate $\omega_a = v_t / R = 0.01$ r/s, of the target aspect angle about the line of sight

$$t_c = \lambda/2\omega_a L_x = 0.03/(2 \times 0.01 \times 15) = 0.1 \text{ s}$$

From Figure 10.10, with $n_e - 1 = t_o / t_c = 0.5$, the rms error is

$$\sigma_s = 0.1\theta_3 = 0.1° = 1.8 \text{ mr}$$

This level of error is almost inevitable in scanning systems. A similar error is found in measurements from frequency-scanning radar.

10.2.3 Monopulse Angle Tracking

Monopulse is a technique in which information concerning the angular location of the target is obtained by comparison of signals received from two or more simultaneous beams. The advantages are a greater efficiency in measurement (as compared to conical or linear scanning), higher data rate, freedom from the effects of target scintillation error, and reduced vulnerability to jamming.

Amplitude-Comparison Monopulse

The conventional amplitude-comparison monopulse radar, using a four-horn feed and reflector, generates four offset beams, and processes them after conversion to sum and difference patterns. Horns are located symmetrically about the axis of the reflector, and are fed in-phase by the transmitter to produce a single pencil beam on the axis. This beam (known as the sum, or Σ beam) is also used in reception for target detection, range and doppler tracking, and as a reference for angle error detection. Up to this point, the monopulse and conical-scan radars are similar. The antenna feed assembly contains a *monopulse comparator*, which forms the sum channel and two difference (or Δ) channels. The additional receiving channels required in monopulse radar for angle error detection are shown in Figure 10.11 as shaded blocks, starting with the four-horn feed assembly, in which the comparator picks up the off-axis components of the echo signals and transforms them into elevation and traverse difference signals. (The traverse coordinate is measured along a great circle passing through the antenna axis at right angles to the elevation coordinate. At zero elevation angle, it is identical to the azimuth coordinate, but at higher angles it represents azimuth times the sine of elevation angle.)

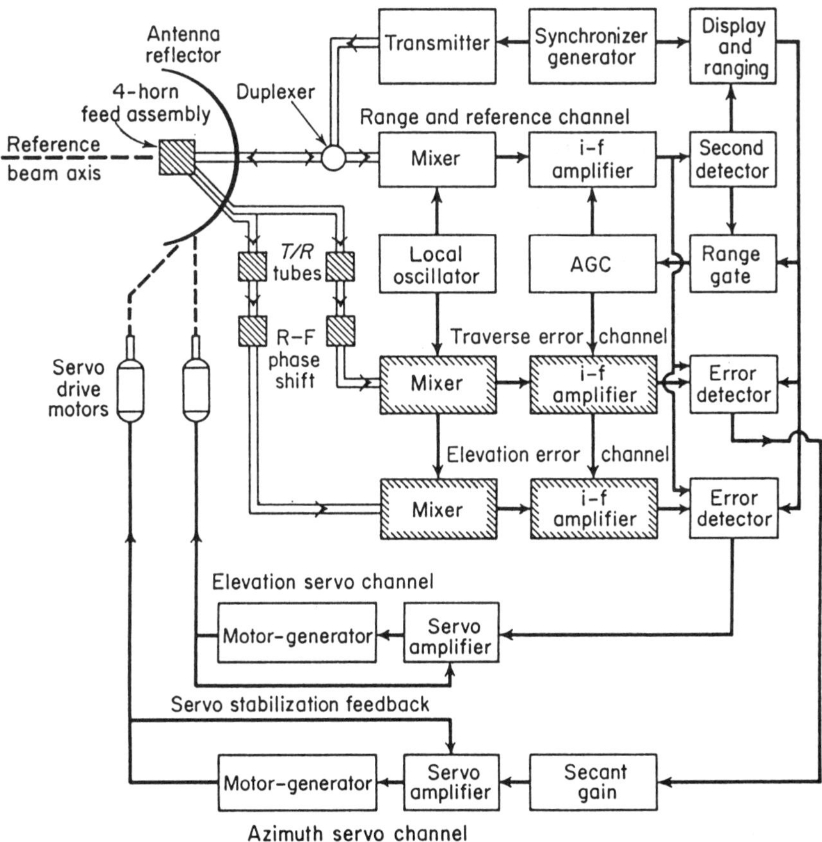

Figure 10.11 Block diagram of conventional monopulse radar.

The two angle difference signals are brought through T/R devices and adjustable phase shifters to mixers, which perform the downconversion to IF using the same local oscillator as is used for the sum channel. After downconversion, the difference signals are amplified in separate IF channels, which deliver all signals to the error detectors with the same phase and amplitude relationships that applied at the feed. The usual method of normalization is to derive an AGC control voltage from the sum channel, and apply it to all three channels to achieve equal gain.

Phase-Comparison Monopulse

The phase-comparison monopulse system also generates four elementary beams, but they originate from phase centers displaced about the antenna axis, pointed in the same direction rather than being squinted. In most cases, the beams are combined into sum and difference patterns, and are processed in the same way as for amplitude-comparison monopulse (except for a 90° phase shift introduced in the feed network). Except for notable differences in antenna sidelobe characteristics, the two types of system can be considered equivalent.

10.3 Monopulse Antennas

10.3.1 Reflector and Lens Antennas

It is convenient to think of an amplitude-comparison monopulse antenna as a reflector or lens fed by four independent horns, each slightly offset from the focal point. However, there is a fundamental incompatibility between the size of the horns needed for good difference patterns, and that needed for a good sum pattern. Hannan [10.5] developed multihorn feeds to correct this illumination problem. In Figure 10.12(a), a 12-horn feed is shown, with an overall dimension twice that normally needed to illuminate the aperture properly. The incompatibility in horn size is overcome by coupling the sum channel only to the center horns, while the difference channels are also coupled to peripheral horns. The multilayer, multimode feed shown in Figure 10.12(b) accomplishes the same result with a simpler structure having greater efficiency.

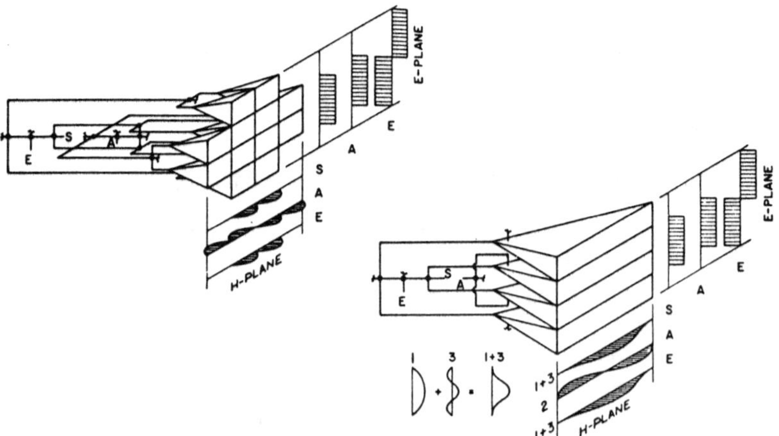

Figure 10.12 Optimized monopulse feed designs. (a) Twelve-horn feed. (b) Four-horn triple-mode feed [10.7].

10.3.2 Planar Array Antennas

Current military requirements have evolved toward systems with multiple-target tracking capabilities. This is one of the requirements that has dictated the use of phased-array antennas in many modern radar systems. Early designs for corporate-fed arrays implemented the monopulse difference patterns simply by dividing the array into quadrants, coupled to the three receiving channels through hybrids to place the difference-channel illuminations of the halves of the antenna in phase opposition. The resulting difference-channel sidelobes are excessively high.

The problem of designing constrained feeds with reduced difference-channel sidelobes was addressed by Lopez [10.6]. Starting from the feed with one central hybrid, Lopez discussed the technique of dividing each line of the array into multiple modules, or subarrays, each with a hybrid and with additional couplers to establish a more gradual stepping of the difference illumination through zero at the center of the array, as in Figure 10.13. This is the approach used, for example, in the naval Aegis phased array radar. While sidelobes are reduced by this approach, the discontinuities in amplitude of the illumination cause significant sidelobe levels at angles far removed from the mainlobe.

A more sophisticated approach to sidelobe reduction in constrained feeds is the Lopez feed, or dual-ladder network (Figure 10.14). In this design, each radiating element is coupled to two feed lines. The primary-line couplers are adjusted to produce the low-sidelobe sum illumination. The secondary-line couplers are adjusted to cancel the discontinuity in the center of the aperture, leaving a smooth, near-sinusoidal difference-channel illumination. The requirement for two precision, wideband microwave couplers per radiating element makes this approach expensive for large arrays.

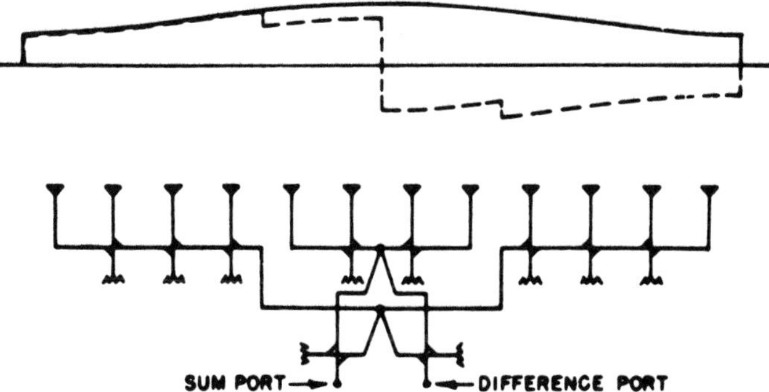

Figure 10.13 Monopulse series-feed network for linear arrays [10.6].

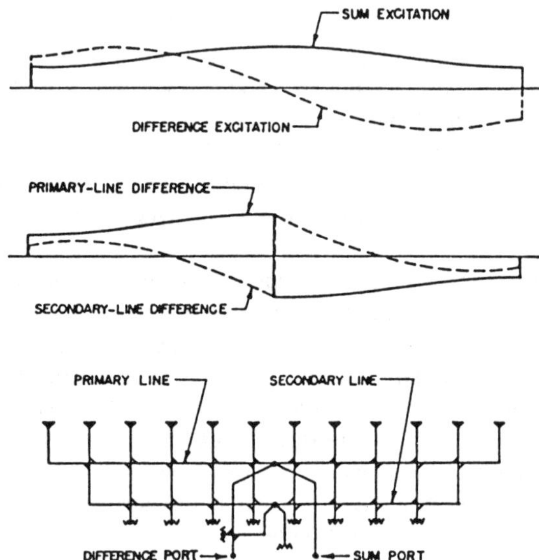

Figure 10.14 The Lopez feed, a dual-ladder network for separate optimization of sum and difference illuminations [10.6].

10.3.3 Space-Fed Array Antennas

The use of a space-fed array, with either a lens or a reflecting array as the main aperture, offers a relatively economical solution to the monopulse feed problem. One of the Hannan feeds (Figure 10.12) can be used to produce near-optimum illumination functions with four to twelve hybrids. The rest of the feed consists of the open space between the horn cluster and the main aperture. An example of this type of array in its lens configuration is found on the Patriot air defense radar, Figure 10.15. The multihorn, multimode feed is located behind a radome panel in the housing which projects from the roof of the vehicle. Illumination from the feed is coupled into dipoles on the rear end of each array element, which in turn are coupled to the front radiators through electronically controlled phase shifters.

10.4 Monopulse Signal Processing

10.4.1 Normalization of Error Signal

The error signal for measurement or tracking of target angle is normalized in the monopulse processor to the form

$$E_\Delta = \left(\frac{\Delta}{\Sigma}\right)\cos\phi \tag{10.4.1}$$

where Δ and Σ are the difference and sum channel voltages and ϕ is the phase angle between them. This equation applies both to amplitude-comparison and phase-comparison antennas where the channels are processed in the sum-and-difference mode. The use of AGC to perform the normalization has been illustrated in Figures 10.3 and 10.11.

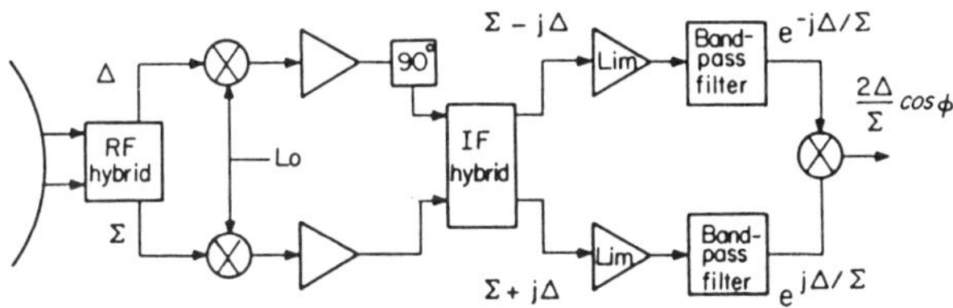

Figure 10.15 Patriot multifunction array radar.

Figure 10.16 shows normalization using **hard limiters**, which operate on channels carrying $\Sigma + j\Delta$ and $\Sigma - j\Delta$. These channels are formed by recombining the Σ and Δ channels in a hybrid, after applying 90° phase shift to one of them. After limiting, the two channels are bandpass filtered to restore the sinusoidal IF, and passed to a phase detector. The phase of $\Sigma + j\Delta$ is $\tan^{-1}(\Delta/\Sigma) \approx \Delta/\Sigma$, and that of $\Sigma - j\Delta$ is $\tan^{-1}(-\Delta/\Sigma) \approx -\Delta/\Sigma$. Hence, the phase detector produces the desired error voltage $2\Delta/\Sigma$ with the sign indicating the direction of the target from the axis.

Figure 10.17 shows normalization using logarithmic amplifiers instead of hard limiters. The Δ and Σ channels are combined in a hybrid without phase shift, to produce $\Sigma + \Delta$ and $\Sigma - \Delta$ channels. These are passed to log amplifiers whose outputs are subtracted to form an error voltage equal to

$$E_\Delta = \log(\Sigma + \Delta) - \log(\Sigma - \Delta) = \log\left(\frac{\Sigma+\Delta}{\Sigma-\Delta}\right) \approx 2\frac{\Delta}{\Sigma} \tag{10.4.2}$$

This is the desired error voltage, multiplied by a constant factor of two. The output sum voltage is produced by summing the two log channels to generate

$$E_\Sigma = \log(\Sigma + \Delta) + \log(\Sigma - \Delta) = \log[(\Sigma + \Delta) \times (\Sigma - \Delta)]$$

$$= \log(\Sigma^2 + \Delta^2) \approx 2\log\Sigma \tag{10.4.3}$$

This is the desired sum signal, again with a constant factor of two.

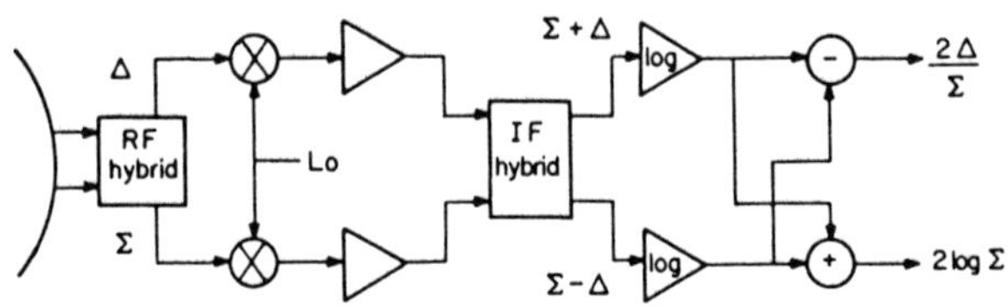

Figure 10.16 Normalization using hard limiters.

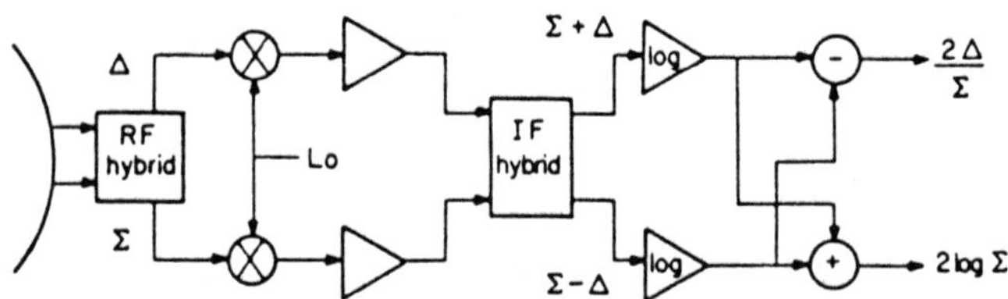

Figure 10.17 Normalization using log receivers.

In each of these two procedures, the normalization is essentially instantaneous. This means that multiple targets, within the same beam but at different ranges, can be measured during the same pulse repetition interval.

10.4.2 Sensitivity to Thermal Noise

The error caused by thermal noise in monopulse angle measurement is

$$\sigma_\theta = \frac{\theta_3}{k_m\sqrt{2E/N_0}} = \frac{\theta_3}{k_m\sqrt{2(S/N)n}} \approx \frac{\theta_3}{2\sqrt{(S/N)n}}$$

where $k_m \approx 1.5$ is the normalized slope of the antenna difference pattern.

10.4.3 Two-Channel Monopulse System

While most monopulse radars use the three-channel (sum-and-difference) processing approach, a slightly simpler variant has been implemented successfully. Figure 10.18 shows a two-channel monopulse system, in which the two difference channels are *time multiplexed* into a combined difference channel Δ, before being amplified and processed in the receiver. In this diagram, a ***microwave resolver*** serves as the multiplexer, and processing uses the log receiver method of Figure 10.17.

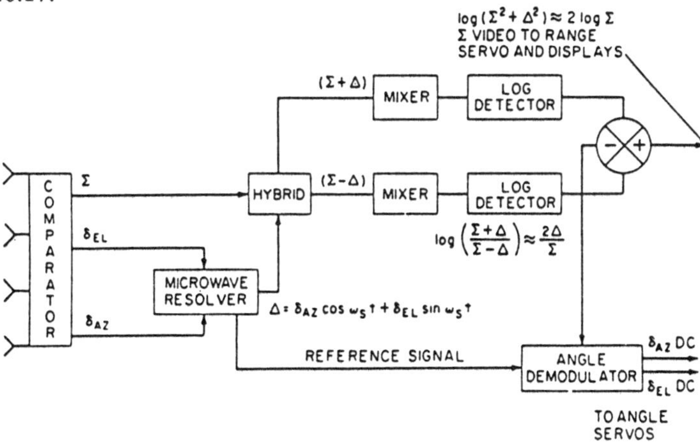

Figure 10.18 Two-channel monopulse system [10.7].

In effect, the receiving system produces two scanning beams, offset on opposite sides of the antenna axis, and scanning at the multiplexing rate. The reliability advantage over three-channel monopulse systems results from the fact that failure of one receiver channel reduces the system to a conical-scan-on-receive (COSRO) system, which can continue to track (but with the scintillation error of that type of tracker). The derivation of COSRO signals from a monopulse feed, in the manner shown in Figure 10.18, is the preferred method of implementing this single-channel sequentially scanned system, because of the elimination of mechanical motion of the feed. The transmitter, in both COSRO and two-channel monopulse, is connected through a duplexer to the sum channel, producing an on-axis beam. In the multiplexing process of a two-channel monopulse radar, and in COSRO, half the difference-channel energy is discarded, reducing the effective *S/N* ratio for measurement by a factor of two.

10.5 Range Measurement and Tracking

Range measurement in radar systems requires estimation of the time delay t_d between transmission and reception of the signal. The range to the target is then given by

$$R = t_d \, c/2$$

In pulsed radar systems, the feature of the signal waveform to which delay is measured can be the leading edge or the centroid of the pulse. Either can be used, as long as the delay is measured between the same reference point of the transmitted and received pulses.

10.5.1 Optimum Range Estimator

The optimum range estimator for a signal whose time delay (and hence phase) is unknown consists of a matched filter, followed by an envelope detector, a differentiator, and a *zero-crossing detector* (see Figure 10.19). This configuration locates the peak of the envelope of the matched-filter output waveform, which will occur at its center. Since there will be many zero crossings in the noise preceding and following the signal, a threshold detector is used to identify the one zero crossing which corresponds to the peak of the echo signal. The threshold must be set high enough to avoid false alarms, which could lead to delay (range) estimates on noise, with resulting gross errors in the true target's range estimate. Hence, the signal energy must be adequate for detection with a high threshold for low false-alarm rate.

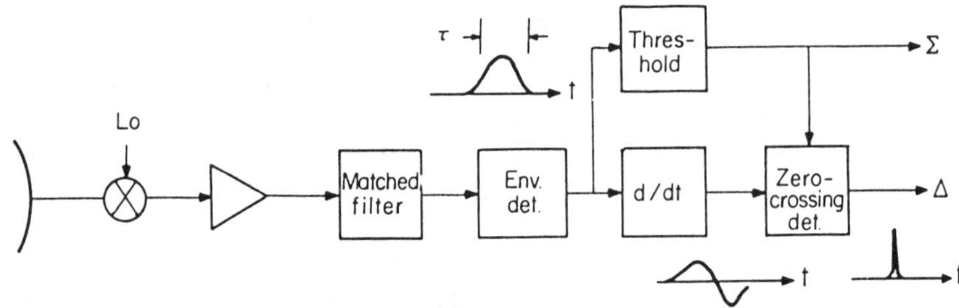

Figure 10.19 Optimum delay estimator.

The accuracy of delay estimation, using the matched filter and differentiator, can be expressed in terms of the signal-to-noise energy ratio and the slope of the differentiated signal output, which is equal to the rms bandwidth β of the transmitted signal spectrum:

$$\sigma_t = \frac{1}{\beta\sqrt{2E/N_0}} \approx \frac{\tau_0}{1.61\sqrt{2E/N_0}} \tag{10.5.1}$$

The last form of this equation makes use of the fact that the half-power width τ_o of the output pulse from a matched filter is determined directly by the rms bandwidth β. The rms bandwidth is defined [10.3] as 2π times the second moment of the transmitted power spectrum about the carrier frequency. The rms bandwidth is approximately twice the half-power bandwidth of the signal, for many spectral shapes.

10.5.2 Range Estimator for a Single Pulse

Correlator Implementation

When all the signal energy is contained in a single pulse, the practical range estimator may closely approach that shown in Figure 10.19. An alternative to the matched filter is the *correlator* implementation shown in Figure 10.20. Here the IF amplifier uses a broadband filter, passing all the spectral components of the pulse to the sum- and difference-channel modulators. The modulator function for the sum channel is approximately the envelope of the pulse, while that for the difference channel is the derivative of the envelope. In many cases, the smooth modulation functions are replaced by rectangular gating signals. Narrowband filters following the modulators perform integration over the pulse, after which an envelope detector produces the sum output for display and detection processing, while a phase-sensitive detector produces the range error voltage. This error voltage is used to control the timing of the modulator functions, forcing these functions to be centered on the signal pulse for zero error output. Unlike the combination of IF filter and differentiator, the correlator produces no useful output unless the correlator functions are adjusted in delay to within one pulsewidth of the input signal. Hence, this approach is appropriate only for closed-loop trackers.

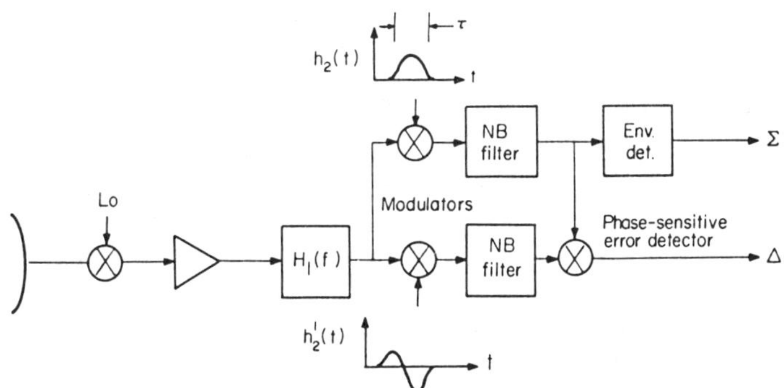

Figure 10.20 Range estimator using correlation technique.

Time-Delay Error for Single Pulse

The time-delay error caused by thermal noise in a system where the filter or correlator is not perfectly matched to the signal can be expressed by using a slope factor $k_t < 1.61$ in the previous equation. This normalized slope constant k_t may be defined, analogous to the monopulse slope constant k_m used in expressing angle accuracy, as

$$k_t = \frac{d(\Delta/\Sigma)}{d(t/\tau_o)} \qquad\qquad (10.5.2)$$

The thermal noise error for a single-pulse time-delay measurement as a function of the actual S/N ratio is

$$\sigma_{t_1} = \frac{\tau_o}{k_t\sqrt{2(S/N)}} \qquad\qquad (10.5.3)$$

As in the case of angle measurement, the normalized slope k_t is dimensionless and will usually be near 1.5, regardless of the signal spectrum.

Video Correlator

If the IF filter does not pass too much noise in spectral regions beyond the significant portions of the signal spectrum, the sum and difference correlators can be implemented at video, with small loss in efficiency. Pulse compression waveforms are normally processed in IF filters in which the phase response is the *conjugate* of the signal spectrum, with the amplitude response weighted to produce low sidelobes. This approximation to the matched filter also permits the video correlator to be used. In either case, it is the *cascaded response* of the filter and correlator which should closely match the signal waveform. The closer the IF filter comes to the matched filter, the narrower should be the width of the correlator functions. In the limit, these functions will become an *impulse function* for the sum channel and a doublet impulse (ideal differentiator) for the difference channel.

10.5.3 Range Estimator for a Pulse Train

The matched filter for a coherent pulse train can be implemented as a *comb filter* at IF, or as a broadband filter followed by a correlator and narrowband filter, as in Figure 10.20, with the correlator function consisting of a train of range gates matched to the arrival times of signal pulses. The equations for time-delay error in terms of energy ratio will remain applicable to the energy ratio of the pulse train, and the output error, expressed as a function of S/N, will be reduced by $\sqrt{n}$ as compared to the single-pulse error.

10.5.4 Split-Gate Range Tracker

In many cases, the sum correlator function is approximated by a rectangular gate, and the difference function by a split gate, as shown in Figure 10.21. These gate functions can be applied at IF, followed by narrowband filters for coherent processing or envelope detectors and low-pass filters for noncoherent processing. If noncoherent processing is to be used, the common practice is to pass the signal through an envelope detector first, and to apply the sum and difference gate functions to the video signal.

10.6 Doppler Tracking

10.6.1 Optimum Doppler Estimator

Measurement of target radial velocity v_r is equivalent to measurement of the doppler shift f_d of the echo signal, since

$$v_r = f_d \lambda/2 \tag{10.6.1}$$

The thermal-noise error in measurement on a single pulse, for optimum processing, is

$$(\sigma_f)_{\min} = \frac{1}{1.81\tau\sqrt{2E_1/N_0}} \tag{10.6.2}$$

where τ is the width of the rectangular pulse being measured.

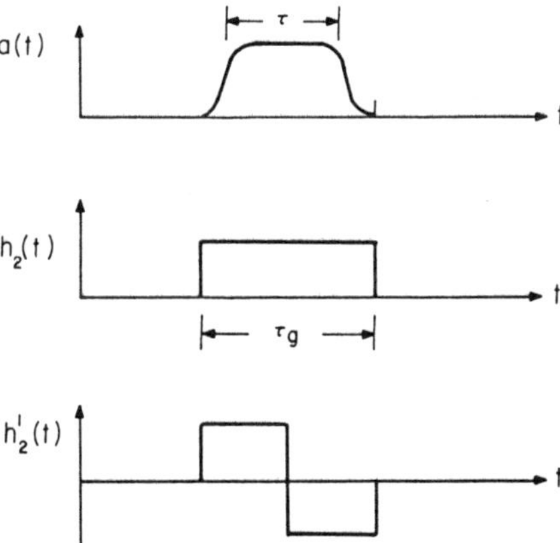

Figure 10.21 Waveforms for split-gate range tracker.

In coherent radar, the signal duration is often controlled by the Gaussian envelope of the two-way antenna pattern as it scans across the target. The corresponding rms duration, replacing 1.81τ in the previous equation, is

$$\alpha_0 = 1.81\,t_o = 1.18/B_3$$

where t_o is the conventional observation time (between the one-way half-power points of the beam) and B_3 is the resulting coherent signal bandwidth (that of the fine lines in a pulsed doppler spectrum).

10.6.2 Coherent Doppler Tracker

In order to track the doppler shift of a coherent pulse train, the doppler filter and discriminator (difference-channel filter) must be approximately matched to the fine lines of the signal spectrum. One convenient method, which avoids the necessity of constructing a comb filter in the frequency domain, is to place a range gate at IF, followed by a narrowband filter (Figure 10.22). The combination of a range gate, matched to the pulse width, and a narrowband filter, matched to the signal duration t_o, constitutes a matched filter for the train of pulses. The discriminator, operating within the bandwidth B_f of the fine-line filter, delivers a DC control voltage to the **voltage-controlled oscillator (VCO)**, which downconverts the input signal to the center frequency of the discriminator.

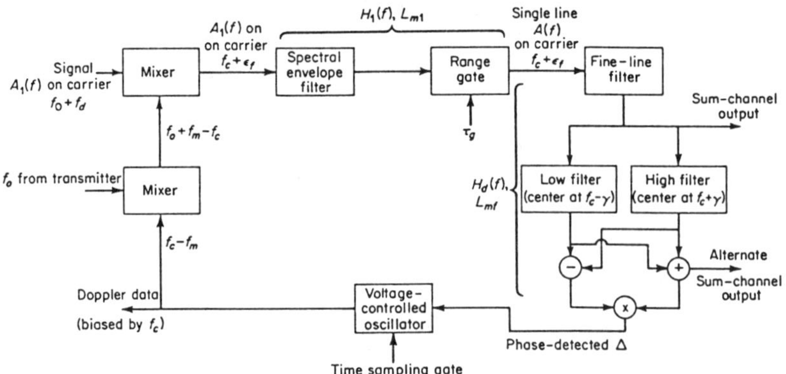

Figure 10.22 Block diagram of pulsed doppler tracker.

10.6.3 Resolution of Doppler Ambiguities

In order to place the narrowband filter of the pulsed doppler tracker on the central line of the signal spectrum, the VCO must be designated to within one PRF interval, $\pm f_r/2$, relative to the doppler-shifted echo signal. There are four methods of obtaining the necessary target data:

(a) The signal frequency may be known *a priori* to lie within a particular PRF interval;

(b) The signal frequency may be measured noncoherently over the pulse train, using a discriminator at IF;

(c) The doppler shift may be calculated from measured range rate; or

(d) The ambiguity may be resolved by measurements at two or more PRFs.

The use of noncoherent frequency measurement to designate the correct line of the spectrum for coherent tracking is only practical if the waveform has a high duty factor (e.g., > 0.03). For lower duty factors, one of the other methods must be used.

10.6.4 Four-Coordinate Tracking Radar

Range and doppler tracking can be combined with monopulse angle tracking, as in Figure 10.23, to implement a four-coordinate tracking radar. The antenna supplies three signals; the sum, elevation difference, and azimuth difference signals. The range and doppler difference channels are formed in the receiver from the sum channel. All five channels are range gated and passed through narrowband IF filters.

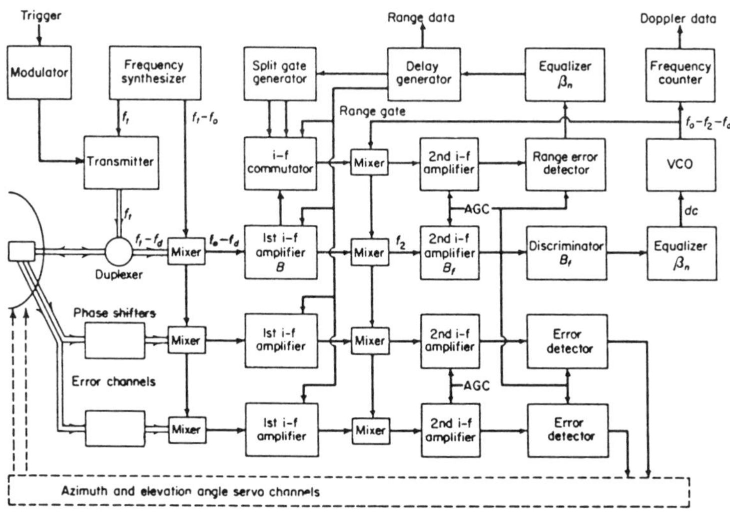

Figure 10.23 Four-coordinate monopulse tracking radar.

10.7 References

[10.1]   D. K. Barton, *Modern Radar System Analysis*, Artech House, 1988.

[10.2]   J. S. Hall, *Radar Aids to Navigation*, MIT Radiation Laboratory Series, Vol. 2, McGraw-Hill, 1949.

[10.3]   D. K. Barton and H. R. Ward, *Handbook of Radar Measurement*, Artech House, 1984.

[10.4]   D. R. Rhodes, *Introduction to Monopulse*, Artech House, 1980.

[10.5]   P. W. Hannan, "Optimum feeds for all three modes of a monopulse antenna," *IRE Trans.* **AP-9**, No. 5, September 1961, pp. 444-461 (reprinted in *Monopulse Radar*, Vol. 1 of *Radars*, Artech House, 1974).

[10.6]   A. R. Lopez, "Monopulse networks for series feeding an array antenna," *IEEE Trans.* **AP-16**, No. 4, July 1968, pp. 436-440 (reprinted in *Monopulse Radar*, Vol. 1 of *Radars*, Artech House, 1974).

[10.7]   R. S. Noblit, "Reliability without redundancy from a radar monopulse receiver," *Microwaves*, December 1967, pp. 56-60.

Chapter 11

RADAR ERROR ANALYSIS

11.1 Mathematical Models of Error

11.1.1 Definitions and Models

The error in a measurement is defined as the difference between the value indicated by the measuring instrument and the true value of the measured quantity:

$$x = U_{\text{measured}} - U_{\text{true}} \tag{11.1.1}$$

The purpose of error analysis is to describe the error in a way that will permit its magnitude to be estimated for operating conditions other than those involved in the testing. This permits limiting the testing program to a few well defined conditions, from which the major error components can be modeled.

Errors are commonly divided into *systematic* (bias) and *random* (accidental or noise) components. The former are characterized by some degree of predictability, and hence may be at least partially corrected by calibration of the instrument. The calibration process will leave a random residual error, which may vary slowly with time or with target coordinates.

Magnitude of Error

Over any limited interval of measurement, the bias may be defined as the mean value of error:

$$\bar{x} = \frac{1}{n} \sum_{i=1}^{i=n} x_i \tag{11.1.2}$$

Measures of the random components are the *standard deviation*, σ_x, and the *variance*, σ_x^2, where

$$\sigma_x^2 = \frac{1}{n} \sum_{i=1}^{i=n} (x_i - \bar{x})^2 \tag{11.1.3}$$

The *rms (root-mean-square) error* includes both bias and random components:

$$(x_{\text{rms}})^2 = \frac{1}{n} \sum_{i=1}^{i=n} x_i^2 = \bar{x}^2 + \frac{1}{n} \sum_{i=1}^{i=n} (x_i - \bar{x})^2 \tag{11.1.4}$$

The term *probable error* is sometimes used, referring to the value which is exceeded 50% of the time. For an error which is *normally distributed* (following the Gaussian curve, as shown for noise in Figure 4.1.2),

$$x_{50} = 0.6745\,\sigma_x \qquad\qquad (11.1.5)$$

A *peak error* x_m is used in some cases, often specified as $\pm x_m$ relative to the true value. The danger in this lies in the fact that normally distributed errors have no finite peak value. A common interpretation is based on

$$x_m \approx 3\,\sigma_x \qquad\qquad (11.1.6)$$

which results in the peak value being exceeded in only 0.3% of the measurements, for a normal distribution. The *peak-to-peak* error is then $2x_m = 6\,\sigma_x$.

Figure 11.1 shows the relationships among the standard deviation, the peak error, and the peak-to-peak error for different error waveforms that may be encountered in radar testing.

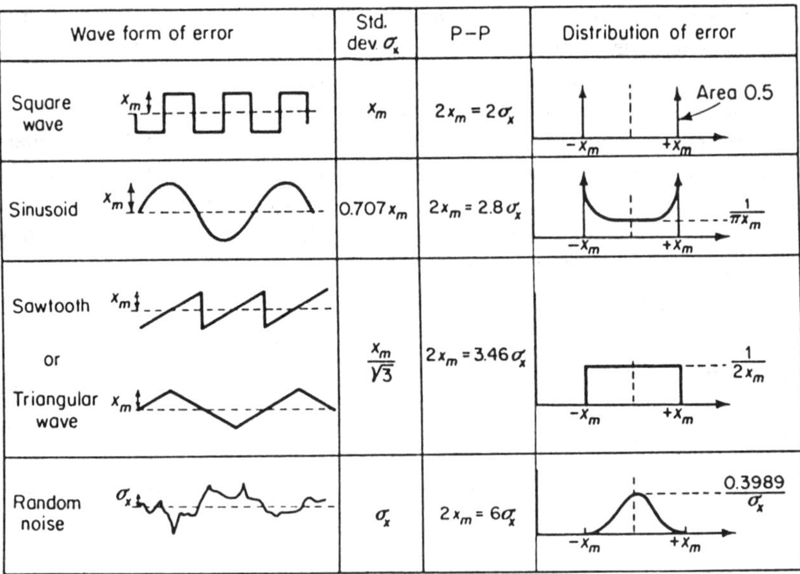

Figure 11.1 Relationships among standard deviation, peak, and peak-to-peak errors [11.1].

Time Functions and Frequency Spectra

The variation of error with time, and the resulting frequency spectrum of error, are important in testing a radar, and in evaluating the effects of subsequent smoothing and prediction of target coordinates. Figure 11.2 shows a typical error spectrum, identifying the several components by frequency:

(a) True bias, an impulse at zero frequency;

(b) Apparent bias, the slowly varying component that remains essentially constant during an observation interval $t_o = 1/2f_1$;

(c) Low-frequency component, extending from zero to a frequency f_a well within the bandwidth of the tracking loop;

(d) Cyclic component, a near-sinusoidal error at frequency f_c; and

(e) White noise component, extending at constant spectral density from zero to the upper limit of the tracking loop frequency response.

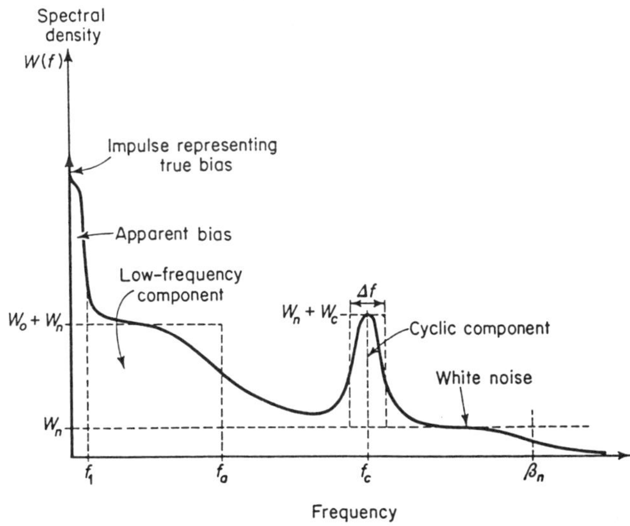

Figure 11.2 Typical radar error spectrum [11.1].

In evaluation of a radar system, different sources of error will often contribute components with different frequency characteristics, permitting each source to be isolated in the data. The frequency spectrum of a particular component can be related to parameters of the radar, the target, or the environment, so that the error may be scaled to anticipated operating conditions other than those used during the test.

11.2 Analysis of Angle Errors

11.2.1 Classification of Error Components

In addition to characterization by amplitude and spectral distributions, radar errors may be classified by their source and by their degree of dependence on target characteristics, propagation conditions, parameters of the tracker itself, and characteristics of the reference instrumentation used in the test. A further classification describes the point at which the error enters into the data system: *tracking errors* cause the tracking axis to deviate from the center of the target, while *translation errors* result from failure to translate the axis position into correct coordinate data at the output. In the case of a mechanical tracking antenna, the tracking errors may be measured directly by a boresight telescope aligned with the tracking axis, while translation errors may be evaluated optically on stars or other points of known position, using the boresight telescope as a *theodolite*.

Table 11.1 lists common sources of angle error in a mechanical tracker, according to these classifications. Radar-dependent errors are governed primarily by the design and construction of the radar itself, although they are influenced by the physical environment (e.g., wind and solar heating). Target-dependent errors vary with different classes of target, although the radar can be designed to minimize these components (e.g., scintillation error is reduced in monopulse radar). Propagation errors apply to a large class of systems, but depend to a minor degree on aperture size (for tropospheric irregularities) and radar frequency (for ionospheric errors). Finally, the apparent, or instrumentation, errors are not related to the radar under test but to the quality and operational integrity of the reference instrumentation against which the test radar data are to be compared.

11.2.2 Description of Errors

Many of the error components of Table 11.1 have been discussed in some detail in other chapters of this handbook:

Thermal noise (Chapter 10);

Multipath error (Chapter 8);

Scintillation (Chapter 10);

Tropospheric propagation error (Chapter 8); and

Jamming (Chapter 12).

Others are peculiar to a specific radar implementation, and will not be discussed further (e.g., boresight drift, servo unbalance and torque effects, mechanical errors). However, the subjects of clutter, target glint, cross-polarization error, and dynamic lag deserve special mention, as they affect almost all radar measurement systems.

Table 11.1 Sources of Angle Error [11.2]

Class of Error	*Bias Components*	*Noise Components*
Radar-dependent tracking errors	Boresight axis setting and drift	Thermal noise
	Torque caused by wind and gravity	Multipath
	Servo unbalance and drift	Clutter and jamming
		Servo noise
		Deflection of antenna caused by wind gusts
Radar-dependent translation errors	Pedestal leveling	Bearing wobble
	Azimuth alignment	Data gear nonlinearity and backlash
	Orthogonality of axes	Data takeoff nonlinearity and granularity
	Pedestal flexure caused by gravity force	Pedestal deflection caused by acceleration
	Pedestal flexure caused by solar heating	Phase shifter error
Target-dependent tracking errors	Dynamic lag	Glint
		Dynamic lag variation
		Scintillation or beacon modulation
		Cross-polarization
Propagation errors	Average refraction of troposphere	Irregularities in refraction of troposphere
	Average refraction of ionosphere	Irregularities in refraction of ionosphere
Apparent or instrumentation errors	Stability of telescope or reference instrument	Vibration or jitter in reference instrument
	Stability of film base or emulsion	Film transport jitter
	Optical parallax	Reading error
		Granularity error
		Variation in parallax

Clutter Errors

The equations of Chapter 10 for thermal noise error may be used also for clutter, by substituting the difference-channel signal-to-clutter ratio S/C_Δ for S/N, provided that this ratio is reasonably high and the clutter return reaching the error detector is random from pulse to pulse. These requirements are usually met in MTI and pulsed doppler radars. In noncoherent radars, the number of independent clutter samples n_c must replace n, and correlation between sum- and difference-channel clutter must be considered [11.2, pp. 531-533].

Target Glint

In the discussion of monopulse tracking, it was shown that the scintillation error component was canceled in the normalization process. However, for targets with multiple scattering centers, an error component known as *glint* remains, regardless of the tracking technique. This component results from the interference between waves from the several scatterers. Typical measured glint errors in angle and range are shown in Figure 11.3. The distribution of glint is seen to extend beyond the physical dimensions of the target, a phenomenon which is predicted by theory and confirmed by experiment. Although the theoretical glint distribution has infinite variance, it is truncated in practical measurement systems by the limited dynamic range and response speed of the instrument. An approximate Gaussian fit to the distribution is obtained with a standard deviation equal to 1/3 to 1/6 of the target span, in any measured coordinate.

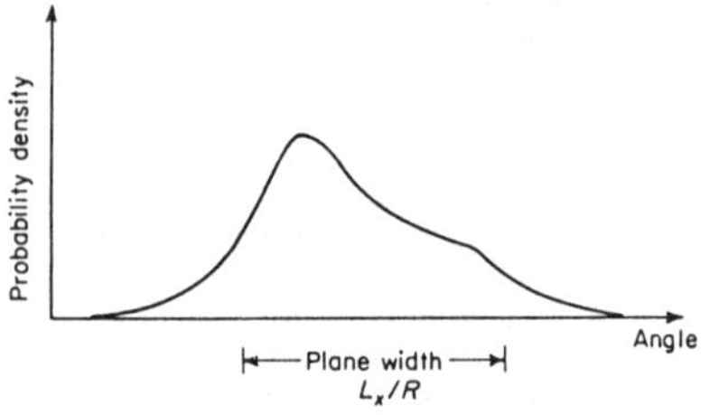

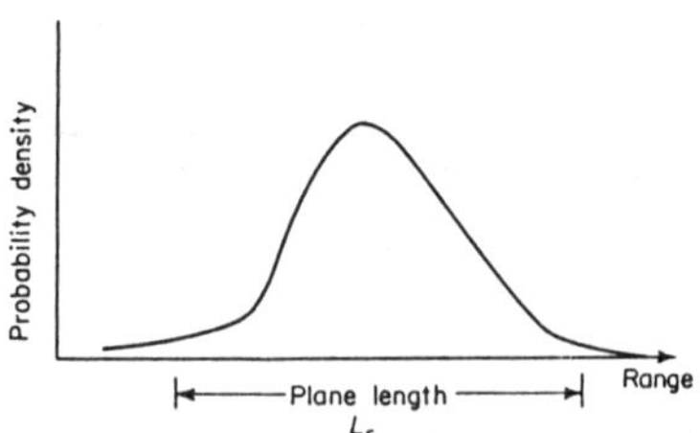

Figure 11.3 Typical aircraft glint distributions in angle and range.

Cross-Polarization Error

In addition to glint and scintillation, a target-dependent error is produced by cross-polarized components of the scattered wave if the radar antenna has response to these components. Figure 11.4 shows typical antenna patterns for monopulse sum and difference channels, where Σ_c and Δ_c represent the response to the polarization orthogonal to that intended.

When the target produces an orthogonal voltage component e_c, in addition to the intended component e, the error signal produced by the antenna will take the form

$$E_\Delta = \frac{e\Delta + e_c\Delta_c}{e\Sigma + e_c\Sigma_c} \approx \frac{e\Delta + e_c\Delta_c}{e\Sigma} \tag{11.2.1}$$

The cross-polarization error can be expressed in terms of the cross-polarized response of the difference pattern and the ratio of cross-polarized to intended target cross section, $\sigma_c/\sigma = (e_c/e)^2$:

$$\sigma_\theta = \frac{\theta_3(\Delta_c/\Sigma)}{k_m\sqrt{2\sigma/\sigma_c}} \tag{11.2.2}$$

In a typical reflector antenna, the ratio $\Delta_c/\Sigma = 0.03$ near the tracking axis, while for aircraft targets the ratio $\sigma_c/\sigma \approx 0.3$. The resulting cross-polarization error will be about $0.008\,\theta_3$, for a monopulse error slope $k_m = 1.4$.

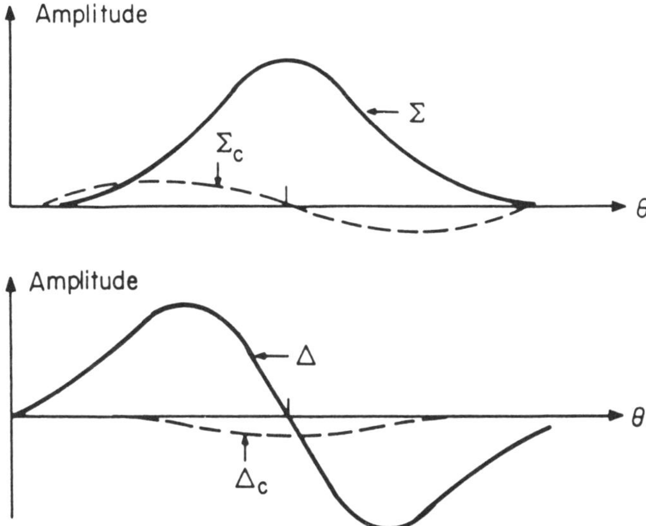

Figure 11.4 Typical antenna patterns for intended and orthogonal polarizations.

Dynamic Lag

The radar measures target position in a spherical coordinate system: e.g., range R, azimuth A, and elevation E. The coordinates of a moving target will vary, and the first derivative of each coordinate represents the target velocity projected onto that coordinate. Higher derivatives will depend not only on the corresponding derivatives of the target motion, but on the geometrical transformation of the trajectory coordinates into the radar's spherical coordinates.

Consider the pass-course situation, Figure 11.5, in which the target travels in a straight line at constant speed, crossing over a point on the surface at ground range R_c from the radar. The angle rates will be very low when the target is tracked at long range, and will increase sharply near crossover. The normalized azimuth rate ω_a and its first three derivatives are plotted in Figure 11.6 as functions of the azimuth angle A. The maximum azimuth rate $\omega_m = v_t/R_c$ occurs at crossover, $A = 90°$. Maxima in azimuth acceleration with amplitudes $\pm 0.65\,\omega_m^2$ occur 30° each side of crossover. The derivatives in elevation and range, shown in Figures 11.7 and 11.8, have similar variations with target azimuth. The parameter $X = h_t/R_c$ affects the elevation derivatives, and the curves are drawn for $X=1$.

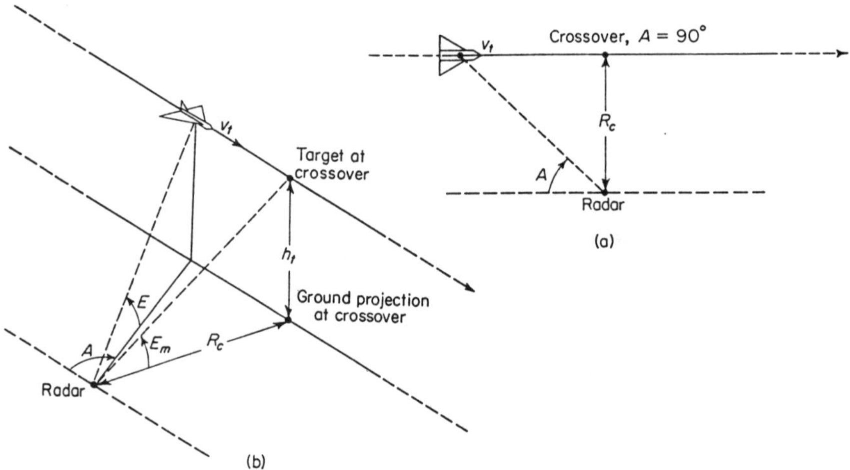

Figure 11.5 Geometry of pass-course problem. (a) Ground projection of flight path. (b) Pass course with target in level flight.

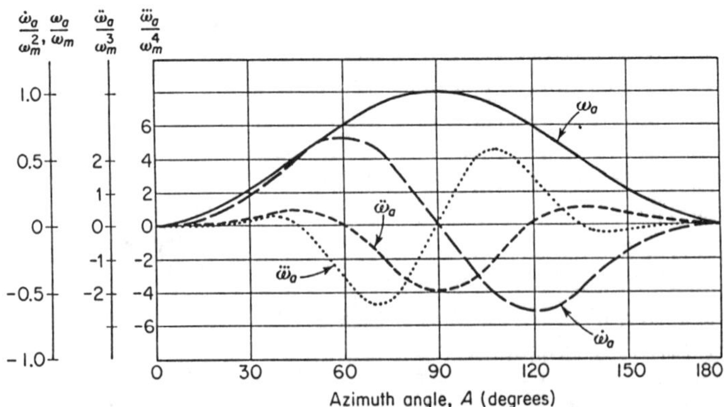

Figure 11.6 Derivatives of azimuth for pass course.

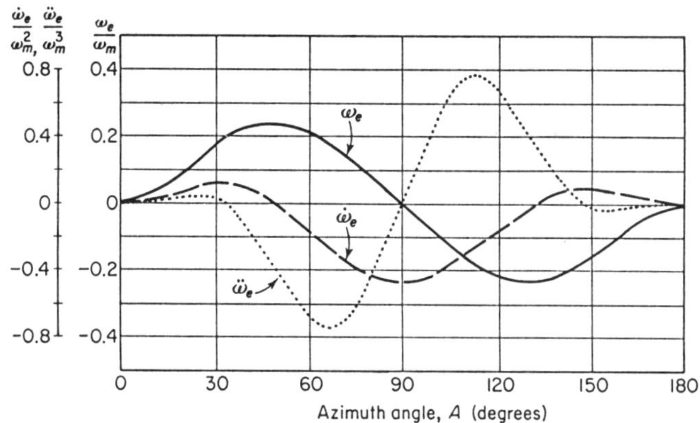

Figure 11.7 Derivatives of elevation angle for pass course ($X=1$).

As an example, consider a subsonic aircraft, $v_t = 300$ m/s, flying at 3 km altitude and crossing at ground range $R_c = 3$ km. The maximum rates and accelerations are:

$$\omega_m = 0.1 \text{ r/s}, \qquad \dot{\omega}_m = 0.0065 \text{ r/s}^2$$

$$\omega_e = 0.024 \text{ r/s}, \qquad \dot{\omega}_e = 0.005 \text{ r/s}^2$$

$$\dot{R}_{max} = 300 \text{ m/s}, \qquad \ddot{R}_{max} = 21 \text{ m/s}^2$$

Thus, the nonaccelerating target introduces in the radar coordinates geometrical accelerations corresponding to about 2g.

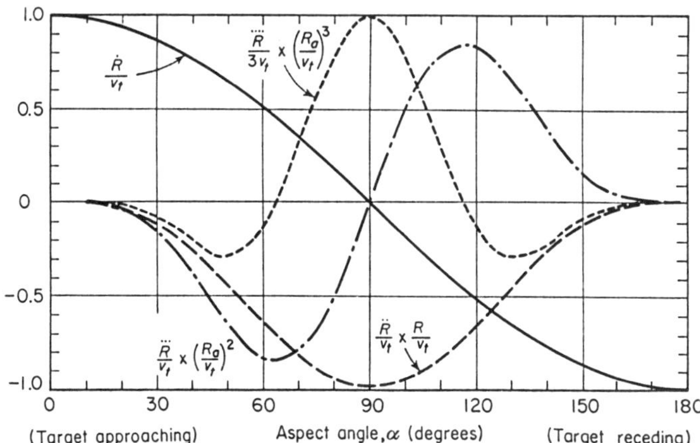

Figure 11.8 Derivatives of range for pass course.

The ability of the radar tracking loops to follow target dynamics can be expressed in terms of the sum of the lag error coefficients; i.e.,

$$\varepsilon_a = \frac{\dot{\omega}_a}{K_v} + \frac{\ddot{\omega}_a}{K_a} + \frac{\dddot{\omega}_a}{K_3} + \dots \tag{11.2.3}$$

The velocity error coefficient K_v can be made arbitrarily large, and the third and higher derivatives of angle will normally be small, leaving the acceleration lag $\dot{\omega}_a/K_\alpha$ as the major dynamic error.

The acceleration error coefficient K_α proportional to the square of the tracking loop bandwidth β_n, or inversely to the square of the smoothing time t_o of the input data, is

$$K_\alpha \approx 2.5\beta_n^2 = 0.63/t_o^2 \tag{11.2.4}$$

Thus, the dynamic lag error can be expressed in terms of the target acceleration and the tracking loop bandwidth or time constant. For azimuth tracking,

$$\varepsilon_a \approx \dot{\omega}_a/K_\alpha = \dot{\omega}_a/2.5\beta_n^2 = 1.6\dot{\omega}_a t_o^2 \tag{11.2.5}$$

The angle acceleration entering into this equation must include the sum of any real target acceleration component, projected into the azimuth coordinate, plus geometrical terms as described above.

The tracking error caused by thermal noise and other random components will increase as the bandwidth of the tracking loop increases, while the dynamic lag error will decrease with increased bandwidth. The total error variance, the sum of the squares of the various components, can be minimized by proper choice of bandwidth. For example, if thermal noise is the dominant random component, the error variance will be

$$\sigma_\theta^2 = \frac{\theta_3^2\beta_n}{k_m^2 f_r B\tau(S/N)} + \frac{a_t^2}{6.3R^2\beta_n^4} \tag{11.2.6}$$

Differentiating this expression with respect to the bandwidth β_n, and setting the result equal to zero, yields the expression for optimum servo bandwidth

$$\beta_n = \left[\frac{a_t^2 k_m^2 f_r B\tau(S/N)}{1.6\theta_3^2 R^2}\right]^{1/5} \tag{11.2.7}$$

As an example, consider a target with acceleration (real or geometrical) $a_t = 20\,\text{m/s}^2 = 2g$, tracked by a radar with error slope $k_m = 1.6$, pulse repetition frequency $f_r = 1000$ Hz, IF bandwidth matched to pulsewidth, $B\tau=1$, signal-to-noise ratio per pulse $S/N = 10$, and half-power beamwidth $\theta_3 = 0.02$ r. The optimum bandwidth is $\beta_n = 2.8$ Hz, and the errors are

$\sigma_\theta = 0.2$ mr for thermal noise,

$\varepsilon_a = 0.1$ mr for acceleration lag,

$\sigma_\theta = 0.22$ mr total error.

The optimum bandwidth in this case is within the capability of a typical antenna servo loop. If the signal-to-noise ratio or target acceleration were higher, the optimum bandwidth might grow so large as to be achievable only with electronic beam steering or with electronic error-correction circuits augmenting the mechanical servo loop.

Surveillance radar data is normally processed in track-while-scan loops using recursive algorithms such as the "α-β" tracker. Multiple-target trackers using phased arrays also use this approach to sampled-data tracking. The tracking loop in this case receives data at discrete data intervals t_t and, on the basis of the observed data $x_o(k)$ received at the kth interval, forms a smoothed estimate $x_s(k)$ given by

$$x_s(k) = x_p(k) + \alpha[x_o(k) - x_p(k)] \tag{11.2.8}$$

where $x_p(k)$ is the previously predicted coordinate for this time. The prediction for the next interval is based on a smoothed estimate of velocity,

$$v_s(k) = v_s(k-1) + (\beta/t_t)[x_o(k) - x_p(k)] \tag{11.2.9}$$

The prediction for the next point is

$$x_p(k+1) = x_s(k) + t_t v_s(k) \tag{11.2.10}$$

It has been shown [11.3] that the acceleration lag error for this recursive tracking loop is

$$\varepsilon_a = (1 - \alpha)t_t^2 a_t/\beta \tag{11.2.11}$$

The error σ_s in the smoothed estimate is related to the error σ_1 of an individual data input by

$$\sigma_s = \sigma_1 \left[\frac{2\alpha^2 + \beta(2 - 3\alpha)}{\alpha(4 - \beta - 2\alpha)} \right]^{1/2} \tag{11.2.12}$$

The equivalent bandwidth of the recursive tracking loop can be expressed as

$$\beta_n = \sigma_s^2/\sigma_1^2 t_t \tag{11.2.13}$$

For example, assume an input data error $\sigma_1 = 0.3$ mr, a data interval $t_t = 0.25$ s, and a recursive loop with $\alpha = 0.6$ and $\beta = 0.43$ (following the optimization criteria of [11.3]). From the equations given above, we can derive the following:

$$\sigma_s = 0.75\sigma_1 = 0.23 \text{ mr}$$
$$\beta_n = 0.56/t_t = 2.2 \text{ Hz}$$
$$K_a = 2.5\beta_n^2 = 12.5 \text{ s}^{-2}$$

This result is intuitively satisfying, since an input data rate of 4 Hz could be expected to support a loop bandwidth of approximately 2 Hz.

Combining of Error Components

The several sources of radar measurement error may be assumed independent, so that the total error is found by adding the individual variances (the squares of the rms values). This procedure is warranted whenever the systematic, or bias, errors have been corrected or calibrated out of the data, leaving all error sources with zero mean value. The total rms error can then be written as

$$\sigma = \sqrt{\sigma_1^2 + \sigma_2^2 + \ldots + \sigma_n^2} \tag{11.2.14}$$

The individual error components will usually vary as a function of range, elevation angle, and target parameters. Figure 11.9 shows a typical curve of error vs. range, for a precision radar tracking an aircraft target. The instrumentation error is 0.1 mr, but this level is approached only at intermediate ranges. At short ranges, the glint error from a target with 25 m wingspan (and hence 5 m rms glint error) will cause relatively large errors in angle tracking. At ranges beyond 200 km, for this system, the thermal noise increases rapidly above 0.1 mr, until the low-signal phenomenon sets in. Loss of track at long range has been discussed earlier.

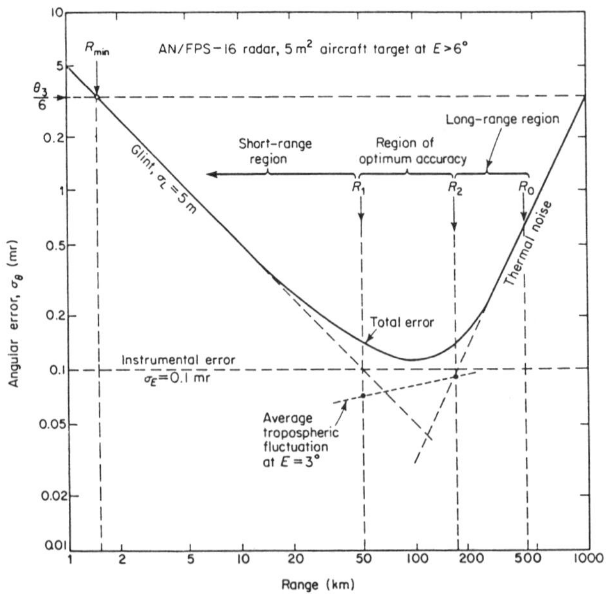

Figure 11.9 Typical monopulse angle tracking error vs. range.

Curves similar to Figure 11.9 can be constructed for different radars and targets, and for low elevations the multipath errors may become dominant over all ranges.

11.2.3 Angle Error in Scanning Radars

The discussions of scintillation error (Chapter 10) and dynamic lag lead to the conclusion that the error of scanning radars will, in most cases, be dominated by these two components. When the radar scans in elevation, as in precision approach radar or nodding-beam height-finders, multipath and refraction errors may also be significant. However, the large magnitudes of these error components usually make it unnecessary to perform detailed evaluations of other sources listed in Table 11.1.

11.2.4 Angle Error in Phased Array Radars

A phased-array tracker may combine the accuracy of monopulse estimation with the advantages of a mechanically fixed (or stable) antenna, minimizing the contributions of mechanical and servo components. Target-dependent errors are still important, especially if the update rate of the beam on each target is low. The discussion of sampled-data tracking systems covers this problem.

The error in beam steering by the array can be estimated if the phase error σ_φ of the individual radiating elements is known. For an array of T elements whose phase errors are independent, the angle error is given by

$$\sigma_\theta = \frac{\theta_3 \sigma_\phi}{1.61\sqrt{T}} \tag{11.2.15}$$

The beamwidth θ_3 varies with scan angle θ from broadside, the error increasing as $\sec\theta$. The rms phase error depends on the number m of phase control bits and the accuracy of phase shifter construction. If quantization error is the only significant source,

$$\sigma_\phi = \frac{2\pi}{2^m\sqrt{12}} \tag{11.2.16}$$

For example, in a typical two-coordinate scanning array with $T = 2000$ elements, $\theta_3 = 0.04$ radian $= 2.3°$ at broadside. When scanned in other than one of the principal planes (such that quantizing errors are minimized), with $m = 3$ bits, we have

$$\sigma_\phi = 2\pi/8 \times 3.5 = 0.23 \text{ radian}$$

$$\sigma_\theta = 0.04 \times 0.23/1.61 \times 45 = 1.27 \times 10^{-4} \text{ radian} = 0.127 \text{ mr}$$

More severe errors occur if a linear array is used, or if the steering is such as to produce correlated quantization errors across rows or columns in a planar array. If the array has $T = 45$ independent linear elements, the error for this example grows to 0.85 mr. The total phase error of an element is often 1.4 to 2 times the quantization limit, further increasing the beam-steering error. However, for properly designed planar arrays of thousands of elements, beam steering is sufficiently accurate that this error source may be neglected.

11.3 Errors

The procedures for describing and combining range error components are similar to those used in angle. Table 11.2 lists some of the errors frequently encountered in range tracking systems. The definitions of the several error classes are the same as in angle, except that the position of the range gate replaces that of the boresight axis.

Table 11.2 Sources of Range Error [11.2]

Class of Error	Bias Components	Noise Components
Radar-dependent tracking errors	Zero range setting	Thermal noise
	Range discriminator shift	Multipath
	Receiver delay	Clutter and jamming
		Variation in receiver delay
Radar-dependent translation errors	Range oscillator (velocity of light)	Range-doppler coupling
		Internal jitter
		Data encoding
		Range oscillator stability
Target-dependent tracking errors	Dynamic lag	Glint
	Beacon delay	Dynamic lag variation
		Scintillation or beacon jitter
Propagation errors	Average refraction of troposphere	Irregularities in refraction of troposphere
	Average refraction of ionosphere	Irregularities in refraction of ionosphere

Two of these factors deserve special mention.

11.3.1 Receiver Delay

The receiver delay results primarily from the use of cascaded filters in the IF amplifier stages. The delay in each stage is approximately the reciprocal of its bandwidth B_1, and if m stages of equal bandwidth B_1 are used the total bandwidth B is

$$B = B_1/\sqrt{m} \tag{11.3.1}$$

Thus, for $m = 10$, the total delay is about $\delta_t = 3/B$, or about twice the pulsewidth for a well-matched system without pulse compression. The receiver delay varies as a function of temperature, signal strength, and tuning of the receiver. Very accurate tuning is usually needed to keep the delay variation within $\delta_t/50 = 1/15B$. When pulse compression is used, the delay is dominated by the compression filter, and is somewhat greater than the width of the transmitted pulse.

11.3.2 Range-Doppler Coupling

In FM pulse compression systems, the dispersive compression line introduces a systematic shift of delay with doppler shift. For a rectangular pulse of width τ and bandwidth B, the delay variation with doppler shift is

$$\delta_t = \tau f_d/B \tag{11.3.2}$$

with the sign of the shift dependent on the upward or downward slope of the transmitted waveform. The errors can be quite large. For example, with $\tau = 100$ μs and $B = 1$ MHz, a doppler shift $f_d = 1$ kHz will cause a delay change $\delta_t = 0.1$ μs, equivalent to a range error of 15 m. The range-doppler coupling error can be corrected in either of two ways:

(1) If range rate is measured over a train of pulses, the doppler shift can be computed and a correction made, based on the above equation; or

(2) The range measurement can be assigned to a time displaced by Δt from the actual time, where

$$\Delta t = f\tau/B \tag{11.3.3}$$

Here, f is the carrier frequency. In the above example, for $f = 3000$ MHz, the time displacement of the data would be 0.3 s, regardless of target velocity.

11.4 Doppler Errors

The procedures for describing and combining doppler error components are also similar to those used in angle. Table 11.3 lists some of the errors frequently encountered in doppler tracking systems.

A particular problem in doppler tracking is the presence of modulated scatterers on the target, such as rotating propellers and jet engine turbines. These sources can produce offset lines in the spectrum whose position has no relationship to the actual vehicle velocity. If one of these lines appears in, or captures, the doppler discriminator, the error can be so large that it cannot be characterized by an rms value.

Table 11.3 Sources of Doppler Error [11.2]

Class of Error	Bias Components	Noise Components
Radar-dependent tracking errors	Discriminator zero setting and drift	Thermal noise
	Gradient of receiver delay	Multipath
		Clutter and jamming
		Variation in receiver delay
Radar-dependent translation errors	Transmitting oscillator frequency	VCO frequency measurement
		Radar frequency stability
Target-dependent tracking errors	Dynamic lag	Glint (target rotation)
		Dynamic lag variation
		Target modulation (propeller or turbine lines)
Propagation errors	Gradient in refraction of troposphere	Irregularities in refraction of troposphere
	Gradient in refraction of ionosphere	Irregularities in refraction of ionosphere

11.5 References

[11.1]   D. K. Barton, *Radar System Analysis*, Prentice-Hall, 1964; Artech House, 1976.

[11.2]   D. K. Barton, *Modern Radar System Analysis*, Artech House, 1988.

[11.3]   S. S. Blackman, *Multiple-Target Tracking with Radar Applications*, Artech House, 1986.

Chapter 12

RADAR ECM VULNERABILITY AND ECCM

The heavy dependence of modern weapons systems on electronic technology has led to the development of sophisticated *electronic warfare (EW)* systems and tactics [12.1, Chapter 1]. Radar systems, in particular, are key elements of many weapons systems. Consequently, they are the object of the continuing development and fielding of *electronic countermeasures (ECM)* systems that are specifically designed to negate or diminish the effectiveness of these radars. To the extent that ECM causes a given radar to be unable to perform its function, the weapons system it supports may become partially or totally ineffective. Thus, radar systems which support military systems should always be designed to include sufficient *electronic counter countermeasures (ECCM)* capabilities to cope with both known and expected ECM threats.

EW began with the development of ECM equipment during World War II, designed to take advantage of the one-way transmission power loss for active jammers compared to the two-way transmission power loss associated with radars. Passive techniques (e.g., chaff) were also developed to create false targets or to screen real targets.

Subsequent radar developments incorporated counter-ECM techniques (ECCM) designed to overcome this transmission power loss imbalance and to minimize the effectiveness of ECM on military radars. The general ECCM philosophy is to minimize the effects of deceptive jamming techniques or to force enemy jamming signals to be spread over extremely broad bandwidths; the latter dilutes the ECM power density levels received at the radar sites. Improvements in radar receiver characteristics (e.g., selectivity, dynamic range, antenna sidelobes, etc.) serve to reduce the effects of active ECM. MTI techniques and improved tracking algorithms reduce the effectiveness of passive ECM (e.g, chaff and decoys).

EW continues to evolve, with more sophisticated ECM techniques being developed to counter advanced radar system designs. In the latter case, the trend is toward ultra-wideband radars utilizing antennas with very low sidelobes (or sidelobe cancellers) plus versatile and sophisticated digital signal processors.

So-called "quiet radars" are also envisioned for future deployment. These will result in *low probability of intercept (LPI)* by hostile receivers that provide the data to direct jammers at specific radar transmitters.

12.1   Radar Vulnerability to ECM

The vulnerability of a given radar to ECM is determined by a number of factors:

(a) The ability of the enemy to know of or to detect the presence of the radar and its approximate location;

(b) The relative positions and sitings of fixed radar and ECM systems, or tactics of airborne radar and EW systems, as well as those deployed on other moving platforms (e.g., missiles, spacecraft, and ships); and

(c) The susceptibility of the radar to ECM signals and techniques.

Clearly, if the enemy has no knowledge of the presence and approximate location of a radar and cannot detect and locate it, he can only use the brute force approaches of broadband noise barrage jamming or sow chaff; either of these measures may restrict his own radar operations. The relative positions and sitings (or tactics), together with the ECM and radar range equations plus the various propagation effects, determine the ability of enemy ECM to generate the jamming signal levels required to exploit a radar's susceptibility. Susceptibility of a radar is characterized by measures of jamming signal power levels required at the radar for various types of ECM to exploit a radar's weaknesses or negate its effectiveness.

The radar designer tries to prevent enemy detection and location of his radars, but is usually forced to provide the required ECCM capabilities on the assumption that he cannot achieve this goal. The best available ECM threat model (relative jammer locations, types of jammers, jammer power levels, ECM techniques, etc.) and overall tactical considerations are used to establish the expected ECM characteristics, power density levels, and tactics (turn-on, turn-off, power management, etc.). These define the ECM environment in which the radar is expected to operate. The radar is then designed to minimize susceptibility to this ECM environment. Radar ECM susceptibility should then be evaluated against this threat environment.

12.1.1 Typical EW System

The typical EW system includes two major elements: a passive receiving subsystem, called *electronic warfare support measures (ESM),* and an active ECM (jamming) or passive ECM (chaff or decoys) subsystem. The ESM subsystem, comprised of wideband receivers (and appropriate antennas), performs the function of analyzing the signal environment and detecting, locating, classifying, and identifying radars. The ECM subsystems then produce the jamming signals and deploy passive devices either in response to the environment or to target a specific victim radar.

The ESM and ECM subsystems typically employ separate antennas to achieve good isolation between transmit and receive functions. These functions are usually time-shared within a given frequency band to achieve the necessary isolation, a technique referred to as "look-through" receiver operation. ESM receivers typically locate radars in angle only by measuring *angle-of-arrival (AOA);* specific radars are classified or identified by a comparative analysis of their frequencies of operation, waveforms, and scanning patterns.

The ECM subsystem can employ various types of jamming, depending upon the tactical situation [12.1, Chapter 2], the type of radar (search and acquisition, tracking, or imaging), and specific knowledge of the victim radar's characteristics (frequency, waveform, etc.).

12.1.2 Radar Vulnerability Analysis

The noise temperature increase at a radar input (antenna) termination caused by a *noise jammer* (see Section 2.2.4) is

$$T_j \;=\; \frac{P_j G_j A_r F_j^2}{4\pi k B_j R_j^2 L_{aj}} \tag{12.1.1}$$

where $P_j G_j$ is the *effective radiated power (ERP)* of the jammer, F_j is the pattern propagation factor of the radar receiving antenna in the jammer direction, B_j is the bandwidth of the jammer spectrum, R_j is the jammer range, and L_{aj} is the one-way propagation loss from the jammer. Vulnerability of a radar to noise jamming is determined by Eq. (2.1.1). The degree of reduction of radar performance results from increases in T_j. For example, a 12 dB increase in total noise temperature of the radar reduces the radar detection range by a factor of two.

Susceptibility factors in Eq. (12.1.1) include A_r (the antenna aperture size), F_j, and B_j. The antenna aperture size is normally dictated by design requirements other than ECCM, and is usually not an ECM/ECCM variable. The factor F_j is important, in that low sidelobe levels (and sidelobe cancellers) may limit jamming effectiveness to the small angular regions of the antenna mainlobes of large aperture radars. The factor B_j can be influenced by the radar design, since wide instantaneous bandwidth or rapid wide band tuning forces the jammer to increase B_j and thus to decrease T_j.

There are no general equations to describe the effects of *deceptive jammers* on a radar other than for calculation of *jamming-to-signal (J/S)* ratios. A general equation for calculating J/S ratios at a radar antenna terminal for single pulses and a monostatic radar with equal transmitting and receiving antenna gains is given by

$$J/S \;=\; \frac{(4\pi)P_j G_j F_j^2 R^4}{\sigma P_t G_r R_j^2} \tag{12.1.2}$$

where R is the radar-to-target range, σ is the target radar cross section, P_t is the radar transmitted pulse power level, and G_r is the radar antenna gain.

The effects of a single-pulse J/S ratio at the radar antenna terminals depend upon (1) the number of pulses processed during a single dwell time, (2) the signal processing gains and losses for both target and jammer pulses, and (3) the J/S ratios after processing required for the jammer to be effective. The latter considerations depend upon the victim radar's target display and tracking methods and logic, since the appearance of a single jammer-induced false target at a display or in a tracking channel may not prevent the radar from performing its function.

Passive ECM, such as decoys and chaff, are designed to present radar cross sections sufficient to be detected and tracked when they are placed within a radar's normal detection range. Thus, they present false targets either to saturate target handling capabilities or to interrupt the radar target tracking capabilities. In the case of chaff, the usual intent is to screen real targets. The susceptibility of radar to passive ECM techniques is determined by the signal and target data processing techniques employed. There is generally no effective radar countermeasure to a well-designed decoy, since by definition the decoy is designed for accurate representation of a valid target.

12.2 ECM Deployment and Tactics

Active ECM deployment typically takes the form of *standoff jammers (SOJ), escort jammers (ESJ),* and *self-protection jammers (SSJ).* Decoys may be either active or passive. They are usually active when deployed by aircraft and frequently passive when deployed with ballistic missile weapons. Chaff is either sown over large areas or in corridors to screen military operations, or dispensed in bundles from an aircraft to disengage radar trackers and missile seekers from their own return signals.

Standoff jammers consist of high power barrage (broadband) noise transmitters on board large aircraft. The ERP of a single transmitter/antenna combination may be in the region of + 50 dBW to + 100 dBW. Several transmitters may be aboard a single aircraft, with one or more of the transmitters dedicated to jam a given type or class of radars (e.g., surveillance, tracking, or imaging radars). Usually, a number of aircraft will patrol a region behind the *forward line of troops (FLOT)* and direct their jamming signals toward the victim radars.

Escort jammers are typically medium to high power emitters on board tactical aircraft which accompany and screen a group of penetrating attack aircraft. ESJs usually employ barrage jammers that cover the frequency bands of the radars and missiles that are a threat to the attack aircraft. ERP levels may be in the region of + 30 to + 40 dBW. Escort jammers may also employ deception techniques to confuse the victim radars with regard to raid size and to saturate radar target data processing subsystems. Ballistic missile weapons may also be accompanied by escort jammer reentry vehicles.

Self-protection jammers (sometimes called "self screening") are employed by both strategic and tactical aircraft, as well as other military vehicles such as ships. These jammers may employ any or all of the known ECM techniques. Self-protection ERP levels are typically not as high as those of SOJs and escorts, since to be effective the ECM levels at the victim radar receiver need only be 10 to 20 dB above the radar skin return signals from the aircraft that employ the jammers. Timing of the turn-on or deployment of ECM during tactical operations is a sophisticated process and an integral part of mission planning and execution. For example, in a low-level aircraft strike mission ECM is not activated until the aircraft is within the detection range of the victim radars. Active ECM may be used only to counter tracking radars and radar-homing missile seekers. Chaff bundles or active decoys may be deployed at the end-game phase of an attacking missile's flight, and sudden maneuvers may be coupled with ECM employment and deployment during this critical phase of missile attack evasion by the penetrating aircraft.

12.3 Active ECM

Active ECM includes noise jamming, a great variety of deception techniques, and active decoys [12.1, p. 13]. Generally, deception and decoy techniques involve the reception and retransmission of radar pulses with time delays, doppler frequency shifts, and other features to simulate target characteristics that are different from those of the real target as seen by the victim radar.

12.3.1 Noise Jammers

The most common type of ECM is noise jamming. ECM noise transmission is either continuous or has a high duty cycle (90% or more). A less than 100% duty cycle indicates the look-through situation, where the quiet time allows for performance of ESM functions. Broadband noise jammers, which cover either the full tuning range of a given type or class of radar or a complete radar band, are known as barrage jammers. Narrowband noise jammers, with bandwidths approximating the instantaneous bandwidth of the radar (determined by the radar pulse width) and tuned to specific victim radar frequencies, are known as spot jammers.

The quality of jamming noise, measured by its effectiveness against a radar, depends upon the methods used for generation or modulation and power amplification [12.2, Section 3.7]. Direct amplification of a noise source (DNA) produces jamming with a nearly Gaussian amplitude distribution, the most effective type of noise. However, with linear amplification the average power output level of the jammer is well below the capability of the jammer output amplifier. Increasing the noise level into the output amplifier and driving it into saturation results in higher average output power at the expense of clipping the noise peaks. Clipped noise is less effective in masking targets and creating false alarms in a radar. Another commonly used approach to noise jamming produces broadband noise by rapidly sweeping over the target frequency band while simultaneously applying a noise FM function to randomize the jamming signal. Thus, full saturated output power level is obtained with some degradation of jamming effectiveness as a result of the departure of this signal from a true Gaussian distribution.

Regardless of the method used for noise generation, modulation, and amplification, the effective ERP of the noise jammer is reduced by a few dB relative to the theoretical Gaussian model. Thus, a noise quality factor (≤ 1.0) can be applied to known ERP levels in evaluating noise jammer effectiveness.

Narrowband spot jammers are more likely employed by escort and self-protection jammers to achieve increased noise power densities in countering either a single radar or (with multiple transmitters) a small number of radars. These jammers usually include look-through capabilities (less than 100% duty cycle) in order to cope with rapid-tuning radars. If 100% duty cycle operation is used, a so-called look-over capability is required. This involves simultaneous transmission and reception by means of extremely high isolation between transmitting and receiving antennas, achieved by signal cancellation schemes employed within the ECM equipment.

Blinking noise jammers with 30% to 60% duty cycles typically employed by the aircraft in a multiple aircraft attack are used to confuse tracking radars and radar missile seekers. These may be either broadband (to counter a large number of radars) or narrowband spot jammers (to counter a single radar or small number of radars).

Frequency swept noise jammers are an older type, designed for barrage jamming with transmitting tubes that have insufficient instantaneous bandwidth to cover the desired band but which can be frequency swept at modest rates. A single jammer of this type is not very effective, since it appears to the radar as a low duty cycle noise jammer. Multiple simultaneous jammers are required to achieve any degree of effectiveness with swept jammers.

12.3.2 Deception Jammers

Deception jammers use special waveforms to produce one or more false targets at coordinates and velocities other than those of the real targets. The jammer receives, modulates, and retransmits each radar pulse (a repeater jammer), or transmits noise pulses or simulated radar pulses upon receipt of a radar pulse (a transponder). Repeater and transponder pulses received at a radar are always delayed in time relative to the reflected signal (skin return) from the jamming vehicle. Thus, in the case of a self-protection jammer the false targets appear at the radar at slightly longer ranges than the real target range, although modern wideband circuitry permits these delays to be in the order of nanoseconds. Separate receiving and transmitting antennas with extremely good isolation are required to prevent ring-around oscillations in the ECM system.

Deception techniques include the production of many false targets to cause *track file saturation* at the radar, *range gate pull-off (RGPO)* by slowly moving a single stronger false target away from the real target, and *velocity gate pull-off (VGPO)* by simulating a stronger target with a different rate of change in position coordinates and corresponding doppler frequency shift. These techniques are all designed to foil tracking radars and missile seekers.

Radar angle deception techniques, also designed to foil tracking radars and missile seekers, include the venerable *inverse scan* technique. Here the false target is amplitude modulated at the scan rate of a conical scanning tracking radar and at a phase angle relative to the conical scanning that cancels the target scanning modulation thus negating the angle measurement capability of the radar. Monopulse tracking radars are immune to this technique, but are susceptible to other angle deception techniques. These include *surface bounce* techniques [12.3, p. 60] and what are sometimes called *phase-front* techniques. Surface bounce techniques are implemented with repeaters using relatively narrow-beam and low-sidelobe antennas to direct the false target signals to reflect off the earth's surface and make the target appear at a different angle than that of the real target. This is similar to the multipath propagation effect discussed in Chapter 8. Phase front techniques include *two-point coherent jamming* [12.3, p. 156] and *cross-polarization jamming* [12.3, p. 153]. Two-point coherent jamming, or *cross eye,* is a repeater technique which utilizes two separated antennas for simultaneous transmission of coherent false target signals with a relative phase relationship that effectively reverses the slope of the radar monopulse angle error curve. This causes the tracking antenna to move its boresite direction away from the target and produce a large angle error or a break-lock condition. *Cross-polarization* jamming is also a repeater process, where the retransmitted signal is polarized opposite to the polarization of the radar transmission. The cross-polarized signal, when received by a parabolic or shaped-reflector radar antenna or when passed through a nonplanar radome, introduces large angle-tracking errors to the radar.

Active decoys are miniaturized repeaters deployed on board small airborne vehicles, rockets, or projectiles, or towed by military vehicles. These devices, operating at locations separated from the real targets, provide beacon signals stronger than the real target echoes, leading to large errors in the tracking radar data.

12.4 Passive ECM

Chaff is the most common and oldest form of passive ECM. Chaff consists of many lightweight strips of metallic or metallized material which are dispersed at an altitude in the vicinity of the targets to be screened or protected. Other forms of passive ECM include decoys, such as drone aircraft equipped with radar reflectors (e.g., corner reflectors and Luneberg lens reflectors) to augment their radar cross sections, and imitation *reentry vehicles (RVs)* that may accompany ballistic missile warheads on their trajectories toward their target areas.

12.4.1 Chaff

The most commonly used chaff is made of either aluminum foil or aluminized glass, with the latter offering the lightest weight for a given number of chaff strips or effective total radar cross section. Each strip of a bundle of chaff designed to be most effective at a given radar frequency is approximately one half wavelength long [12.3, p. 185]. These appear to a radar as a short-circuited dipole and resonate at the frequency $f = c/2d$, where $d = \lambda/2$ is the dipole length. A plot of dipole cross section as a function of frequency shows resonances at $f = nc/2d$, where $n = 1, 2, 3,\dots$. However, for values of n greater than one, the resonant peaks are much lower than for the case of $n = 1$. Typically, a chaff bundle is made up of dipoles of several lengths, giving resonances over a sufficient bandwidth to cover several radars operating at different frequencies or capable of rapidly turning over a broad frequency band [12.3, p. 187].

Chaff is generally deployed either to screen a number of targets (within a chaff cloud) or to decoy a tracking radar away from its target. In a screening operation, one or more aircraft will sow a corridor of chaff by sequentially dropping chaff bundles from high altitudes. Decoying for self-protection usually consists of dropping relatively small bundles of chaff, spaced to provide discrete chaff targets, to distract the radar (or radar missile), or dispensing chaff bundles by means of rockets. In the latter case, the object is to lure a tracking radar or missile seeker away from the target (e.g., a ship) or to place the chaff closer to the threat (radar or missile) with the intention of making the chaff appear as a higher priority target.

Immediately following deployment, the chaff bundle appears to a radar as large cross-section discrete target. However, depending upon the dispersing method and wind conditions (as well as turbulence created by the dispensing vehicle), the bundle quickly spreads and begins to fall. Spreading results in the chaff becoming a form of volumetric clutter; closely spaced bundles blend into a more or less continuous cloud of chaff. Wind may modify the shape of the cloud due to shearing, provide turbulent motion to the dipoles, and spread the chaff horizontally. Vertical spreading may result from wind, but more likely is determined by varying fall rates caused by the different sizes and physical orientations of the dipoles. The latter effect may cause a chaff cloud to divide into two vertically spaced portions having oppositely polarized (vertical and horizontal) radar reflections.

Chaff fall rates are typically between 0.3 and 0.55 m/s at sea level, and 0.6 and 1.1 m/s at 12 km altitude [12.3, p. 183]. Hence, chaff clouds sown at high altitudes may remain at relatively high altitudes for from one to three hours.

The average RCS of an individual dipole of chaff, when averaged over all aspect angles, is 0.15 λ^2 (Chapter 5, Section 5.5). Modern bundles of chaff can achieve total RCS values of $\sigma \approx 2 \times 10^4 \, \lambda W m^2$ over a two-octave band, where W is the bundle weight in kg.

The velocity spectrum of chaff is similar to that of rain (see Section 5.5). The standard deviation of the velocity spectrum is $\sigma_v = 0.3 \, k_{sh} \, R\theta_e \cos \alpha$, where R is in km, α is the azimuth angle between the beam and the wind direction, and the one-way elevation angle θ_e is in radians. Values of k_{sh}, the shear coefficient, between 2 and 4 m/s per km are common, with a total velocity change of up to 40 m/s between the surface and the highest (chaff) clutter.

Since chaff is essentially artificial volume clutter, a radar is designed to cope with chaff, in the same way as with rain clutter, through the application of doppler signal processing and MTI techniques (Chapter 7). Chaff may be used to decoy tracking radars temporarily, but sophisticated tracking techniques can recognize and defeat most such attempts.

12.4.2 Passive Decoys

Non-radiating decoys may be employed to increase the number of targets seen by a radar. For example, drones with enhanced radar cross sections can be flown in company with penetrating aircraft to increase the apparent raid size in order to dilute the effectiveness of defending weapons or even to saturate defending radar tracking capabilities. Radar reflectors (corner reflectors, Luneberg lens reflectors, etc.) may be configured to simulate high-value fixed targets as seen by airborne radars, thus diluting the effectiveness of air-to-ground weapons delivery systems. However, these decoy techniques are expensive and present substantial logistics problems, and are not likely to be used extensively over any extended period of time.

The more likely use of passive decoys is with ballistic missile reentry systems, where a large number of simply configured lightweight replicas of a nuclear warhead reentry body may accompany it in a nuclear attack. These decoys would be erected (or inflated) and ejected from the final rocket stage of the weapon after it is out of the denser atmosphere, and would continue on trajectories similar to that of the warhead. Again, the intent would be to saturate the radar tracking facilities and to dilute the effectiveness of defending missiles.

Radar techniques designed to reject decoys and to concentrate tracking resources on real targets may make use of target signature analysis (e.g., polarization properties, scintillation, frequency dependence, etc.) or may (in the case of moving targets) rely on differences in target kinematics. An example of the latter method is the use of sophisticated tracking filters that sense the different acceleration coefficients of heavy ballistic missile warheads and lightweight decoys. The decoys are affected by the higher density at the lower levels of the atmosphere as they reenter the atmosphere.

12.5 Electronic Counter Countermeasures

The technology of electronic warfare is becoming increasingly sophisticated. Thus, there is no given set of ECCM techniques that can be incorporated into a radar design that will render it immune to all present and future electronic countermeasures. However, a number of general radar design concepts have evolved that provide a measure of success in limiting the degradation inflicted by ECM. These ECCM techniques function either by reducing the sensitivity of the victim radar to the ECM, or by discriminating ECM signals from legitimate target signals.

12.5.1 Techniques to Reduce Sensitivity to ECM

Antenna Techniques

Low Sidelobe Antennas — For those situations in which the jammer operates through the victim radar's receiving antenna sidelobes, reduction of the sidelobe gain will directly reduce the effectiveness of the jamming. Significant reduction of the sidelobe level can be achieved through the use of a suitably tapered (and accurately controlled) illumination function across the receiving antenna aperture.

Narrow Beam Antennas — Increasing the angular resolution capability of the radar enhances its capability to counter ground bounce jammers and to reject false targets generated by SSJs and ESJs.

Sidelobe Cancellation — Coherent adaptive sidelobe cancellers can be used to counter one or more significant jammers. An auxiliary antenna with a broad pattern covering the sidelobes of the main antenna is utilized (see Figure 12.5.1). A portion of the signal from the auxiliary antenna (predominantly the jamming signal) is coherently subtracted from the main channel. The weighting coefficient utilized in this subtraction is derived by a correlator comparing the auxiliary channel with the corrected main channel. This extracts the gain and phase of the residual jamming signal in the main channel relative to the auxiliary channel. The correction loop reduces the jamming signal component in the main channel until there is no longer any correlated output. The combined pattern of the two antennas will have a null at the jammer location that will move automatically to remain on the jammer as the main antenna beam scans. Multiple cancellation loops allow simultaneous cancellation of several jammers. The cancellation can be effective for a jammer located within the mainbeam or in the sidelobes of the main antenna.

Some limitations to this technique are:

(1) It is primarily effective against high duty factor jamming;

(2) Since a portion of the auxiliary antenna pattern is added to the main pattern to create a null in a specific desired location, the general sidelobe level of the antenna can be increased, with a resultant increase in the vulnerability of the system to sidelobe jamming; and

(3) Similarly, cancellation of mainlobe jamming can result in a distortion of the main beam shape so that monopulse tracking can be seriously affected.

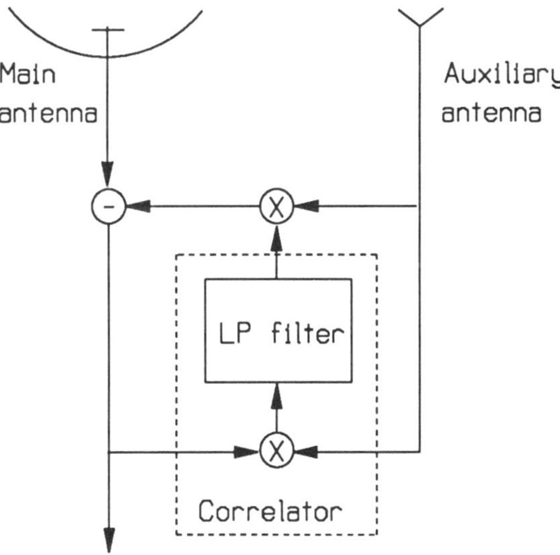

Figure 12.5.1 Coherent sidelobe cancellation.

Sidelobe Blanking — Signals arriving through the sidelobes of the receiving antenna can be recognized as such by comparing the signal strength as received through the main antenna with the signal strength received through a small auxiliary antenna. A signal arriving in the main beam of the main antenna will be much stronger than in the auxiliary channel, whereas a signal arriving in the sidelobes will be stronger in the auxiliary channel. A blanking control signal can be generated by comparing the ratio of the main and auxiliary channel signal strengths. All returns from those sectors of the main antenna having gain less than the auxiliary antenna are blanked in the radar receiver. The blanking is instrumented at a rate corresponding to the signal bandwidth of the radar, so that the receiver is blanked only for those instants when individual jamming pulses are received through the sidelobes.

This technique is effective only for low duty factor jamming. Since a continuous jammer would continuously turn off the radar receiver, suitable threshold circuits must be included to prevent this. For a radar that utilizes a coherent burst waveform, the entire burst must be blanked for the affected range cell to avoid spreading of clutter into the doppler passband.

Planar Antennas and Radome — Cross-polarization repeater jamming can introduce large angle-tracking errors if the radar receiving antenna has a significant cross-polarized response, as discussed in Section 12.3. Since control of the cross-polarization response characteristics of a radar antenna is difficult for parabolic or shaped reflectors, or a nonplanar radome, an appropriate counter countermeasure is to utilize a planar antenna and a more planar radome.

Polarization Diversity — Another appropriate counter countermeasure to cross-polarization repeater jamming is to use polarization diversity on reception, and to track on the stronger polarization signal produced by the jammer. Alternatively, an arbitrary polarization, unknown to the jammer, can be used on reception.

Multiple Phase Center Antennas — Cross-eye jamming attempts to generate large tracking error signals in the victim radar by retransmitting the radar signal through an interferometer antenna pattern so that a null is positioned at the receiving antenna aperture. Since the jammer signals are retransmitted back toward the source of the radar signals (i.e., the radar transmitting antenna) a counter countermeasure technique consists in displacing the radar receiving antenna from the transmitting antenna. This can be accomplished either by using separate transmit and receive antennas, or by displacing the phase center of the receiving antenna relative to the transmitting antenna in a common antenna aperture.

Receiver and Signal Processor Techniques

Coherent Integration — Coherent integration, by reducing the vulnerable receiver bandwidth, directly reduces the effectiveness of wideband barrage jamming.

Large Dynamic Range — Intermittent or blinking jammers that attempt to drive the radar receiver alternately into saturation and then into a desensitized condition can be partly countered by increasing the dynamic range in the receiver.

Fast AGC — A further counter to intermittent or blinking jammers is the use of fast AGC that reduces the time spent by the radar receiver either in saturation or desensitization below the noise level.

Improved AFC — Skirt-frequency jamming attempts to exploit the mismatch in phase of the receiver sum and difference channels as a function of frequency across the IF passband. The counter for this technique is to use automatic frequency control in the radar to position the dominant signal (the jamming signal) in the center of the IF passband, and operate in a ***track-on-jam (TOJ)*** mode.

Preselection Filtering — Image-frequency jamming injects a signal separated from the radar signal by twice the receiver IF (i.e., at the image frequency). Two-frequency jamming injects two signals separated by the radar IF frequency. Both attempt to create large monopulse angle errors, and both can be countered through the use of RF preselection filtering in the radar receiver.

Monopulse Channel Phase Matching — Careful matching of the phase vs. frequency characteristics of the monopulse sum and difference channels can be an effective counter to both skirt-frequency jamming and image-frequency jamming.

Guardband Controlled Blanking — Swept frequency jamming can be countered by using guardband filters adjacent to the radar operating frequency band to control the blanking of the main receiver channel during those times when the jamming signals are within the radar bandwidth.

Doppler Filtering and MTI — Doppler filtering and MTI processing often can provide very effective rejection of chaff on the basis of velocity, since chaff velocities drop quickly to the local wind speed after deployment.

Frequency Agility

Rapidly varying the radar frequency of operation over a wide tunable bandwidth can force a barrage jammer to broaden its radiated bandwidth accordingly. This provides a frequency domain filtering advantage for the radar, since the bandwidth of the radar receiver need be only as wide as that of the instantaneous pulsed radar signal, while the jamming signal is forced to spread over a much wider bandwidth. The effective jamming power is thus reduced after the receiver filtering by an amount equal to the ratio of the frequency agile bandwidth to the radar signal bandwidth. Frequency hopping by the radar is usually performed on a pulse-to-pulse basis for non-coherent modes, and on a dwell-to-dwell basis when dwells are coherently processed.

12.5.2 Techniques to Discriminate against ECM Signals

A number of techniques can be used by the radar either to exploit or to discriminate against jamming signals and false target returns in favor of true target returns. Included in this category are target tracking techniques and logic that permit discrimination between true and false targets, and techniques that employ specific target characteristics or signature information.

Passive Angle Tracking — It is possible to make use of the ECM signals themselves to obtain target data for those cases (self-protection jamming, for example) in which the jamming is carried on the target vehicle. The strong jamming signals permit highly accurate angle-of-arrival measurements by the radar operating in the passive (receive only) mode. This defines a line of position to the target, which of itself may be sufficient to launch a weapon. Further, it is possible under some circumstances to obtain additional intersecting lines of position through maneuvers of the radar platform, thus permitting a range estimate to be made. When a number of radar systems can share their data in a netted manner it is possible to combine, by triangulation, the various measured lines of position into a more detailed definition of the target location.

Multiple Track File Maintenance — In cases in which the jamming generates a false target to lead the radar tracking circuits away from the true target, and hence to break track, it can be advantageous for the radar to recognize when the target under track appears to separate into two or more pieces, and to establish and maintain separate track files on each. This allows continued tracking of the true target without the necessity for time-consuming reacquisition when the false target is recognized as such, perhaps through the dynamic filtering discussed below.

Dynamic Target Filtering — Range gate and velocity gate stealers are examples of a class of deceptive jammers that attempt to disrupt radar performance by generating false targets which capture the radar's tracking gates and lead them away from the true target. The radar track processor can provide some measure of rejection of these false targets by recognition of any kinematic inconsistencies, i.e., the change of range with time should be consistent with the doppler velocity measurement, and the acceleration should be within the bounds of known target capabilities.

Target Signature Analysis — When detailed target signature data are available to the radar they can be utilized as discriminants against false targets. Given appropriately fine radar range resolution, target range profile measurements can be obtained and target length can be estimated and compared with typical target dimensions. At a finer level of detail, the distribution and locations of specular reflection points can be observed. If a polarization diversity capability is provided in the radar, the polarization scattering matrix characteristics of the target can be measured. In each of these cases the detailed measurements obtained by the radar can be used as discriminants to separate true targets from false targets.

12.6 Summary of ECM Effects and ECCM Responses

12.6.1 Search Radar ECM and ECCM

The primary goals of ECM directed against search radars are (a) to prevent target detection and acquisition and (b) to generate false targets. Table 12.6.1 summarizes the various types of ECM directed against search radars, along with their deployment methods and primary effects on the victim radar. ECCM techniques that can reduce the radar vulnerability to each type of ECM are also summarized in this table.

Barrage noise generated by a *stand-off jammer (SOJ)* can deny detection and acquisition of a quiet target within the radar mainbeam, and can significantly degrade detection if directed at the radar sidelobe region. Instrumenting for low sidelobes on receive can reduce the radar sensitivity to the sidelobe jamming. Coherent sidelobe cancellation can also be used. Barrage noise produced by *self-screening jammers (SSJ)* or *escort jammers (ESJ)* can also prevent target detection and acquisition. However, these jammers provide strong beacon-like signals that betray their angular locations, which then can be exploited by the radar in weapon designation. If the radar is netted with other radars, the jamming target can be located through triangulation. Coherent integration narrows the receiving bandwidth and thus decreases the radar vulnerability to the jamming.

Spot noise may be used by various types of jammers in attempting to deny target acquisition. A highly frequency-agile radar, however, usually forces the jammer to revert to barrage jamming. In the case of SSJs and ESJs, spot noise at short range may drive some victim radar receivers into saturation. This can be countered by widening the dynamic range in the radar receiver.

Frequency-swept noise jamming can generate false target detections when the jamming frequency sweeps through the radar receiver bandwidth. An effective radar counter is the use of guardband-controlled blanking in the receiver.

Repeaters can generate false targets in the search radar, and can saturate the radar track file capacity. A useful countermeasure to this technique is to provide an extended track file capacity, and to establish and maintain track on each detected object until each is evaluated. Sidelobe blanking can also be used to reject false targets injected into the sidelobes via repeater jamming.

Decoys carrying active repeaters can create the same effects in the radar as the repeater jammers discussed above. Similar ECCM techniques can be effective against them. Passive decoys can be countered through the use of target signature analysis, exploiting as discriminants any available differences between the desired targets and the decoys (such as length, dynamic motion, and polarization characteristics). However, there is no counter to a "perfect" decoy, which must be handled as another target.

<table>
<tr><td colspan="3" align="center">Table 12.6.1
ECM against Search Radar</td></tr>
<tr><td>ECM Type</td><td>Effects of ECM on Radars</td><td>ECCM Techniques</td></tr>
<tr><td>Barrage noise
• SOJ</td><td>Prevent target detection and acquisition</td><td>Low sidelobe antennas
Sidelobe cancellation
Coherent integration</td></tr>
<tr><td>• SSJ and Escort J.</td><td>Prevent target detection and acquisition</td><td>Jammer strobe processing
Coherent integration</td></tr>
<tr><td>Spot noise
• SOJ</td><td>Prevent target detection and acquisition</td><td>Same as Barrage noise (SOJ), plus Frequency agility</td></tr>
<tr><td>• SSJ and Escort J.</td><td>As above, plus receiver saturation</td><td>Same as Barrage noise (SSJ), plus Frequency agility
Wide dynamic range receivers</td></tr>
<tr><td>Swept noise</td><td>Same as Barrage noise, plus False target generation</td><td>Guardband controlled blanking</td></tr>
<tr><td>Repeater (SSJ and Escort J.)</td><td>False target generation
Track file saturation[1]</td><td>Multiple track file maintenance [1]
Sidelobe blanking (ESJ)</td></tr>
<tr><td>Decoys
• Active repeater
• Passive</td><td>False target generation
Track file saturation [1]</td><td>Multiple track file maintenance [1]
Target analysis</td></tr>
<tr><td>Chaff</td><td>Prevent target detection and acquisition
False target generation</td><td>Doppler filtering and MTI</td></tr>
</table>

---

[1] Assumes track-while-scan (TWS) capabilities.

Chaff can be used to prevent or delay target acquisition throughout a region, as well as to generate false targets. Doppler filtering and MTI processing can usually reject chaff based on its velocity characteristics.

12.6.2 Track Radar ECM and ECCM

The primary objectives of ECM against a tracking radar fall in the categories of (a) preventing or delaying tracker acquisition of the target, (b) disrupting or introducing large errors in the range, doppler and/or angle tracking function of the radar, and (c) generating false targets. In Table 12.6.2 the principal types of ECM used against tracking radars are summarized, along with their primary effects on the victim radar. The table also indicates the appropriate ECCM responses to each of these ECM techniques.

Barrage noise jamming not originating at the target location (i.e., SOJ or ESJ platforms) can prevent target acquisition by masking the quiet target return. The combination of a low-sidelobe receiving antenna and sidelobe cancellation can counter this type of jammer. Noise jamming originating at the target itself (SSJ) denies target range to the radar, but generates an angle strobe which can be utilized by the radar to track-on-jam and initiate a weapon launch, and for missile homing. Mainlobe cancellation can be used to counter barrage jamming originating from nearby platforms (ESJ).

Narrow band (or spot) noise jamming can also be used to deny target range to the radar, and is countered in similar fashion.

Simple repeater jammers generate false targets that can delay acquisition of the desired targets and saturate the track file capacity of the radar data processor. Multiple track file maintenance and the use of target signature analysis to reject false targets are useful countermeasures.

More sophisticated repeater jammers can program the range and/or doppler of the false targets to pull the radar tracking gates away from the true target and thus disrupt the radar tracking function. Multiple track file maintenance allows the tracking radar to continue tracking on the true target as well as on the false targets, and kinematic analysis on each track file (dynamic filtering) provides false target discrimination.

Cross-polarization repeater jamming attempts to disrupt angle tracking by redirecting a signal with orthogonal polarization back to the radar. The radar sensitivity to cross-polarization jamming can be greatly reduced through use of planar antennas and radomes which provide significant rejection of cross-polarized returns. If the radar incorporates polarization diversity on receive, it can select the stronger signal, and thus achieve an angle track on the jammer. If the radar uses an arbitrary polarization on receive, the cross-polarization jammer may be rendered ineffective.

Table 12.6.2 ECM against Tracking Radar		
ECM Type	*Effects of ECM on Radars*	*ECCM Techniques*
Barrage noise • Noise	Prevent target acquisition	Low sidelobe antennas
• SSJ and Escort J.	Deny target range	Sidelobe cancellation Coherent integration Jammer strobe processing (track on jamming) Mainlobe canceller (Escort J.)
Spot noise (SSJ and Escort J.)	Deny target range Receiver saturation	As above, plus Wide dynamic range receivers.
Repeater (SSJ and Escort J.) • Simple type	Generate false targets (delay acquisition) Track file saturation	Multiple track file maintenance Target signature analysis
• Range gate pull-off (RGPO)	Break track	As above, plus Dynamic track filtering
• Velocity gate pull-off (VGPO)	Break track	As above, plus Dynamic track filtering
• Cross polarization jammer	Generate large tracking errors or break track	Planar antennas and radomes Polarization diversity
• Cross eye jammer	Generate large tracking errors or break track	Multiple phase center antennas Separate transmit and receive antennas.
• Surface bounce	Generate large tracking errors or break track	Narrow beam antennas Target signature analysis.

<table>
<tr><td colspan="3" align="center">Table 12.6.2
ECM against Tracking Radar (Continued)</td></tr>
<tr><td>ECM Type</td><td>Effects of ECM on Radars</td><td>ECCM Techniques</td></tr>
<tr><td>Special jammers
• Blinking (noise)
  jammer</td><td>Upset AGC circuits
Break track (multiple SSJs)</td><td>Fast AGC
Wide dynamic range receivers</td></tr>
<tr><td>• Skirt frequency jammer</td><td>Saturate receiver</td><td>Multiple track file maintenance</td></tr>
<tr><td>• Image frequency
  jammer</td><td>Generate large tracking errors or break track</td><td>Improved AFC
Σ and Δ channel phase matching</td></tr>
<tr><td>• Two frequency
  jammers</td><td>Generate large tracking errors or break track</td><td>RF preselection filter</td></tr>
<tr><td>• Chaff</td><td>Prevent target acquisition
Generate false targets</td><td>Doppler filtering and MTI
Multiple track file maintenance</td></tr>
<tr><td>Passive decoys</td><td>Generate false targets
Track file saturation</td><td>Multiple track file maintenance</td></tr>
</table>

Cross-eye repeater jamming attempts to disrupt angle tracking by redirecting the signal back to the radar through an interferometer type of antenna, or through the use of separate transmit and receive antennas.

Surface bounce jamming generates large elevation angle errors by illuminating the ground between the radar and the target with the repeater signal, thus forcing the radar to track the ground-imaged signal rather than the true target. The use of a narrow beamwidth in elevation can resolve the true target from the false target, and target signature analysis can provide a discrimination capability.

Blinking jammers can upset the AGC circuits in the radar, causing the receiver to be alternatively saturated and desensitized below the noise level. Use of fast AGC permits the receiver to recover from saturation quickly, and increasing the dynamic range of the receiver will reduce its vulnerability to saturation effects. Multiple blinking jammers can cause the radar to move from one jammer location to another. Maintenance of multiple track files can be a counter to this tactic.

Skirt-frequency jamming exploits phase mismatch errors in the receiver monopulse channels across the IF passband. An improved AFC that centers the signals in the IF passband permits angle tracking to be maintained on this type of jammer, and these jamming signals can be reduced or rejected through use of RF preselection filtering. Image frequency jamming and two-frequency jamming are also countered by RF preselection filtering.

Chaff is most often countered through the use of doppler filtering and MTI which permit rejection based on velocity. Both chaff and passive decoys can generate false targets that can be managed by establishing track files which will allow their ultimate rejection through trajectory or other signature analyses.

12.7 References

[12.1]   Robert J. Schlesinger, *Principles of Electronic Warfare*, Prentice Hall Space Technology Services, 1961.

[12.2]   David K. Barton, *Modern Radar System Analysis*, Artech House, 1988.

[12.3]   D. Curtis Schleher, *Introduction to Electronic Warfare*, Artech House, 1986.

[12.4]   M. V. Maksimov, et. al, *Radar Anti-Jamming Techniques*, Artech House, 1979.

Chapter 13

RADAR EVALUATION PRACTICES

This chapter discusses the radar evaluation process through a combination of analyses, simulation, subsystem testing, and both laboratory and field system testing. Analyses and simulations are based upon the radar system's design characteristics, and the measured performance of the radar is evaluated by comparing the results to the requirements of the system performance specification.

The first step in the development of a radar system is to derive from the performance specification the required subsystem design parameters; e.g., transmitter power and waveform, antenna gain and beamwidths, receiver noise factor and frequency response, and signal processor characteristics. This derivation of subsystem design requirements and parameters is performed by the radar system engineer, normally during the contract definition phase, as part of the radar contractor's initial study effort. During development and assembly of the prototype radar, testing is conducted on critical components, circuits, and subsystems to establish that these portions of the system and associated software will satisfactorily support the roles assigned to them by the interpretation of the system specification.

Following successful testing of individual subsystems, the radar is assembled into an operating prototype system, on which limited testing will normally be performed at the development laboratory or contractor's facility. These tests may include operation in the physical environments covered by the specification (e.g., temperature, moisture, vibration), and limited tests of system performance. However, the ability of the radar to meet its overall system performance specification can only be evaluated by testing the entire system in the field, which includes a realistic operational environment (of moving targets, clutter reflections, weather, jamming and other interfering signals, and possibly attempts at ESM intercept of the radar signals).

Before a radar system design or model is evaluated, the requirements for its different modes of operation and performance in each mode must be established. This means that a system-level specification must be available that describes all of the radar functions, how well each function must be performed, the regions over which these functions are to be performed, the targets to be detected, measured, or identified, and the background environment in which this is to take place.

Radar evaluation is normally made in several steps, depending on the status of the program and the resources available to the evaluator. These steps typically follow the sequence:

(1) Analysis and simulation;
(2) Subsystem tests;
(3) Laboratory tests;
(4) Field tests; and
(5) Extrapolation from tests, using analysis, simulation, and field test results.

The flow of the radar system design process and system testing is illustrated in Figure 13.1.

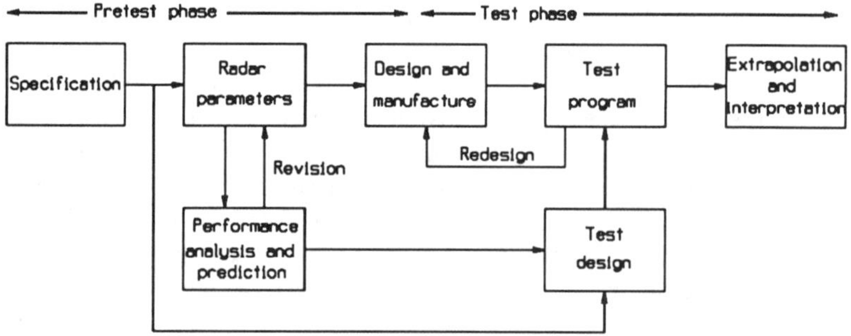

Figure 13.1 Flow of radar system design and testing program.

13.1 Requirements Analysis

A different statement of radar system requirements may be necessary for each type of evaluation. In the analysis phase, for example, the target cross section is specified by its mean value in square meters, with a fluctuation model usually selected from the Swerling cases. Atmospheric conditions will be calculated from validated models and used to adjust the signal strength in the simulator. In field tests, a specific aircraft is used, its cross section (possibly augmented) having been measured in advance. Weather conditions are observed by standard meteorological instruments, and rainfall estimated over the radar-target path. Extrapolation of radar performance to other targets is made either by scaling to their physical sizes or by using actual measurements from a radar cross-section range. Extrapolation to other weather environments relies on models or direct measurement of attenuation on communication circuits or radars similar to the radar under evaluation.

One difficulty in this evaluation process is that the specification, against which the radar system may have been proposed or developed, may have stated the requirements in only one way; e.g., detect a 1.0 m^2 aircraft target (Swerling Case 1), flying at 1 km altitude in 4 mm/h rain, with 0.9 single-scan probability of detection at a range of 100 km with a stated probability of false alarm. The procedures for analysis and simulation in this case are well defined, but the translation to a specific test target and the establishment of the actual rainfall rate under test conditions are more difficult. The evaluator must be able to translate the numerical specifications to the actual target and environment, and back again, if the field evaluation is to be meaningful.

The basic purpose of the test program is to relate the actual performance of the radar to the requirements imposed by the specification. These requirements may either be spelled out explicitly in the specification, or they may be derived from overall performance requirements or goals.

13.2 Performance Prediction

Neither the hardware nor the software is available for test in the early stages of a radar development or procurement program, and evaluation must be made on paper using analytical techniques. The necessary analyses may start from fundamental theoretical models of radar performance, or from available test data on similar radar equipment which may be adapted or improved to meet the new requirement. Some areas of radar performance are well understood, and accurate calculations of system performance can be made from known radar parameters and models of the external environment in which the radar is intended to operate. In other areas, reliable theoretical procedures have not been developed that permit accurate prediction of radar performance, and simulation or field test will be required. Even in those areas where adequate theory exists, there still remains considerable uncertainty as to the validity of models used to represent target and environmental effects, and key aspects of performance can only be validated by test. A thorough analytical evaluation is required, however, to identify the specific critical areas in which tests are necessary to resolve existing uncertainties.

The need for analytical evaluation prior to testing is based on the limited test resources available and on the statistical nature of most radar performance measures. If the test resources are spread over the entire range of conditions in which the radar is expected to operate, there is little chance that the number of data points in any one region will be adequate to arrive at statistically valid conclusions. Analysis, if properly carried out, can identify the most critical regions in which the radar design is marginal or the analytical models are weak, as well as broad regions in which satisfactory performance can be expected with high confidence. A relatively few tests in these regions, repeated to give valid statistical measures of performance, can serve to validate or correct the analytical and simulation models near their boundaries of uncertainty, and give confidence in their use to extrapolate to other cases. The purpose of testing is to resolve uncertainties. Little new information is obtained when the tests merely confirm that which was already known with high confidence.

An example (shown in Figure 13.2.1) in which the desired system parameter (e.g., detection probability, acquisition time, or target discrimination probability) is plotted as a function of target range. The range region in which the system performance parameter changes rapidly can be identified through analysis, and the test plan can be written to explore this area thoroughly. In the figure, it can be seen that testing in the vicinity of R_{50} is most likely to be productive and to validate the radar system design with minimum resources. Once the predicted performance in this region is validated, extrapolation to higher and lower levels of performance can be carried out with confidence. If affordable, three series of tests can be carried out, covering the upper knee and the lower knee of the performance curve, as well as the steep slope near the center. It must be recognized, however, that the number of tests necessary to confirm probabilities near 99% or near 1% will be far higher than those needed near 50%. On the other hand, tests scattered over the entire curve, with few samples at each level, may be entirely inconclusive, giving no confidence at any level of radar system performance.

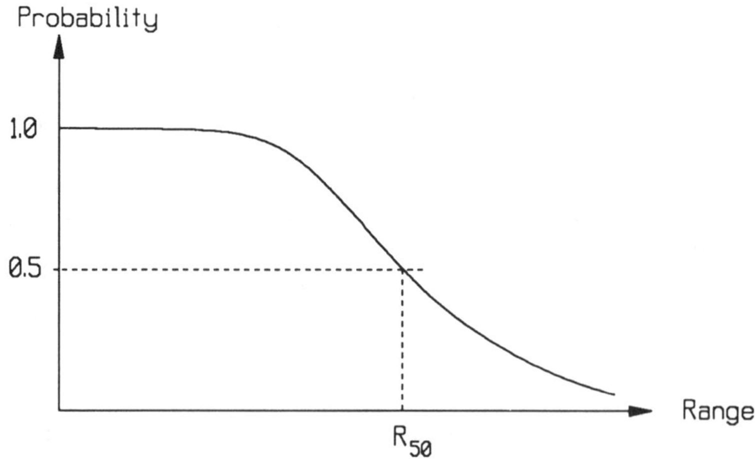

Figure 13.2.1 System parameter as function of target range.

13.3 Identification of Critical Performance Parameters

For each level of radar testing, certain performance parameter measurements are critical to achieving a thorough system performance evaluation. For example, receiver noise factor and transmitter output power are important at the subsystem level, while measurements of detection range and subclutter visibility are more important in system-level testing.

13.3.1 Subsystem Performance Parameters

The *antenna* subsystem is defined here to include the aperture, feed, and beam-control equipment. The latter would include beam pointing or scanning mechanisms, the beam steering computer of an electronically scanned array, and any radomes, shrouds, or polarization mechanisms.

The critical performance parameters for antenna evaluation include:
(a) Beamwidths and beam shapes in both azimuth and elevation planes;
(b) Scan angle coverage;
(c) Scan rates or beam position switching times;
(d) Power gain;
(e) Polarization;
(f) Sidelobe levels; and
(g) Impedance and VSWR.

Beamwidths and shapes, power gain, polarization, sidelobe levels, and VSWR typically do not vary with scan angle changes for mechanically scanned antennas. However, electronically scanned arrays are characterized by significant variations in these parameters as functions of the electronic scan angles. Consequently, electronically scanned arrays should be evaluated at a number of scan angles in each dimension (azimuth and elevation) and for combinations of scan angles in both dimensions.

The *transmitter* subsystem is defined here to include the power amplifier (or final power output device), modulator, waveform generator, and master oscillator/exciter, together with associated power supplies.

The critical performance parameters for transmitter subsystem evaluation include:
(a) Peak output power;
(b) Average output power;
(c) Bandwidth (tunable);
(d) Waveform (PRF, pulsewidth, coding);
(e) Stability (amplitude and frequency) and noise sidebands; and
(f) Electromagnetic interference (EMI).

Amplitude and frequency instabilities which are manifested as amplitude and phase modulation sidebands (both sinusoidal and noiselike) have the effect of degrading the performance of coherent radars to achieve specified subclutter visibility. *Electromagnetic interference (EMI)* is a measure of the signal levels radiated by the transmitter subsystem at frequencies other than those associated with the radar waveform within the design operating frequency band.

The *receiver* subsystem is defined here to include all of the receiver components and circuits between the receiver input (at either the antenna or T/R switch) and output (either the video signal level after final detection or the A/D converter).

The critical performance parameters for receiver subsystem evaluation include:

(a) Noise factor;
(b) Instantaneous bandwidth;
(c) Tunability; and
(d) Linearity or signal compression characteristics.

Linearity measurements are important in establishing the receiver's ability to support coherent signal processing, particularly where high SCV performance is required. In many cases, signal limiting characteristics at specified signal amplitude levels or other signal compression characteristics (e.g., output signal level proportional to the logarithm of the input signal level) are required.

The critical performance parameters for *signal processor* evaluation include:

(a) Improvement factor;
(b) Visibility factor;
(c) Dynamic range;
(d) Dwell times;
(e) Coherent bandwidths;
(f) Non-coherent integration; and
(g) ECCM.

The improvement factor and visibility factor are critical to determining the capabilities of a radar to detect targets in clutter. The dynamic range, typically determined by the number of bits carried through a digital signal processor, is also critical to determine performance in clutter and in establishing the radar's ability to accommodate large input signals without saturating.

The coherent bandwidth of a digital signal processor is determined by the processor sampling rates; dwell times relate to the amount of memory in the processor. Non-coherent integration usually places demands for modest amounts of memory, but is important in determining the visibility factor.

The ECCM capabilities of a modern radar are determined by the characteristics of many of its subsystems. The characteristics of the signal processor and the data processing/display subsystems are of special importance, however; in particular the logic associated with thresholding and alarm processing. In addition, the "post processor" or data processor are key elements in coping with deceptive types of ECM. The complexity of the circuitry and software in this subsystem generally reflects the ECM threat defined for the radar.

The *data processor* that is typically used for post processing and radar control, together with its software, is usually tested in the same manner as a general purpose computer, by means of exercise and the use of diagnostic software routines. Tests should be conducted not only to establish functionality for all radar operating modes, but also to establish proper operation of the control processor when operating in concert with subsystem control units (interface boxes).

Radar displays can be evaluated by critical observation during operation using either simulated or live targets. However, a thorough technical evaluation of the displays would require the measurement of several factors, including resolution, contrast, dynamic range, persistence (or refresh rate), stability, and brightness.

Tests and measurements for evaluating the above subsystem performance parameters are discussed in Section 13.5.

13.3.2 System-Level Performance Parameters

The critical system-level performance parameters consist of:
(a) Detection range and volumetric coverage;
(b) Subclutter visibility and velocity response;
(c) Target acquisition time;
(d) Number of targets handled simultaneously;
(e) Resolution in four radar coordinates (angles, range, radial velocity);
(f) Accuracy in four radar coordinates and in output coordinates;
(g) Electromagnetic compatibility;
(h) Vulnerability to ECM; and
(i) Target discrimination capability.

Tests and measurements for evaluating these parameters are discussed in detail in Sections 13.6 and 13.7.

13.4 Evaluation by Simulation

In many areas of radar performance the processes applied to the signals and the resulting output data are complex, involving nonlinear operations and logical paths which may not lend themselves to any reasonable mathematical analysis. While simplified models can be created to obtain approximate analytical results, simulation may be the only way to predict the performance of these processes for multiple target inputs and interference. This is especially true in areas such as target identification and multiple-target tracking.

The techniques of analysis and simulation are seldom reliable enough to permit positive decisions on radar production without validation through actual field test. On the other hand, radar designs or action features which have fundamental flaws or limitations can often be rejected on the basis of analysis alone. When particular areas of concern are identified by analysis or simulation, it is usually possible to design test programs to determine whether these areas are adequately addressed by the radar design.

13.5 Subsystem Tests

Subsystem testing is an integral part of the radar development process. Subsystem specifications are derived from the overall system performance specification, and individual subsystems are separately designed, fabricated, and tested before the system is integrated into a single operational entity. Some or all of the subsystems may be tested in the radar system evaluation phase to verify the integrity of the design or to provide additional input data to the analytical performance predictions where the desired data are not otherwise available from the designer. On the other hand, subsystem tests may be conducted for diagnostic purposes only, or to investigate the causes of overall system performance deficiencies.

The following sections briefly describe experimental methods for measuring the performance of the radar antenna, transmitter, receiver, and signal processor. Testing of the radar control subsystem (typically digital) is conducted in the same manner as any special purpose digital hardware operating in conjunction with a general purpose control processor (possibly emulated with a PC) together with the pertinent radar control, test, and exercise software.

Digital displays are considered integral elements of the digital subsystem. Analog displays (e.g., PPIs) are usually evaluated qualitatively in the course of field testing. Quantitative testing of analog displays is more complex and requires a special display evaluation facility, equipped to allow precise photometric measurements.

13.5.1 Antenna Measurements

Two methods for measuring antenna patterns are described here. The first approach utilizes a conventional far-field antenna range. Such a facility should have an unobstructed view over a large area, where bridges, towers, and other objects are sufficiently distant to minimize reflections that might limit the depth of antenna sidelobes that can be measured. An antenna range of this type requires a precisely calibrated antenna positioner in order to permit pattern cuts at any angle.

The compact range is a newer type of facility for antenna pattern measurements [13.1]. The compact range resembles the configuration of a shaped reflector antenna with an offset feed, as shown in the sketch of Figure 13.5.1. The test signal source is connected to the feed, and plane-wave propagation is obtained within the "quiet zone." The antenna under test is placed within the "quiet zone" on a positioner which moves the pointing direction of the antenna in order to make the required pattern measurements.

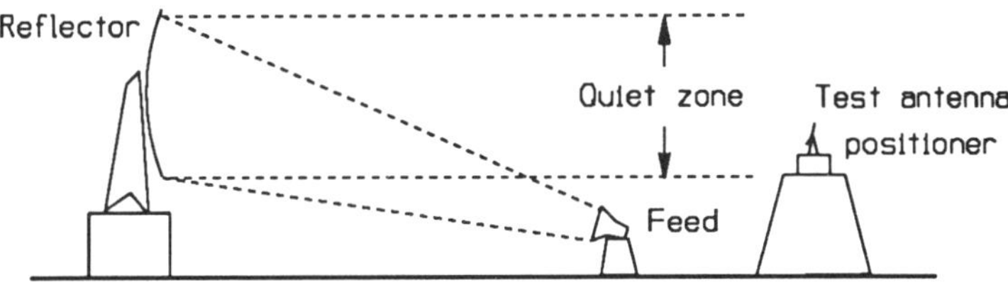

Figure 13.5.1 Compact antenna range configuration.

Modern antenna pattern test ranges (both far-field and compact types) are usually operated under computer control, including the antenna scanning regimen and antenna pattern plotting routines.

The equipment configuration for measuring antenna patterns is illustrated in Figure 13.5.2, where the option of analog or digital recording is shown. This configuration can also be used for measuring scan-angle coverage and scan rates or beam switching times. Power gains are measured by adding a standard-gain antenna near the antenna under test together with a microwave switch and a calibrated variable attenuator that are switched to the receiver, disconnecting the test antenna. Polarization measurements can be made with appropriately polarized transmitting antennas (or feeds) and a means for rotating the polarization.

Impedance and VSWR measurements can be made either at an antenna pattern range or at any location where the antenna will not be affected by nearby objects or structures (e.g., metallic objects near the aperture), or where spurious radiation of in-band signals will be a problem. These measurements are made in the same way as with any passive microwave device, such as with a slotted line.

Procedures for antenna testing are described in detail in [13.2].

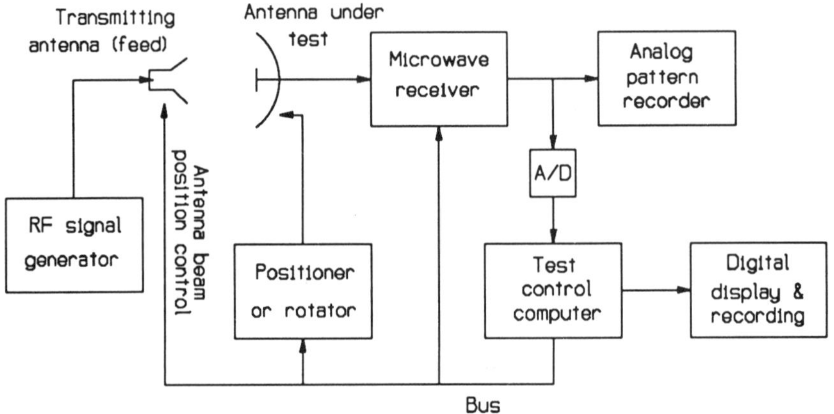

Figure 13.5.2 Antenna pattern measurement configuration.

13.5.2 Transmitter Tests

The test configuration for measuring the critical parameters of a transmitter subsystem is shown in Figure 13.5.3. For output power measurements, the instrumentation is a power meter. Tunable bandwidth measurements include output power measurements across the specified band. Frequency measurements may be made by direct connection of frequency measurement equipment to the master oscillator/exciter.

Waveform measurements require oscilloscopes [13.3], electronic counters [13.4], and spectrum analyzers. Transmitter stability and noise sideband measurements involve spectrum analyzer instrumentation, with the addition of special circuitry to null out the carrier and PRF spectral lines to accommodate the dynamic range limitations of available spectrum analyzers.

Electromagnetic interference (EMI) measurements are preferably made with the radar antenna connected, in order to include the filtering and impedance matching effects of the antenna. These measurements are part of the system laboratory tests (Section 13.6.2) and are included under the more general category of electromagnetic compatibility (EMC) testing.

Transmitter testing procedures are further described in [13.5].

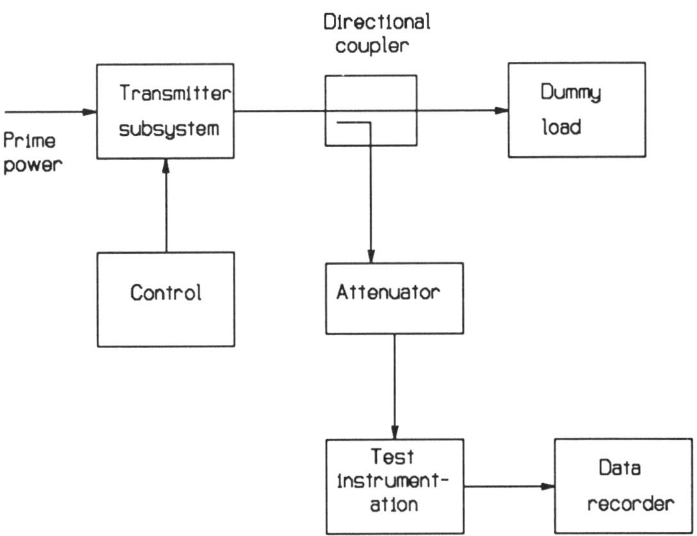

Figure 13.5.3 Transmitter test configuration.

13.5.3 Receiver Tests

The generalized equipment configuration for measuring receiver performance is shown in Figure 13.5.4. The signal generation equipment would typically consist of combinations of RF or IF signal generators, pulse generators, and waveform synthesizers. Application of typical test equipment to receiver testing is described in [13.6], [13.7], and [13.8].

Procedures for conducting receiver tests and measurements are described in detail in [13.9]. This reference also describes test procedures for measuring MTI canceller performance. These procedures are useful when the MTI is part of the analog receiver subsystem.

When pulse compression circuits are incorporated within the analog receiver subsystem, performance measurements of these circuits can be made with waveform synthesizers such as described in [13.6].

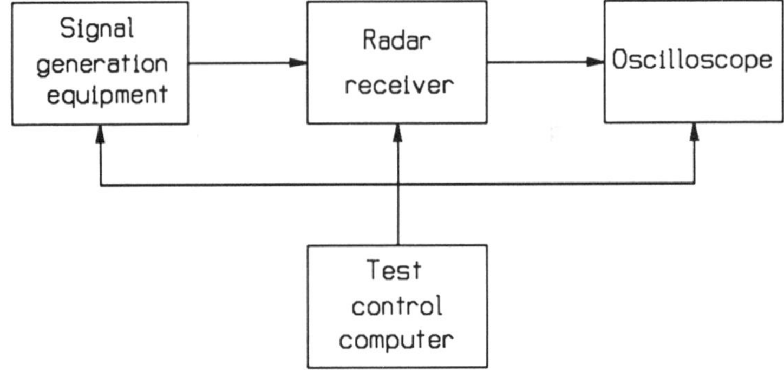

Figure 13.5.4　Receiver test configuration.

13.5.4 Signal Processor Tests

A typical equipment configuration for digital signal processor testing is shown in Figure 13.5.5. A digital computer (generally a PC) is used to control the test operation and to process the output signal data. Waveform synthesis for the simulation of targets, noise, clutter, and ECM is provided by equipment described in [13.6] and [13.7]. The A/D converter shown is the same as that incorporated in the radar (actually, there may be two or more A/Ds, to accommodate quadrature channels plus sum and difference channels). The effects of quantization and bit truncation or round-off that typically occur as digital signals propagate through the signal processor are of special importance in the evaluation of digital signal processor dynamic range and visibility factor.

The more important performance measures of the combined receiver-and-signal processor are described in Section 13.6.2.

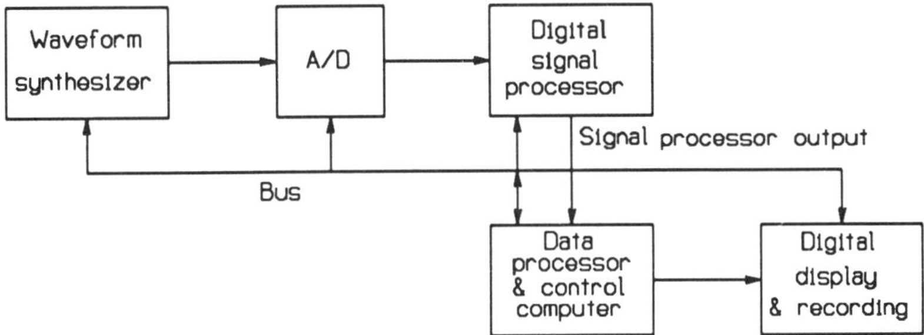

Figure 13.5.5 Digital signal processor test configuration.

13.6 System Laboratory Tests

A number of system-level parameters are best isolated and evaluated in the laboratory or at special test facilities, such as antenna test ranges. These tests involve individual subsystems or combinations of a partial set of subsystems (e.g., the receiver and signal processor). Signals representing targets, clutter, noise, interference, and ECM are generated with equipment similar to that described in Section 13.5.3.

Conducting these tests in the laboratory, rather than in the field, is more cost effective. However, such tests alone are not sufficient to fully characterize the radar system performance. The field tests of Section 13.7 are still needed to obtain the fundamental radar performance benchmarks.

Step-by-step procedures for system-level laboratory tests, sometimes called closed-loop tests, are described in [13.10].

13.6.1 Computer-Aided Testing

Radars testing involves a set of tasks that are ideally suited to ***computer-aided testing (CAT)*** [13.11]. Such testing is characterized by a well-defined set of repetitive tasks where timing, accuracy, and repeatability are important. Modern radars operate under computer control and significant portions of the receiver are implemented in digital processors. Thus, it should be recognized that access points at the input and output of each subsystem must be provided either in the original design or in the form of special interface equipment.

Waveform Simulation

In using computer-aided testing, receiver input waveforms must be simulated that have specific characteristics related to realistic target situations. The "threat model" defines the number of targets and, for each target, provides a waveform sequence with parametric variations (appropriate sum and difference channel amplitude, phase, and frequency) that correspond to the trajectory of that particular target in the threat model. Threat models can be constructed to cover the entire operating environment of the radar. The CAT computer can be programmed to record and display a wide array of subsystem data as well as system-level mission results. Since the CAT computer establishes the threat, it is a straightforward matter to display and score the resulting radar report. The threat generator should include the ability to incorporate both clutter and ECM waveforms, and to vary these inputs to cover the expected range of battlefield conditions. The CAT computer should also be capable of varying the key receiver parameters over a range of values commensurate with the variations expected as a function of environmental conditions.

The general test approach is diagrammed in Figure 13.6.1. Simulated signals are injected into the receiver and digital subsystems. The performance of the radar is determined by monitoring the digital subsystem outputs. The tests are controlled and managed by means of a *test control computer (TCC)*, a small PC-type desktop computer with the usual peripherals. The test computer is programmed to control both the simulated signal generator and the radar under test through programs written for each specific test to be conducted.

The test control computer inserts test scenario data into the radar system and extracts performance data from it by means of a *direct memory access (DMA)* interface with the radar control processor. Interrupt signals from the TCC can control the sequencing of the radar system functions. By this means the radar system operation can be halted to allow examination or modification of data in memory, mode changes can be forced to take place, and target returns and track histories can be inserted or removed.

The TCC also exerts control over the test signal simulator equipment. The timing and sequencing of target returns is defined in order to simulate target returns at desired ranges and angular coordinates. Detailed signal parameters, such as pulse width, pulse shape, and doppler frequency, are also established by this computer. Console keyboards, displays, and recording equipment associated with the TCC provide for overall test control and documentation. A completely flexible method of control of system testing can be achieved in this way.

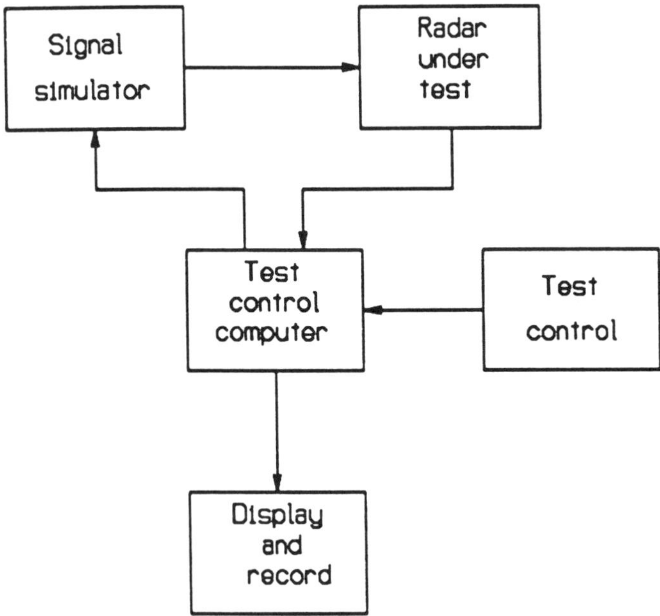

Figure 13.6.1 Laboratory testing by computer control.

13.6.2 System Laboratory Test Procedures

Specific system-level parameters that are best evaluated in the laboratory include:

    (a)   Velocity response;
    (b)   Target acquisition time;
    (c)   Simultaneous target handling capacity;
    (d)   Doppler resolution; and
    (e)   Angular resolution.

Velocity Response

The velocity response could be measured by flying live targets with known radar cross sections toward the radar at all (achievable) velocities of interest, but doing so would constitute a very inefficient use of resources. Laboratory measurements can completely characterize the relative velocity response function. A relatively small number of field measurements of SCV and detection range are necessary to calibrate the velocity response curves obtained in the laboratory.

The technique for measuring the radar's velocity response utilizing the computer assisted testing setup shown in Figure 13.6.1 is as follows. The signal simulator generates target echo pulses with varying amounts of doppler shift. These pulses are triggered by the timing signals generated in the radar control processor, so that they properly represent any staggered PRF scheme utilized by the radar. These pulses can be injected into the radar under test at either the IF or the video portion of the receiver. The velocity response is obtained by monitoring the output of the signal processor where the doppler filtering occurs.

Target Acquisition Time

Target acquisition time limits in modern radars are determined by signal processor filtering and signal integration delays, plus the delays resulting from other data processing functions associated with detection, target reporting, classification, and track initiation. If target tracking is not implemented with electromechanical servos, but with an electronically scanned array antenna, the delay times associated with antenna beam steering are negligible compared with signal and data processing delays. Track-while-scan systems introduce delays associated with the scan rates, coverage volumes, and revisit times. Target acquisition times are best measured in the laboratory, involving only the digital subsystem.

Measurement of target acquisition time with the CAT configuration of Figure 13.6.1 is as follows. The signal simulator generates a train of pulses representing the radar returns from a target. These pulses are injected into the IF or video portion of the receiver, or directly into the signal processor, in digital form. The test computer sequences the radar system through its normal operational steps of signal filtering, detection, post-detection integration, target verification, classification, and track initiation. The effects of scan coverage procedures and revisit times are included through coordinated control by the test control computer of the pulse train timing of the signal simulator and of the mode sequencing of the radar control computer. The progress of the simulated target signals through the various steps leading to acquisition is monitored and reported by the test control computer.

Simultaneous Target Handling Capacity

The number of resolvable targets that can be handled simultaneously by a given radar is determined partly by signal processor delays and, more particularly, by the speed and memory capacity of the data processor that performs target reporting, target verification and classification, track initiation, and track continuation. The limitations imposed by switching electronically scanned antenna beams from target to target are established by the dwell time requirements (integration delays) and the data processor scheduling function, rather than by antenna beam switching times (which are typically measured in microseconds).

Evaluation of a radar's target handling capability is accomplished by programming the signal simulator to generate a number of simultaneous target returns at resolvable ranges and doppler frequencies. These pulses are injected digitally at the input to the digital subsystem. Separate tests can be conducted with all simultaneous targets within one antenna beam, or distributed throughout the scan volume. The test control computer monitors the operation of the radar in filtering, detecting, and tracking these targets. Additional new target returns can be added to determine the capability of the radar system to establish new track files while maintaining existing target tracks.

Target Resolution

Target resolution in three (doppler and two angles) of the four radar coordinates is most efficiently accomplished by laboratory testing (antenna range measurements are treated as laboratory tests here), since controlling field tests with live targets spaced at barely resolvable separations is extremely difficult (the preferred method of measuring range resolution is described as a field test in Section 13.7.4). Mapping radars, including airborne synthetic aperture radars, are the exceptions to this argument, where resolution and map quality are significantly affected by the accuracy achieved by the radar in compensating for nonideal motion of the radar platform. In these cases radar maps of known scenes (including test patterns of reflectors) must be acquired in field tests and subsequently analyzed to determine resolution and map quality. These latter cases are discussed in Section 13.7.4.

Although the preferred method for measuring range resolution is described in Section 13.7.4 as a field test, an alternative approach to measuring range resolution and range sidelobe levels (assuming pulse compression is employed) with real point target returns is to inject the transmitted pulse-waveform (at either an IF level or an RF level) available from the waveform generator into the IF or RF amplifier. The timing of receiver and signal processing operations relative to pulse initiation (for each repetition interval) must be adjusted to avoid having the simulated target appear at zero range. A disadvantage of this method is that possible effects of the transmitter on the receiver response function (usually on range sidelobes) are not included in the test.

Figure 13.6.2 represents the general nature of the radar output response function in a particular dimension (x). The 3 dB width of the function is given by the dimension A, and the peak sidelobe level is given by the dimension B. The dimension A is typically taken to be the primary measure of resolution, since this value is (approximately) the minimum spacing for which two equal magnitude target signals can be resolved. Resolving a small target signal in the presence of a large one puts the small target in competition with the skirts of the response function for the larger signal, and thus requires greater target spacing before resolution is achieved. In the limit, the sidelobe levels determine the magnitude of the small-target cross section that can be detected in the presence of a large target. When the ratio of the target signal levels is greater than the mainlobe-to-sidelobe ratio (dimension B), the smaller target cannot be detected in the sidelobe region of the larger target.

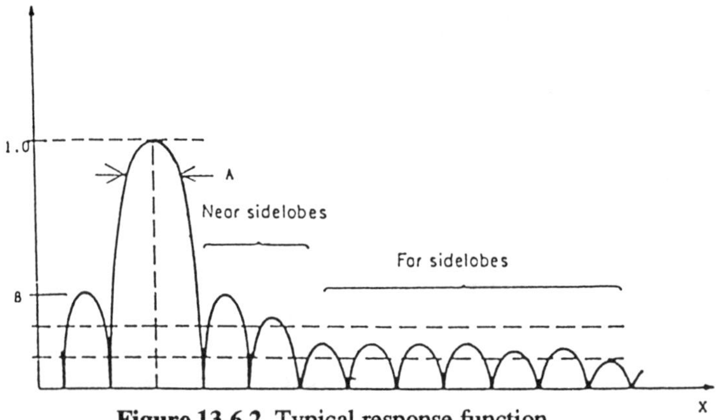

Figure 13.6.2 Typical response function.

In general, radar requirements focus on resolution of targets of equal or nearly equal sizes or signal strengths, although peak range sidelobes may be specified for pulse compression systems in order to accommodate small targets in the presence of large targets. Synthetic aperture mapping radars typically require low sidelobes in both the range and doppler frequency dimensions in order to accurately map areas of low reflectivity adjacent to areas of high reflectivity (e.g., land-water boundaries).

Doppler Resolution

The doppler response function can be directly evaluated using the measurement setup of Figure 13.6.1. The simulated target signal is injected at the digital signal processor input. The output of the signal processor is recorded as the doppler shift of the input signal is varied. A second type of test consists of simultaneously applying two signals closely spaced in doppler to the signal processor input to verify that they are resolvable.

Angle Resolution

In the azimuth and elevation dimensions, the radar response function corresponds to the antenna patterns in the two planes. Measurements of azimuth and elevation angle resolutions are obtained by antenna subsystem pattern measurements as described in Section 13.5.1. These pattern measurements can also provide the data required to assess elevation coverage for search radars which employ shaped beams in elevation (typically csc^2 shaping), and ECM vulnerability through sidelobes.

Electromagnetic Compatibility (EMC)

Electromagnetic interference caused by the radar is of concern for a number of reasons. These include interference with other radars of the same type operating within the same band but on different channels, interference with other RF systems such as other radar types, communications and data transmission systems, and the possibility of the radar being detected and located through its own spurious radiation. Determination of the radiated power levels within the radar operating band, but outside the instantaneous transmission bandwidth of the radar under test, can be made using the transmitter subsystem alone and operating into a dummy load, as shown in Figure 13.6.3.

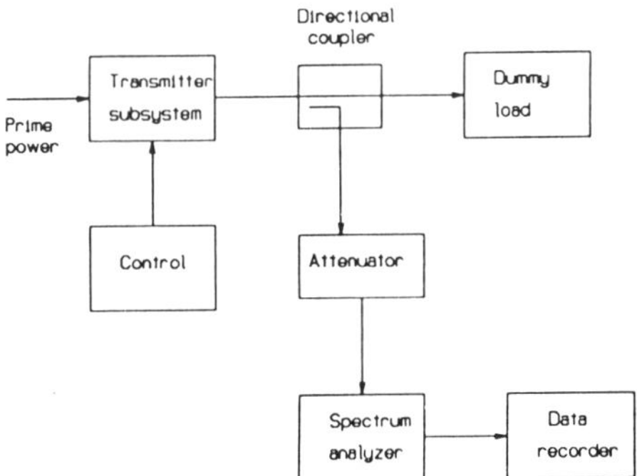

Figure 13.6.3 Transmitter test for in-band EMC.

Measurement of radiated power levels outside the radar operating band should be made with the transmitter feeding the antenna and with the spectrum analyzer connected to a receiving antenna as shown in Figure 13.6.4. The antenna is included in these tests, because it acts as a filter to out-of-band signals. Measurements should be made in the frequency regions where sum and difference plus harmonic combinations of exciter internal frequencies can produce signals outside the radar band. Measurement at the second and possibly (depending on how high the fundamental frequency is) the third harmonic of the radar (fundamental) carrier frequency should also be made. The receiving antenna can either be in the near field or far field of the radar antenna beam.

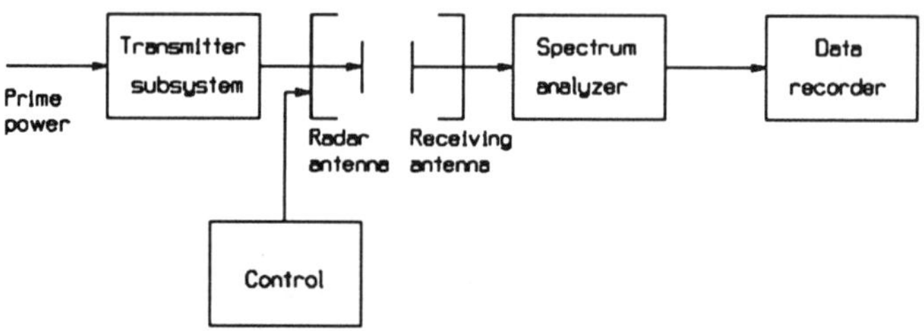

Figure 13.6.4 Transmitter test for out-of-band EMC.

Spurious radiation also includes radiation from the low frequency oscillator and counters in the exciter and waveform generator and from the digital subsystem. In addition, both in-band and out-of-band low-level radiation at the frequencies of interest may be of concern when the radar transmitter is turned off to avoid detection, location, and *anti-radiation missile (ARM)* attack. Measurements for this case can be a repetition of the previously described tests, with the transmitter final amplifiers turned off.

ECM Vulnerability

Testing radar vulnerability to ECM can be accomplished in the laboratory by coupling simulated ECM signals directly to the radar receiver (at the antenna or RF amplifier) and injecting simulated targets at the same point. This technique allows testing without exposing the threat scenarios to possible compromise, but it has the disadvantage that live target characteristics and actual clutter cannot be included in the tests without special simulators which may be unduly complex. Hence, field testing for ECM is preferred if intercept of ECM signals by a potential enemy is considered to be an acceptable risk.

Target Classification

Although field testing for establishing target classification effectiveness is preferable, it is also possible here to use simulated or recorded target signals, typically injected at IF or video. Target classification, depending upon types of targets and radar functions, may be made by analysis of phase or amplitude modulations (e.g., turbine compressor blade modulations) or of extremely fine resolution (usually in range). Simulations require faithful modelling of such characteristics, and recordings of live target returns (if available) need to be excellent reproductions in phase and amplitude and possibly have very wide bandwidths. The advantage of the simulation approach is that it can yield representative data when live targets of interest are not available for discrimination evaluation.

13.7 System Field Testing

Radar field testing requires that a complete operating radar system be placed in an open environment in which targets, clutter, and other sources of interference can be coupled through the antenna to the transmitter and receiver. Because the costs for field testing are usually much greater than for laboratory tests, and actual test conditions are more difficult to control, field testing should be reserved for evaluation of major radar performance parameters in which the electromagnetic test environment is a critical factor.

In general, field test conditions are difficult to describe accurately, especially those related to clutter and propagation effects. Given the limited resources usually available for testing, it is preferable to perform repeated tests under a few well defined conditions. This will permit adequate statistical validation of system performance for those conditions. This avoids spreading the test resources over all possible combinations of operating conditions and ending up with insufficient data and questionable results. Utilizing the results of a limited set of field test conditions, it is possible to use analytical models to extrapolate radar performance to conditions not too distant from those covered by the field test data.

13.7.1 Design of Field Tests

After suitable analysis, tests may be designed to measure each major system performance parameter. The test design includes specifying such factors as:

(a) Target type, size, velocity, flight path;
(b) Radar location, mode, waveform, scan procedure;
(c) Environment — weather, ECM, clutter (determined by location);
(d) Presence of multiple targets, real or simulated; and
(e) Types of data to be recorded.

The test design is not carried out to the level of step-by-step procedure, but rather is a description of the test objectives, general methods to be used, and conditions of the radar and its environment. Step-by-step procedures are described in [13.10].

13.7.2 Evaluation of Field Test Data

The type, range, and quality of test data expected should be specified during the preparation of the test plan and procedure as well as methods to be employed for evaluating the data. For example, in the case of target position, it may be specified that target position data from the radar under test will be transformed to Cartesian coordinates, compared with range instrumentation radar (e.g., AN/FPS-16) data, and differences calculated. The mean and standard deviations of these differences would be the output. If the desired data are errors in radar spherical coordinates, the AN/FPS-16 angle data would be converted to spherical coordinates at the test radar, differences obtained, and the mean and standard deviations of those differences would be the outputs.

13.7.3 Interpretation of Test Results

The final step in the system field test process is the interpretation of the test results in terms of subsystem parameters, such as transmitter power, receiver noise factor, or antenna gain and pattern. This interpretation can lead to decisions as to the next lower level of radar testing, including the detailed evaluation of subsystems, circuits, and components to determine reasons for any departure from the specified or expected system performance. It can also lead to discovery of defects in the test procedure or environment of the test, providing a guide to retesting under more accurately controlled conditions.

13.7.4 System Field Test Procedures

System level parameters that are best evaluated in the field include:
(a)   Detection range and volumetric coverage;
(b)   Subclutter visibility (SCV);
(c)   Range resolution;
(d)   Angle-tracking response;
(e)   Angle-tracking accuracy;
(f)   Susceptibility to electromagnetic interference;
(g)   Vulnerability to ECM;
(h)   Target discrimination; and
(i)   Vulnerability to interception and location.

Detection Range and Coverage Measurements

The best system-level test for radar detection range is conducted with the radar in full operation and with live targets having calibrated radar cross sections. The radar should be sited to permit an unobstructed view of the region where the single scan probability of detection of the test target is expected to be 50% as well as to provide safety of (aircraft) target operation and accurate position determination. Clutter levels within this region should be well below the level for which the detection performance becomes clutter limited (i.e., the clutter-to-noise level should be at least 10 dB less than the SCV capability of the radar).

The radar should be sited at a location where the foreground terrain is moderately rough (such as grassy, scrubby, or plowed fields), in order to avoid forward scattering from more reflective ground that may introduce lobing in elevation coverage patterns. The latter effect will result in increased detection ranges at the lobe peaks and reduced ranges midway between peaks.

The best calibrated targets for these purposes are metallic spheres, corner reflectors, or Luneberg lenses. In most cases, the reflector will be mounted on an aircraft, so that it can be easily positioned and moved through the range interval of interest, and its approach speeds controlled either to position the target velocity or to know the location of the target within the velocity response function of the radar. Ideally, the target speed is controlled so that its approach velocity occurs at the optimum velocity point on the velocity response function.

Preferably, a small aircraft will carry the reflector, in order to minimize the contribution of the aircraft radar cross section to the overall target cross section, since aircraft cross sections vary significantly with small changes in aircraft aspect angle (toward the radar). Also, a calibrated constant cross section point target (over a wide aspect angle), avoids the issue of selecting the appropriate Swerling fluctuation model. Of the three examples listed, the corner reflector and Luneberg lens are especially appealing, in that each provides a large cross section as a result of its effective antenna gain. The cross section of such a reflector will dominate that of a small aircraft. A reflector-to-target ratio of at least 10 dB is desired.

One problem that a large cross section test target may present is that the radar detection range against the target, in the region of interest (Pd = 50%), may occur beyond the unambiguous range of the radar if it has been designed to detect very low cross section targets (e.g., an artillery locator). In this case, the radar's data processing software may have to be modified to overcome any technique employed to reject ambiguous range targets, and the range ambiguities would have to be accounted for in data evaluation.

In order to measure the volumetric coverage of the radar, a number of test runs should be made at each of three or more target altitudes to obtain elevation coverage profiles. Target position data can usually be obtained from the aircraft navigation equipment, but can also be obtained from other measurement systems, such as a nearby tracking radar. Target velocity data can be obtained either from the aircraft on board instruments or from the radar itself (if velocity measurement is a function within the digital subsystem).

In determining the volumetric coverage of a radar with an electronically scanned antenna, three or more azimuth scan angles should be evaluated, since antenna gain varies with scan angle. In this case, the antenna (or radar) may be mechanically rotated if necessary to keep the test target operations within the assigned flight zone.

Subclutter Visibility (SCV) Measurements

Although the most realistic measurement method is to operate the radar against an actual target and a live-clutter background, this approach presents some difficulties based on the uncertainty of the clutter signal level within the target resolution cell at any specific time during the test. The preferred method is to use actual clutter signals which are phase modulated to simulate a moving target [13.12]. In this procedure, as shown in Figure 13.7.1, a calibrated phase modulator is inserted in series with the first local oscillator the IF COHO. By shifting the phase between alternate pulses by the same amount, the received clutter signal is modulated to achieve the effect of a target at optimum velocity (doppler frequency equals 1/2 PRF). This optimum phase is superimposed upon all the received clutter (i.e., within all range gates). The simulated target-to-clutter ratio is dependent on the degree of phase modulation, and the optimum phase condition results in pure phase modulation without accompanying amplitude modulation.

Radar siting for SCV tests can be identical to that used in the detection range measurements, provided that one or more strong single points of clutter (e.g., towers, water tanks, mountain cliffs) are within radar view and at sufficiently close range to provide clutter return signals equal to or greater than the ratio of SCV over the receiver noise level.

An alternate SCV test procedure, which has the advantage of not requiring a special phase modulator, is to superimpose (at RF or IF) a signal-generator synthesized target on live clutter. If pulse compression waveforms are used by the radar, this approach requires elaborate waveform generation equipment. Also, the test-target signal should be gated in range and angle to place it coincident with the point clutter. Measurement errors may result from misalignment between the two signals.

Depending on the emphasis placed on radar operation in rain, snow, and chaff, as well as the design of the velocity response function to reduce their effects, it may be necessary to repeat SCV measurements under these conditions. This results in added difficulties of test calibration and reduced measurement accuracy caused by the time variations of the clutter characteristics (cross sections, spectra).

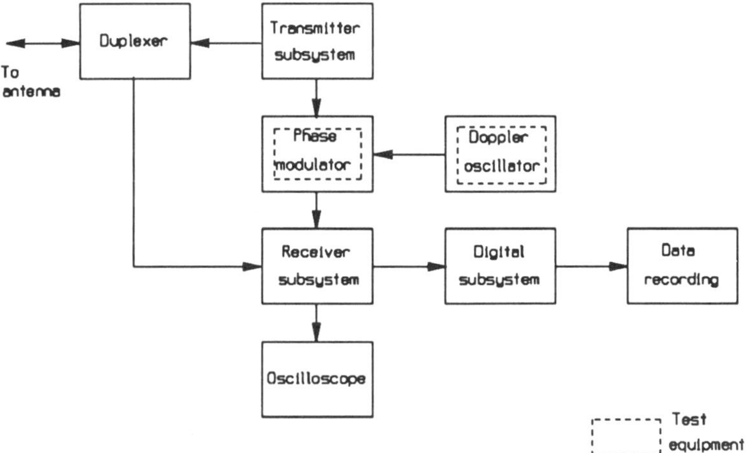

Figure 13.7.1 Subclutter visibility measurement.

These forms of clutter have similar characteristics (see Section 2.2.3). Their doppler spectral widths are dependent upon the wind shear and turbulence. Unlike ground clutter, the clutter return from rain, snow, and chaff is offset in doppler frequency as a result of the average wind speed component in the direction of the radar.

Range Resolution Measurement

The preferred method for measuring range resolution and range sidelobes with a radar employing pulse compression is to use an isolated point clutter source similar to the clutter signal used for the SCV tests. This approach offers an advantage over the laboratory test described in Section 13.6.2 in that it includes the effects of the transmitter power amplifier, such as amplitude limiting and phase ripple, particularly as they affect range sidelobes. Figure 13.6.2 applies to the range dimension (x = range), where the sidelobes are the time sidelobes of the pulse compression process.

Range resolution measurements can be made using the test configuration for the SCV tests, but without moving-target simulation. With a sufficiently strong point-source return, typically more than 40 dB above the receiver noise, the range gate outputs are sampled by the signal processor over a range interval centered at the target range and having a width twice that of the transmitted pulsewidth. A test point within the signal processor may provide these data for observation on an oscilloscope, or the data may be available to the radar display for diagnostic purposes.

Angle Tracking Response

The closed-loop response of an angle tracking servo, be it analog or digital, including the antenna and receiver, is best evaluated on a target signal radiated from a boresight tower. These tests may be conducted with either a single-function tracking radar, or with an electronically scanned radar operating in a tracking mode. Two types of response should be measured — step response and frequency response.

To conduct step response measurements, a target signal is radiated at various levels above noise, and for each signal level the radar is locked onto this signal in all tracking coordinates. The angle-tracking loop is then switched to manual, and the beam is displaced a small fraction of a beamwidth in azimuth from the target. The azimuth data are recorded while the loop is switched to automatic track, and the step response is observed in the recorded data. The test is repeated in the elevation coordinate.

For frequency response measurements, the angle loop is locked on the boresight tower source as in the step-response test. A low-frequency function generator signal is added into the error signal amplifier, and adjusted to produce a small fraction of a beamwidth in angle motion at a single low frequency (e.g., 0.1 Hz). Peak beam deflection is recorded as the frequency is varied up to and beyond the nominal loop bandwidth. Azimuth and elevation loops are both tested in this manner.

Servo and Receiver Noise Measurements

To measure servo and receiver noise, angle output data are recorded while the radar is tracking the boresight tower signal, without externally generated error signals. Samples of data are taken in each coordinate as the signal level is reduced in steps to the minimum at which tracking is possible. The rms value of beam position error, relative to the source, is determined for each value of signal-to-noise ratio. The limiting value for high snr will indicate the level of noise in the servo loop, while the curve of position error vs. snr will measure the antenna and error detector slope constant for thermal noise.

Angle Tracking Accuracy Measurement

The purpose of this test is to establish the absolute angle accuracy of the radar system under field operating conditions. Previous analysis will have established the expected contributions of the following major error sources:

(a)   Thermal noise;

(b)   Target noise (glint and scintillation);

(c)   Clutter;

(d)   Jamming;

(e)   Multipath; and

(f)   Instrumental error in the radar.

These components of error will vary with the scenario, and a limited number of realistic scenarios must be defined which will permit evaluation of the radar angle response to each component.

Thermal Noise Angle Error Measurement

To conduct tests of the effects of thermal noise on tracking accuracy, a well-defined point target is provided by mounting a corner reflector, Luneberg lens, or repeater on a small target aircraft or drone. The target is flown at a range such that S/N ratios near + 10 dB are observed at elevations high enough to avoid ground clutter and multipath errors. The standard deviation of the angle data from a smoothed target trajectory provides a measure of thermal noise, at low S/N ratio, and the limit as S/N ratio is increased represents the instrumentation error of the radar system.

Clutter and Multipath Error Measurement

For this test, the well-defined target of the thermal noise test is flown at reduced elevation angles to produce errors from clutter and multipath. The target trajectory is selected to pass from a region in which strong clutter is coincident in range with the target to a region in which the target is relatively free of clutter. The terrain contributing multipath errors should be as uniform as possible, to permit the isolation of the clutter error component as clutter varies.

Recorded target data are compared with reference data from optical tracking systems or instrumentation radars, keeping in mind that the instrumentation radar may also have multipath errors at low elevation angle. If the siting is chosen to permit the instrumentation radar to track at shorter range (and higher elevation angle) than the radar under test, this problem may be minimized. The differences between recorded target data and the true target position represent the combined effects of the several sources listed above, plus possible atmospheric refraction, which can be calculated and removed from the data. The thermal noise and target glint contributions can be minimized by use of a large reflector or other source. The target trajectory is selected to produce clutter and multipath signal levels which exceed the radar instrumentation error.

The clutter and multipath error components can also be separated from each other by running the same test trajectory with larger reflectors; this has the effect of reducing the relative importance of clutter. Small reflectors cannot be used to increase relative clutter unless the physical size of the target vehicle is small enough to avoid significant glint and scintillation components.

Measurement of Errors Due to Target Noise

For these measurements, a normal target flown without cross-section enhancements yields realistic glint and scintillation effects. The trajectory is selected to provide a strong signal from an elevation high enough to minimize clutter and multipath. Recorded data are analyzed to obtain rms deviations from the true trajectory.

Flights at different ranges can be scheduled to isolate the glint errors from scintillation errors. Glint is represented by a constant deflection of target position in meters, and hence the angle error varies inversely with range. Scintillation produces a constant angle error, independent of range, if the target scintillation spectrum (i.e., rate of rotation about the radar line of sight) remains constant. This rate can be held essentially constant by flying the target with a rate of turn which greatly exceeds the azimuth angle rate observed by the radar.

Measurement of Susceptibility to Electromagnetic Interference (EMI)

The radar specification may cite EMI levels over the full electromagnetic spectrum (DC to millimeter wavelengths) to which the radar may be subjected without impairment of radar performance, as a result of excessive spurious signal levels entering the receiver, exciter, waveform generator, digital signal processor, data processor, or display.

The EMI test consists of the generation and radiation of CW and pulse signals at the specified frequencies (or frequency bands) and power levels measured at the outside of the radar shelter. For these tests, the radar is in operation, except for the final power amplifier. As the test signal frequencies, power levels, and wave forms (CW, pulses of various widths, and PRFs) are varied, the IF receiver and radar output is monitored and recorded.

Future radars may be subjected to intentional disruption by directed high-powered electromagnetic energy. Typically, the frequencies of concern range from the low end of the radar band into the millimeter wave region (above the waveguide cutoff frequency). The effects of such radiation may include receiver saturation and burnout. Testing for these effects is essentially identical to other EMI tests, with the far-field generation of the specified energy levels measured at the radar antenna. In this case, simulated targets should be injected at RF, either by a signal generator located in the antenna near field or from a boresight test tower. The radar output data and display is monitored for saturation effects (desensitization), front-end burnout (total loss of signal), or excessive false alarms. An alternative to simulating target signals is to operate the radar at full power and monitor returns from strong fixed targets (clutter).

Tests for Vulnerability to ECM

A variety of ECM techniques may be employed by attack aircraft to defeat certain types of battlefield radars, or by ground vehicles to defeat airborne (and RPV) radars. ECM techniques that may be used against search and acquisition radars include repeaters or noise pulses to saturate target handling circuits, and noise jammers for the same purpose or to desensitize detection circuits. Techniques used against tracking radars include repeaters which effect range-gate and velocity-gate pulloff, the so-called phase-front techniques (cross polarization, cross-eye, and ground bounce), and noise jammers. High-power standoff jammers are typically deployed well behind the FLOT in a number of large aircraft at high altitude. They generally produce barrage jamming with wideband noise to counter all of the opposing radar types within the battle area, and attempt to introduce strong enough signals through the radars' sidelobes to greatly reduce radar detection against small cross section penetrators regardless of the radar's mainbeam orientation.

The procedures for testing the radar vulnerability to ECM are essentially the same as those employed for detection range measurements and live target tests for tracking accuracy. Actual "live" jammers on board test aircraft can be used, or simulated ECM signals can be injected into the radar antenna by means of a small antenna (e.g., a horn antenna) in front of the radar antenna. The simulator signal generator may consist of a set of waveform generators modulating an RF power source or it may be a set of actual jammers.

In conducting ECM vulnerability tests of search and acquisition radars or search and acquisition modes, detection range limitations and target acquisition probabilities should be determined as a function of the type of ECM and the ECM power level. An alternative is to establish the ECM power levels at the antenna corresponding to the specified threat and then determine if the radar performance requirements, such as detection range, are met.

Evaluation of the effects of ECM on tracking radars should be based on the recorded tracking errors as a function of ECM levels compared to the specification, and on ECM levels at which tracks are broken or lost.

Radars designed to cope with strong ECM environments may incorporate antenna sidelobe cancellation techniques or adaptive antenna arrays to negate the effects of a few powerful jammers that would otherwise upset radar operation by entering the receiver through antenna sidelobes. Standoff jammers are of major concern in this regard.

Sidelobe cancellers and adaptive arrays can be effectively evaluated only by testing on an antenna range, or in conjunction with field measurements of detection range and tracking accuracy, with the simulated jammer sources in the radar antenna far field. Thus, radars incorporating these ECCM techniques would be tested as previously described using live targets, with simulated jammers or real jammers located at diverse points in the far-field regions of the radars. The number of jammer sources used in each case should equal the number called out in the radar performance specification, since cancellation performance and other radar performance measures (e.g., tracking accuracies) are dependent upon the number of jammers and number of canceller loops employed. Noise jamming sources are usually preferred for these tests, since SOJs are often the principal threat.

Testing Vulnerability of Radars to Interception and Location

Radars used for battlefield reconnaissance, intrusion detection, navigation, air defense, and hostile weapon location have long been the target of intercept and warning receivers, and more recently of *anti-radiation missiles (ARMs)*. Their vulnerability to these EW devices results from the relatively large two-way path loss involved in the basic radar-target geometry compared to the one-way loss for the intercept receiver. This one-way advantage of an intercept receiver located on the target allows the intercept receiver to operate over an entire radar band (e.g., $B = 1$ GHz) or even over an octave band (e.g., 8 to 16 GHz), while retaining high probability of intercept of the radar signal, despite a resulting mismatch to the radar transmission.

Some modern radars are designed to use special frequencies, waveforms, low-sidelobe antenna patterns, and operational procedures to reduce the vulnerability to intercept. Such radars are described as *low probability of intercept (LPI)* radars. Options in the design of these radars include the following:
  (a) Reduced average power, or "power management";
  (b) Reduced peak power (e.g., CW or modulated CW);
  (c) Wide signal spectra (e.g., wideband modulation, frequency hopping);
  (d) Low transmitting antenna sidelobes;
  (e) Sporadic transmission;
  (f) Selection of frequencies in densely occupied bands;
  (g) Selection of frequency and waveform to simulate communication signals;
  (h) Selection of frequencies in atmospheric absorption bands.

The process of evaluating vulnerability to signal interception of both conventional and LPI radars is closely related to the evaluation of the intercept receivers. Indeed, many LPI radar analyses depend on postulation of intercept-receiver characteristics which are mismatched to the radar signal. In the absence of a set of standard intercept receiver characteristics against which to evaluate a radar, any radar with a relatively wideband waveform can be claimed to have "LPI" properties.

An initial step in the evaluation of the vulnerability of a radar to signal intercept is to establish the intercept receiver threat models against which the radar is to be evaluated. This can be by analysis, computer simulation, laboratory test, or field test. The definition of these models must be made by an organization familiar with the existing intercept receiver technology, with future trends and forecasts in this technology, and with its probable deployment. Different threat models are needed for strategic reconnaissance, tactical reconnaissance, radar warning receivers, look-through receivers of ECM systems, and ARM seekers.

Most LPI techniques are valid only in dense signal environments, where the intercept receiver is, in effect, jammed by a multiplicity of irrelevant signals. A realistic evaluation of radar vulnerability to intercept should be made with the radar signal immersed in a realistic set of such interfering signals. Again, models for this environment are best produced by those organizations which deal with complex EW scenarios.

The vulnerability of the radar to signal intercept depends, in many cases, on the physical surroundings of the radar. Factors include the following:
    (a)   Atmospheric and weather attenuation of the signal;
    (b)   Scattering of the signal by terrain or cultural features;
    (c)   Scattering of the signal by rain or chaff; and
    (d)   Masking by terrain.

For example, if low-sidelobe antenna features are used to prevent the signal from reaching the intercept receiver, scattering from sources within the main beam can provide unwanted paths to the receiver. Accuracy of location of the victim radar in this case may not be adequate for the weapon system, e.g., ARM seeker, but the receiver may be cued to the correct frequency and waveform by the scattered signal.

Models are available for analysis and simulation of these environmental effects. These should be supplemented by field and laboratory tests to validate the models. The procedure for testing conventional radars and LPI radars for vulnerability to ARM attack consists of operating the radar at the field test site, with signal generator and simulators used to represent additional nearby radars and other signal sources of the battlefield environment. Generic intercept receivers attempt to lock on the radar under test. Before initiating full-scale field tests, laboratory tests should be conducted to explore the wide range of variables, such as number of emitters and location, in a non-real-world environment to limit the scope of the field-test effort. These laboratory tests involve special LPI waveform generators to simulate signals from certain new radar designs, antenna pattern generators to simulate low sidelobes of new antenna designs, and multipath scattering models to simulate the effects of different standard terrain situations and radar siting options.

Target Discrimination Tests

Future radars may incorporate techniques for discriminating among different types of targets for purposes of classifying or identifying them and establishing threat priorities. These techniques typically exploit unique phase, amplitude modulation, or extremely high radar range resolution effects of given target types.

The preferred method for evaluating target classification techniques is to use live targets of the types of interest, including both desired and false targets. These tests may be conducted in the same manner as the radar detection range measurement tests, with the evaluation criteria based on target discrimination or classification specifications.

Section 13.6.2 describes an alternate approach to this test, where simulation or recorded target signals would be used when live targets are not available. Another possibility is to conduct field tests with available targets and augment these tests with recorded target signals if they are available to cover the cases where live targets cannot be used.

Target Acquisition Time

An alternate method to laboratory measurement of target acquisition time of tracking radars, electronically scanned systems that perform both search and track functions, or track-while-scan systems, is to use a live target. This test can be accomplished by recording the data processor outputs (including a time tag on each data report, if the output data do not already include time) and observing the time difference between the first reported detection of a target and the report of target track being established.

13.7.5 Airborne Radar Field Tests

The normal operating conditions for airborne radars establish special requirements for conducting field tests, since these include in-flight operation. However, many field tests can and should be run on the ground in order to avoid undue expense of flight operations.

Essentially, all of the previously described laboratory and field tests can be conducted on the ground, with the airborne radar preferably located on a high mount, such as a rooftop or at the edge of a cliff. In such a situation, ground effects, multipath propagation, and clutter are similar to those observed when the airborne radar is installed in an aircraft flying at low altitude.

The SCV measurements effected in a "rooftop" environment represent the inherent capabilities of the radar and must be interpreted in the light of typical flight conditions. For example, SCV in actual flight is lower than that measured in a fixed or rooftop environment as a result of clutter spectral spreading caused by aircraft motion and imperfect own-ship velocity compensation by the *inertial measurement unit (IMU)*. However, the rooftop measurement can represent an important benchmark. For a given set of radar characteristics, including velocity response, only the velocity compensation subsystem will contribute to SCV deterioration below that which can be analytically derived from the above benchmark measurements and the radar characteristics.

Airborne flight tests should be conducted under representative clutter conditions with live targets to obtain qualitative and semi-quantitative (e.g., knowing the type of clutter and approximate moving target cross section) measurements. These observations and test results can then be compared with the expected performance measures derived from benchmark tests.

Of particular interest are the cases of airborne mapping radars, including those deployed on RPVs. For these radars, non-coherent mapping can be evaluated by flying over representative target areas and recording maps with whatever recording equipment is contained within the system or with additional recording equipment.

Synthetic aperture radars (SAR) can only be fully tested for map quality, resolution, and accuracy by flight testing. The mapping and recording procedures apply here as well; however, the scenes selected should provide the opportunity to observe and measure the finest resolutions specified. Special SAR resolution and accuracy test facilities are available for such measurements [13.10, p. 107].

13.8 References

[13.1]   R. C. Johnson, H. A. Ecker, R. A. Moore, "Compact range techniques and measurements," *IEEE Trans.* **AP-17,** 1969, pp. 568-576.

[13.2]   International Test Operations Procedure (ITOP) 6-2-020 Radar Antenna Tests, U.S. Army Test and Evaluation Command, 15 April 1988.

[13.3]   Radar System Characterizing and Testing, HP5185A Waveform Recorder, HP5185T Digitizing Oscilloscope, Application Note 313-10, Hewlett Packard Company 02-5952-7905, March 1988.

[13.4]   Radar System Characterization and Testing, Using the HP5345A Electronic Counters, Application Note 174-14, Hewlett Packard Company May 1987.

[13.5] International Test Operations Procedure (ITOP) 6-2-530, Radar Transmitter Procedures, U.S. Army Test and Evaluation Command, 1 May 1988.

[13.6] Receiver Testing with the HP8770S Arbitrary Waveform Synthesizer System, Hewlett Packard Application Note 314-1.

[13.7] Vector Modulative Measurements, Coherent Pulsed Tests of Radar and Electronic Warfare Systems, Hewlett Packard Application Note 343-3, November 1, 1986.

[13.8] Applications and Performance of the 8671A and 8672A Microwave Synthesizers, Hewlett Packard Application Note 218-1, September 1979.

[13.9] International Test Operations Procedure (ITOP) 6-2-529, Radar Receiver Procedure, U.S. Army Test and Evaluation Command, 15 April 1988.

[13.10] Advanced Radar Testing Methodology Investigation II, TECOM Project No. 7-CO-R88-EPO-007, ANRO Engineering Consultants, Inc., September 30, 1988.

[13.11] Introduction to Computer Aided Test, Hewlett Packard Application Note 316-0, May 1983.

[13.12] H. R. Ward, "Clutter Filter Performance Measures," IEEE International Radar Conference, April 1980.

Appendix A

GLOSSARY OF RADAR TERMS

This glossary is intended as a guide to terminology for the working engineer, rather than as a formal standard of usage. Except when indicated as having been derived from one of the standard dictionaries [A.1]-[A.3], the terms have been given simple definitions which should be adequate to promote understanding by users of the *Radar Evaluation Handbook*.

acceleration error constant. A servomechanism or tracker constant which relates the dynamic lag error component to the acceleration of the tracked target. The lag error component is the target acceleration divided by the acceleration error constant.

accumulator. A digital circuit or process which adds successive binary inputs over an integration interval.

active decoys. A technique in which jamming is radiated from a point removed from the target, either towed behind the target or dropped from the target and expended.

adaptive cancellation. The process by which unwanted interference is cancelled in a radar antenna or circuit, based on real-time estimation of the interfering environment, leaving the desired signal with minimum degradation.

aliasing. The conversion of frequency components of a sampled signal, from frequencies above one-half the sampling rate to frequencies between zero and one-half the sampling rate.

alpha-beta (α-β) tracker. A recursive, sampled-data tracking algorithm in which predictions are made from past data and updated with present data at the time of reception of a new sample.

ambiguity function. A function representing the response of a matched filter to a particular radar waveform, as a function of the time delay and doppler offset of the received signal.

analog-to-digital converter. A device which converts analog signals (voltages, in radar applications) to digital form.

aperture. A surface, near or on an antenna, on which it is convenient to make assumptions regarding field values for the purpose of computing the field at external points [A.2].

aperture distribution. The variable intensity of radiation from different portions of a transmitting aperture, and of weighting applied to the received wave.

aperture efficiency. The gain of an antenna, relative to that which would have been achieved with uniform illumination of the same aperture.

array pattern factor. The antenna pattern effect due to the illumination function across the array, as opposed to the element factor.

aspect angle (of a target). The angle between the radar line of sight and the principal axis of the target object.

atmospheric attenuation. The loss in RF energy of a signal caused by transfer of electrical energy to heat in the molecules or ions of the atmosphere.

atmospheric lens loss. Reduction in the power of the radar signal caused by variable refraction in the atmosphere, which spreads in elevation the energy leaving the antenna at low angles.

backscattering coefficient. Monostatic radar cross section of a target.

barrage noise jamming. Noise jamming which is spread across the tuning range of the radar, or the entire band occupied by several radars.

baseband. The band near zero frequency to which RF or IF signals may be converted.

Beaufort wind scale number. A number from 1 to 11 which indicates the wind condition. Beaufort 1 is "light air," while 10 is a "whole gale."

beam pattern. A curve representing antenna (voltage or power) gain as a function of angle in a selected coordinate.

beam pattern constant. A constant relating the antenna gain to the product of beamwidths.

beamshape loss. A loss factor included in the radar equation to account for the use of the peak antenna gain in the radar equation instead of the effective gain that results when the received train of pulses is modulated by the two-way pattern of the scanning antenna [A.2].

binary integration. A video integration process in which the individual pulse signals are first reduced to binary form, before being added in a digital accumulator.

binary integration loss. The ratio of the S/N ratio required at the input to a binary integrator to the S/N ratio required at the input of the envelope detector of a video integrator, to achieve a given detection performance.

bistatic. Referring to a system in which the transmitter and receiver are at different locations.

blind speed. A target radial velocity at which an MTI or doppler system has a null response.

blinking noise jammer. A noise jammer which is turned on and off with timing to enhance its effectiveness or to prevent track-on-jam or home-on-jam.

blip-scan ratio. The single-scan probability of detection of a target.

blockage. The shadowing of portions of the antenna aperture by structure (e.g., the feed system in a reflector).

blockage efficiency. The ratio of antenna gain to that which would be achieved by the same aperture in the absence of blockage.

Boltzmann's constant. The number k that relates the average energy of a molecule to the absolute temperature of the environment; k is approximately 1.38×10^{-23} joules/kelvin [A.2].

Brewster angle. The grazing angle at which the real part of the Fresnel reflection coefficient of a particular surface material passes through zero.

bright-band effect. The increase in reflectivity of precipitation clutter at altitudes where melting of the precipitation occurs.

burn-through range. The range at which the radar echo signal becomes detectable against the competing noise jamming.

carrier. The sinusoidal waveform to which amplitude or frequency (phase) modulation is applied.

cardinal plane. A plane containing the beam axis and one of the major aperture axes.

Case 1 target. A target model characterized by a Rayleigh distribution of reflected signal amplitude (exponential distribution of power), fluctuating slowly enough to provide complete correlation of pulse amplitudes over a single observation time of the radar.

cell-averaging CFAR. A CFAR process in which the threshold or channel gain is controlled by averaging the amplitudes of cells (range gates or doppler filters) surrounding the detection cell.

CFAR loss. The increase in the S/N required to overcome the higher threshold established by a CFAR system, as compared to a system using a fixed threshold giving the same detection performance with fixed level of Rayleigh-distributed input interference.

Chebyshev filter. A filter using a particular polynomial approximation to approach a rectangular frequency response function.

chi-square distribution. The distribution of a quantity which represents the sum of M independent quantities, each having Gaussian distribution. M represents the number of degrees of freedom of the resulting chi-square distribution.

circular polarization. Wave polarization in which the E-field oscillates in a plane which rotates about the direction of propagation. In right-handed circular polarization the rotation is clockwise, as viewed by an observer looking in the direction of propagation.

clutter. Unwanted radar echoes, typically from ground, sea, rain or other precipitation, chaff, birds, insects, and aurora [A.1].

clutter detectability factor. The S/C ratio required at the input to an envelope detector to obtain a given level of detection performance.

clutter distribution loss. The increase in the S/C ratio required for non-Rayleigh clutter, as compared to that required for a given detection performance in Rayleigh-distributed clutter.

coherent. Having a predetermined phase relationship over some time period.

coherent integration. The process in which signal pulses or samples are combined prior to envelope detection, to increase signal-to-noise ratio and resolution in the frequency domain.

coherent integration time. The interval over which input signals are combined in a coherent process to increase their signal-to-noise ratio and to obtain resolution in the frequency domain.

coherent pulse radar. A radar in which the successive pulses have a controlled phase relationship, permitting processing to combine the signals prior to envelope detection.

collapsing loss. The increase in input signal-to-noise ratio required to overcome the effects of adding extraneous noise samples, not associated with signals pulses, in the video integration process.

collapsing ratio. The ratio of the total number of envelope-detected noise samples to the number of signal samples integrated at video.

comb filter. A filter whose otherwise uniform response has many sharp peaks or nulls, usually at intervals of the radar PRF for clutter cancellation or coherent integration of specific target velocities.

cone of silence. The conical region centered directly above a rotating search radar, not reached by the upper portion of the radar beam.

conical-scan tracker. A tracking radar system in which the beam axis is offset from the tracking axis and rotated to produce an amplitude-modulated signal indicative of tracking error.

conical scan on receive only (COSRO). A conical scan system in which the transmitting beam is on the tracking axis and only the receiving beam is scanned.

constant false-alarm rate (CFAR). Having the property of maintaining a preset false-alarm rate under variable input conditions (e.g., change in noise or interference level).

constant-γ (clutter model). A model in which the clutter reflectivity σ^0 is modeled as a constant times the sine of the grazing angle.

continuous-wave (CW) radar. A radar using continuous transmission, as opposed to keyed or pulsed transmissions.

correlation gate. A gate in range and angle coordinates, formed in track-while-scan processing, in which a further target detection is expected for a given target track.

correlation frequency. The spacing of samples in frequency which gives essentially independent values from the underlying amplitude distribution.

correlation time. The spacing of samples in time which gives essentially independent values from the underlying amplitude distribution.

corrugated horn. A feed horn with ridged walls, producing a nonuniform field at its output.

cosecant-squared gain function. The elevation pattern of an antenna for which the gain, above or below the mainlobe, varies as the $\csc^2$ of the elevation angle to produce a coverage pattern having constant performance along lines of constant target altitude.

cosecant-squared loss. The reduction in antenna gain caused by extension of coverage beyond the main elevation beam, using a cosecant-squared gain function.

coverage diagram. A plot of radar signal strength or detection performance in the vertical plane (range vs. height or elevation angle).

cross-eye jamming. A technique in which jamming is radiated from two sources with 180° phase difference, generating artificially large glint errors in a tracking radar.

cross-modulation. The process by which two or more signals interact in nonlinear circuits, causing generation of spurious or additional components of interference.

crossover loss. The loss in two-way signal power which occurs as a result of squinting the beam axis of a conical-scan radar from the tracking axis.

cross-polarization jamming. A jamming technique in which the jammer radiates a large component at a polarization orthogonal to the radar receiving antenna polarization, to induce large tracking errors.

cumulative detection. A detection process in which successive pulses (or scans) are tested individually for threshold crossing, and a target is declared if at least one pulse (or scan) crosses the threshold.

cumulative detection loss. The ratio of the S/N ratio required to achieve a given cumulative probability of detection to that required with an optimum video integration process having the same detection probability and false-alarm time.

cumulative probability of detection. The probability that a target will have been detected on at least one out of many scans of the radar.

cut-off wavelength (of waveguide). The wavelength above which transmission through the guide is severely attenuated.

deceptive jammer. An active jammer which transmits an approximate replica of the radar signal to generate false target indications.

decibel. $10 \log_{10}(P/P_0)$, where P/P_0 is a power ratio.

delay-line canceller. An MTI canceller in which signals from one PRI are stored for one or more PRIs and added to the current signal with weights which lead to clutter cancellation.

design illumination function. The intended illumination function of the radar antenna, excluding errors and imperfections in antenna construction.

detectability factor. In pulsed radar, the ratio of single-pulse signal energy to noise power per unit bandwidth that provides stated probabilities of detection and false alarm, measured in the IF amplifier and using an IF filter matched to the single pulse, followed by optimum video integration [A.1].

detection probability. The probability that a target signal will be detected (produce an output alarm) in competition with noise and other interference.

diffraction. The deviation of the direction of energy flow of a wave when it passes an obstacle, a restricted aperture, or other inhomogeneities in a medium (excluding effects of reflection) [A.2].

diffuse reflection. Reflections from a rough surface, which are spread over a large angular sector about the specularly reflected ray.

direct wave. The wave which passes from the radar to the target and back over a line-of-sight path through the atmosphere.

directive gain (directivity). The ratio of the radiation intensity in a given direction from the antenna to the radiation intensity averaged over all directions [A.2]. The directive gain exceeds the power gain by the dissipative loss in the antenna.

discrete clutter. Strong clutter resulting from scatterers which are concentrated in a region smaller than the radar resolution cell (e.g., water tanks or radio towers, as observed by radars with usual resolution properties).

diversity gain. The amount by which fluctuation loss is reduced through use of diversity in time, space, or frequency.

doppler shift. The shift in frequency of a wave emitted or reflected from a moving target.

ducting. Confinement of near-horizontally directed electromagnetic waves to a restricted horizontal layer in the atmosphere, resulting from a sufficiently steep negative vertical gradient of the refractive index in a limited altitude region.

duplexer. A microwave switch which connects the antenna alternately to the transmitter and the receiver.

eclipsing loss. The increase in input signal-to-noise ratio required to restore detection probability which is lost in range-ambiguous systems on targets whose delay causes them to overlap a subsequent transmitted pulse.

effective aperture area (of an antenna). In a given direction, the ratio of the available power at the terminals of a receiving antenna to the power flux density of a plane wave incident on the antenna from that direction, the wave being polarization matched to the antenna [A.2].

effective earth's radius. The radius of a sphere over which straight ray paths will represent the actual curved paths caused by atmospheric refraction.

effective echoing area. Radar cross section of a target.

electromagnetic (EM) field. A state of the region in which charged bodies are subject to forces by virtue of their charges and velocities.

electromagnetic interference(EMI). Interference with the radar from equipment other than enemy jammers, or interference caused by radar emissions entering other portions of the radar equipment or co-located friendly systems.

electromagnetic spectrum. The distribution of the radiated energy over various frequencies from about 1 Hz to 10^{21} Hz.

envelope detection. The process of removing the carrier of the signal, retaining its envelope amplitude.

envelope detector loss. The ratio of the S/N ratio required at the input to an envelope detector to the S/N ratio required at the input to a synchronous detector with reference phase matched to the signal phase for a given detection performance.

error slope. The slope of the curve of error-detector voltage vs. target deviation from the tracking point (e.g., the tracking axis, in angle measurement).

escort jamming (ESJ). Jamming from a platform which accompanies other penetrating aircraft into a defended area.

exponential distribution. A probability density function following an exponential relationship, for values of the variable which are positive (or negative, but not both).

false alarm. An erroneous radar target detection decision caused by noise or other interference exceeding the detection threshold [A.1].

false-alarm rate. The average rate at which false alarms are produced at the output of a detection system (the reciprocal of false-alarm time).

false-alarm time. The average time between false alarms at the output of a detection process.

Faraday rotation. The process of rotation of the polarization ellipse of an electromagnetic wave when it propagates in a magneto-ionic medium [A.2].

far field. The region, distant from the antenna, in which the radiated power density varies inversely with the square of distance, and in which the antenna pattern in angle remains constant.

fast Fourier transform. A process of combining a sequence in input signals (in time or across an antenna array) to yield the Fourier transform (in frequency or angle), with reduced processing burden.

ferrite phase shifter. A phase shifting device for phased arrays in which a ferromagnetic material is exposed to a magnetic control field to vary transmission phase.

flat-plate area. The area of a flat plate, normal to the radar beam, which would produce the same RCS as does the actual target.

fluctuation loss. The increase in input signal power required to preserve a given level of radar performance on a fluctuating target, as compared to a steady target of the same average RCS.

fluctuation rate. The rate at which target RCS lobes pass across the radar line of sight.

forward-scattering. Scattering of a radar signal in the same direction as the incident wave (away from the radar).

free-space propagation. Propagation equivalent to that in a vacuum, without interfering particles or surfaces affecting the wave.

frequency density. The relative frequency of occurrence of different values of a signal or other variable.

frequency distribution. A curve or function representing the relative frequency of occurrence of different values of a signal or other variable.

frequency scanning antenna. A type of phased array in which the phases of the radiating elements are changed systematically by varying the operating frequency of the radar.

frequency swept noise jammer. A noise jammer in which the carrier frequency is swept rapidly across the jammed band.

Fresnel reflection coefficient. The ratio of amplitude of the wave reflected from a smooth surface to the amplitude of the incident wave.

Fresnel zone. A region of the surface within which the phases of reflected rays change by 180°. The first Fresnel zone is the region about the point of specular reflection, bounded by an ellipse corresponding to 180° phase shift relative to the phase of the specularly reflected ray.

front face mismatch. The ratio of the gain of an array antenna to that which would be achieved if the elements were perfectly matched to the arriving wave.

Gaussian (distribution function, antenna pattern, or waveform). Having the form of the normal distribution.

glint. The inherent component of error in measurement of position and/or doppler frequency of a complex target due to interference of the reflections from different elements of the target [A.1].

grating lobes. Array sidelobes caused by periodic errors in phase or amplitude across the array, and resulting either from excessive spacing between elements or other errors having a regular period.

grazing angle. The angle between the radar wave and the local surface of the earth.

ground clutter. Clutter resulting from the ground or objects on the ground [A.1].

ground plane. A metallic surface on which antenna elements are mounted, preventing radiation on the rear side of the plane.

ground wave. From a source in the vicinity of a planetary surface, that wave which would exist in the absence of an ionosphere [A.2].

half-wave dipole. A dipole antenna element whose length is half the radar wavelength.

high-PRF (HPRF) radar. A radar with blind speed higher than the velocity region over which targets and clutter are distributed.

histogram. A representation of a frequency distribution by means of rectangles whose widths represent class intervals and whose heights represent the corresponding frequencies [A.3].

hits per scan. The number of pulses received during the observation time of a scanning radar.

home-on-jam (HOJ). The process of homing, with a radar seeker, on the source of jamming.

hybrid junction. A waveguide or transmission line arrangement with four ports which, when the ports have reflectionless termination, has the property that energy entering at any one port is transferred (usually equally) to two of the remaining three [A.2].

identity friend or foe (IFF). Equipment used for transmitting radio signals between two stations located on ships, aircraft, or ground, for automatic identification [A.2].

illumination efficiency. The ratio of the directive gain for an aperture with a given illumination function to the gain of the same aperture with uniform illumination.

illumination function. The function describing intensity of radiation from different portions of the transmitting antenna aperture, and the corresponding weighting applied to the received wave.

improvement factor (MTI). The signal-to-clutter ratio at the output of the clutter filter divided by that at the input, averaged uniformly over target radial velocities of interest [A.1].

impulse function. A function which is zero at all points except for one unity value.

impulse radar. A radar whose transmission is generated by applying a short video impulse to the antenna. The transmitted waveform is a sinusoidal wave with a duration in the order of one or a few cycles.

instrumented range. The range over which a display or other detection device is permitted to detect target echoes.

integrating filter. A filter at IF, baseband, or after envelope detection which combines signals for integration.

integration. The combination of multiple signals of a radar to enhance detectability, reduce measurement errors, improve angular resolution, or improve performance in some other way [A.1].

integration loss. The loss in the video integration process as compared to optimum coherent integration of the same number of signal samples.

integration weighting loss. The increased S/N ratio required to overcome the effect of nonoptimum weighting of samples in a video integrator.

intermodulation. The modulation of the components of a complex wave by each other, as a result of which waves are produced that have frequencies equal to the sums and differences of integral multiples of those of the components of the original complex wave [A.2].

interrupted CW. A waveform in which extended bursts of CW transmission are keyed on and off.

inverse-gain jammer. An amplitude modulated jammer whose amplitude is varied in inverse relationship with the incident radar signal amplitude, to induce error or beak-lock in conically scanned radars and other scanning systems.

inverse synthetic aperture radar. A radar technique in which coherent signals from a moving (rotating) target are processed to form an image of the target.

jammer strobe. A line formed on a display at the angle of a strong jammer.

jamming. A form of electronic countermeasure (ECM) in which interfering signals, typically noise-like, are transmitted at frequencies in the receiving band to obscure or distort the signal.

Kalman filter. A filter whose parameters are adjusted to minimize output error (or noise), making optimum use of *a priori* data on the nature of the desired signal and the competing noise or other interference.

kelvin. A unit of temperature equal to 1/273.16 of the Kelvin scale triple point of water [A.3].

knife-edge diffraction. Diffraction of a wave as it passes over an obstacle above a surface. The diffracted component modifies the field in the region just clear of the knife edge, and causes the wave to propagate at reduced amplitude into the shadow behind the obstacle.

limiter. A circuit which prevents the signal from exceeding a given peak level.

linear circuit. A circuit whose output is the sum of the input waves without intermodulation products.

linear envelope detector. An envelope detector from which the output voltage is directly proportional to the peak carrier voltage of the input signal.

linear polarization. Wave polarization in which the E-field oscillates in a single plane (e.g., vertical or horizontal).

lobing. The effect of multipath reflection on the elevation coverage pattern of the radar that creates lobes when time interference is constructive and nulls when the interference is destructive.

logarithmic amplifiers. An amplifier whose output amplitude is proportional to the logarithm of the input.

log-normal distribution. A probability density function in which the logarithm of the quantity has a normal (Gaussian) distribution.

loss budget. A list of loss factors, describing the departure of a radar from idealized conditions.

low-PRF (LPRF) radar. A radar with unambiguous range in excess of the greatest target range of interest.

Marx generator. A network in which high voltages are produced by charging capacitors in parallel and discharging them in series.

masking angle. The angle below which the radar path to distant targets and clutter is obstructed by a surface obstacle (e.g., hill or tree).

matched filter. A filter that maximizes the output ratio of peak signal power to mean noise power. For white Gaussian noise, it has a frequency response that is the complex conjugate of the transmitted spectrum, or, equivalently, has an impulse response that is the time inverse of the transmitted waveform [A.1].

matched point. A point in four-dimensional radar space to which the antenna and receiving system have been matched (tuned or pointed) for optimum reception of the signal.

matching loss. The ratio of the signal-to-noise ratio at the output of a matched filter to that at the output of an unmatched filter, given the same input conditions.

maximal length (m) sequence. A pseudorandom binary phase code generated by an *n*-stage shift register.

maximum unambiguous range. The range at which the round-trip signal delay equals the radar pulse repetition interval.

medium-PRF (MPRF) radar. A radar in which the PRF is higher than that required for LPRF and lower than that required for HPRF operation.

microwave resolver. A device which combines two orthogonal components of a microwave signal (e.g., two polarizations) into a composite signal representing time-multiplexed samples of the two input components.

modulation. The process by which some characteristic of a carrier is varied in accordance with a modulating wave [A.2].

monopulse comparator. A network (usually at RF) in which the beams of a monopulse radar are combined to give a sum channel and one or two difference channels at the output.

monopulse radar. A radar which obtains angular data on targets by comparison of signals received simultaneously in two or more antenna beams, as distinguished from comparison of a single beam output at different times.

monostatic. A system using the same location for transmitting and receiving, although not necessarily the same antenna.

Monte Carlo. Involving the use of random sampling techniques [A.3].

MTI canceller. The circuit in which signals from the current PRI are combined with those from previous PRIs to cancel clutter.

MTI loss. The increase in the S/N ratio required at the input to an MTI system, as compared to non-doppler IF processing of the same input pulse train to obtain a given level of detection performance.

multipath. Propagation of a wave over more than one path. This causes the pattern-propagation factor to vary in amplitude (multipath lobing), and also introduces measurement error in all radar coordinates.

multipath error. The error in radar measurement of target position or velocity caused by multipath reflection from the surface beneath the path.

noise factor (figure). The ratio of (A) the total noise power per unit bandwidth delivered by the system into an output termination to (B) the portion thereof engendered by the input termination, whose noise temperature is 290 K (kelvins) [A.2].

noise jammer. An active jammer which transmits an approximation of random noise to mask radar targets.

noncoherent integration. The process of combining signals after envelope detection to increase the signal-to-noise ratio (also known as video integration).

nonlinear circuit. A circuit whose output cannot be expressed as a constant factor times the input.

nonparametric CFAR. A CFAR process in which amplitude properties of reference cells other than the mean or standard deviation are used to establish the detection criterion. For example, the amplitudes of several cells may be placed in rank order, and a target declared if the detection cell exceeds the median (or other chosen rank) by a given amount.

normal probability distribution. A probability density function that approximates the distribution of many random variables (see Eq. 4.1.4 for the mathematical representation of this function).

observation time. The time over which a target signal can be observed and integrated in the radar processor (sometimes called "time-on-target"). It is measured by convention as the time during which the target is within the one-way, half power points of a scanning beam.

obstacle gain. The increase in intensity of waves in the shadow of an obstacle, as compared to the intensity which would have existed over a smooth earth without the obstacle.

operator loss. The ratio of the input S/N ratio required for detection by an operator to that required by an optimum automatic detection process for the same detection performance.

optical region. The frequency region above the resonant region, in which the RCS is not significantly affected by resonance phenomena.

passive ECM. A countermeasures technique in which reflecting objects (e.g., chaff or decoy targets) are deployed to confuse the radar.

peak error. The maximum deviation of a measured quantity from its true value. For a normally distributed random error, the peak is usually taken as three times the standard deviation, a value exceeded only 2.6% of the time.

peak power. The rms power of the transmitted carrier when the pulse envelope is at its peak.

peak-to-peak error. The difference between the maximum positive and negative deviations of a measured quantity from its true value. For a normally distributed random error, the peak-to-peak is usually taken as six times the standard deviation.

pencil beam. An antenna pattern which has a single, narrow lobe in both angle coordinates.

phased array. An antenna composed of radiating elements which are controlled electronically to scan the beam in one or two angular coordinates.

phase-front jamming. Jamming techniques which induce large glint errors in a tracking radar by distorting the angle of arrival of the true wave front.

phase shifter. A device used to control the phase of a wave (e.g., the wave emitted by an element of an array antenna).

phase-shifter efficiency. The ratio of output power of a phase shifter to input power.

polarization. That property of a radiated electromagnetic wave describing the time-varying direction and amplitude of the electric field vector [A.2].

polarization loss. The ratio of the received power of an echo to that which would be received with antenna polarizations properly matched to the polarizations for which target RCS was measured.

polyphase pulse compression codes. Discrete phase codes in which the phase varies in steps less than 180°.

post-detection integration. The process of combining signals after envelope detection to increase the signal-to-noise ratio.

power-aperture product. The product of average transmitter power and effective aperture of the receiving antenna, used in the search radar equation to determine maximum range of coverage.

power gain (of an antenna). The ratio of the power flux per unit area from an antenna to the power flux per unit area from an isotropic radiator at the same location with the same power input as the subject antenna [A.2].

predetection filter. A filter located in the IF stages of the receiver, or using in-phase and quadrature baseband signals, prior to an envelope detection process.

probable error. The error which is exceeded 50% of the time.

probability density function. The first derivative of the probability distribution function; it represents the probability of obtaining a given value [A.2].

pulse compression. The processing of a wideband, coded signal pulse, of initially long time duration and low range resolution, to result in an output pulse of time duration corresponding to the reciprocal of the bandwidth and, hence, higher range resolution, and with approximately the same pulse energy [A.1].

pulse repetition frequency. The rate at which pulses are transmitted.

pulse repetition interval. The interval between leading edges of successive pulses.

radar. A device for transmitting electromagnetic signals and receiving echoes from objects of interest (targets) within its volume of coverage [A.1].

radar cross section (RCS). A measure of the reflective strength of a radar target; usually represented by the symbol σ, measured in square meters, and defined as 4π times the ratio of the power per unit solid angle scattered in a specific direction to the power per unit area in a plane wave incident on the scatterer from a specified direction [A.1]. For monostatic radar, the two directions coincide but are opposite in sense.

radar horizon. The point at which the radar wave, refracted by the atmosphere, is tangent to the earth's surface.

random error. An error which varies in an unpredictable way, as a result of input interference, system noise, or accidental variation in system parameters.

range-gate pull-off. An active jamming technique in which a repeater signal is generated and varied in time delay to capture and pull the gate off the range of the target.

range sidelobes. Sidelobe responses at the output of a pulse compression filter, preceding and following the main response.

Rayleigh criterion. A criterion defining as a "smooth surface" one which reflects substantial power as a specular reflection.

Rayleigh distribution. The distribution of the amplitude of a random noise or target echo envelope, for which the in-phase and quadrature components are independent Gaussian variables (see Eq. 4.1.5 a mathematical representation of this distribution).

Rayleigh region. The frequency region in which the RCS varies inversely with the fourth power of frequency, as a result of scatterer dimensions which are small relative to radar wavelength.

receiving line loss. The ratio of the input power from the antenna to the receiving RF components (waveguide or other, including duplexer) to the power delivered to the receiver.

reciprocity. The property of passing signals in either direction with equal response.

reflection. The bending back of a wave when it reaches the surface of a medium having refractive index different from the medium in which the wave initially travels.

reflector antenna. An antenna consisting of one or more reflecting surfaces and a radiating (receiving) feed system [A.2].

refraction. Bending of the radar beam in the atmosphere.

resonant region (of radar cross section). The frequency region within which the RCS oscillates because of interaction of the creeping wave, circulating around the object, with the reflection from the point of first contact.

response (of a radar receiver or circuit). The output voltage or power, measured as a function of an input variable such as time, frequency, or other signal characteristic.

Rician distribution. The distribution of a sinusoidal signal plus bandlimited Gaussian noise, as derived by S. O. Rice.

root-mean-square (rms) value. The square root of the mean-square value of a variable. The rms value includes any bias or DC component of the variable.

scan distribution loss. The increase in input signal energy needed to overcome the inefficiency in the cumulative detection process over several scans, as compared to optimum video integration of the same signals.

scatterer. An object or portion of an object which scatters or reflects the radar wave, back toward the radar or in some other direction.

scintillation error. Error in radar-derived target position or doppler frequency caused by interaction of the scintillation spectrum with frequencies used in sequential measurement techniques.

sea state number. A number from 1 to 8 indicating the height of the waves. Sea state 1 is "smooth," and 8 is "precipitous."

search frame time. The time required for a scan by the radar over the entire search volume.

search loss factor. A factor including the effect of all departures from ideal assumptions in derivation of the search radar equation.

search radar. A radar used primarily for the detection of targets in a particular volume of interest [A.1].

search radar equation. An equation relating radar power-aperture product to the target RCS, the volume of search coverage, and other radar requirements.

self-screening jamming (SSJ). Jamming that originates at the radar target.

single-scan detection probability. The detection probability of a target during a single scan of the radar.

slotted waveguide. An antenna component in which radiating elements are formed by cutting slots in the wall of a waveguide.

small-signal suppression. The process in which noise causes a reduction in signal information, when both pass through a nonlinear circuit under conditions of small signal-to-noise ratio.

smooth-sphere diffraction. The diffraction effect which reduces the amplitude of a signal propagating over a path near the smooth surface of the earth, and which causes propagation of the wave with reduced amplitude beyond the radar horizon.

speckle. The random variation in level of an image formed by coherent radar processing.

specular reflection. Reflection of radar waves from a smooth surface, where the angle of reflection equals the angle of incidence.

specular scattering coefficient. The ratio of amplitude of the specularly reflected wave from a roughened surface to the amplitude for a smooth surface of the same material.

split angle gate. A processing gate used in scanning radar, in which signals from the early portion of the observation time are integrated with one polarity, and signals from the later portion with opposite polarity, producing a null output when the received pulse train envelope is centered in the gate.

spot noise jamming. Noise jamming which is tuned to the radar frequency, occupying a relatively narrow band about the radar signal spectrum.

spotlight mapping. The process of forming a synthetic aperture radar map of a limited region, using signals from a narrow real radar beam which is pointed at that region as the radar platform moves.

square-law detector. An envelope detector from which the output voltage is directly proportional to the square of the peak input carrier voltage.

stacked-beam radar. A 3D radar system in which multiple receiving beams are stacked vertically, within the broad beamwidth of the transmitting antenna. All beams receive in parallel, using multiple receiving channels.

standard deviation. The square root of the mean square deviation of a value from its mean. When the mean value is zero, the standard deviation is equal to the rms value.

stand-off jamming (SOJ). Jamming from a platform which stays at a constant range, generally beyond the reach of defensive systems.

steady target. A target whose signal, as observed by a given radar, does not fluctuate.

straddling loss. The increase in the S/N ratio required to overcome the effect of gates, filters, or processing windows not centered on the target signal.

superrefraction. Confinement of near-horizontally directed electromagnetic waves to a restricted horizontal layer in the atmosphere, resulting from a sufficiently steep negative vertical gradient of the refractive index in a limited altitude region.

surface-bounce jamming. A jamming technique in which strong jamming is directed at the surface beneath the path from the radar, causing increased multipath error in the radar and possibly leading the radar to track the reflected signal rather than the actual target.

surface tolerance efficiency. The ratio of antenna gain to that which would be achieved with the same aperture in the absence of dimensional errors in construction.

synthetic aperture beamwidth factor. The ratio of the ideal cross-range resolution cell width to that of an SAR process in which weighting is applied to reduce sidelobe levels.

synthetic aperture radar (SAR). A radar which coherently combines signals obtained sequentially from different locations of a moving antenna, in order to form high-resolution beams for mapping or target identification purposes.

synthetic video. A video signal generated in a processor, and having characteristics determined by digital logic.

systematic error. An error which varies in a regular way, as a function of particular environmental or operating conditions.

tapering. The process of adjusting illumination across the aperture, with reduced amplitude at the edges, to reduce sidelobes.

theodolite. An optical instrument for measuring angles to a distant target.

three-dimensional (3D) radar. A search radar whose output data include range and two angle coordinates of each detected target.

time-averaging CFAR. A CFAR process in which the threshold for the detection cell is established on the basis of the average output of that cell over several scans of the radar.

track-file saturation. The condition of having more targets or apparent targets than can be processed by the tracking system.

tracking error. An error which causes the tracking axis or range gate of the radar to deviate from the target centroid.

tracking radar. A radar whose primary function is the automatic tracking of targets.

track-on-jam (TOJ). The process of angle tracking on the received jamming, giving the angle to the jammer platform.

track-while-scan. An automatic tracking process in which the radar antenna and receiver provide periodic video data from a search scan, together with interpolation measurements, as inputs to computer channels that follow individual targets [A.1].

translation error. An error in converting the position of the beam axis or range gate into true target coordinates.

transmission line loss. The ratio of the input from the transmitter to the RF components (waveguide or other transmission line and duplexer) to the power delivered to the antenna.

two-dimensional (2D) radar. A search radar whose output position data consist of target range and position in one angle coordinate (usually azimuth or bearing).

two-parameter CFAR. A CFAR process in which the threshold for the detection cell is determined by combining the mean and standard deviation of the amplitudes in the reference cells.

two-point jamming. A jamming technique which induces large glint errors in a tracking radar by creating two strong point sources within the radar beam.

unambiguous range. The range at which the time delay of the target signal equals the pulse repetition interval of the radar.

vegetation coefficient. The ratio of the reflected specular or diffuse wave amplitude from a surface covered with vegetation to that which would have been reflected in the absence of vegetation.

velocity-gate pull-off. An active jamming technique in which a repeater signal is generated and varied in frequency, relative to the incident radar frequency, to capture and pull the velocity gate (doppler tracking filter) off the target.

video integration. The process of combining signals after envelope detection to increase the signal-to-noise ratio.

voltage-controlled oscillator (VCO). An oscillator whose frequency can be varied by application of a variable control voltage.

Weibull distribution. A probability density function, often used for clutter amplitudes, which is characterized by a mean value and a spread parameter a which may be greater than the value $a = 1$ for the more usual Rayleigh distribution.
white (noise or spectrum). Having uniform spectral density over frequencies of interest.
wind shear. The vertical gradient of wind velocity.

zero-crossing detector. A circuit which produces an output pulse when the input voltage passes through zero.

References for Appendix A

[A.1]　　IEEE Standard Radar Definitions, IEEE Std 686-1990.
[A.2]　　IEEE Standard Dictionary of Electrical and Electronics Terms, ANSI/IEEE Std 100-1988.
[A.3]　　Webster's New Collegiate Dictionary.

Appendix B

LIST OF SYMBOLS USED

The section number is the location of the first use. Symbols using subscripts such as $_{max}$ (indicating maximum) are not separately defined, as their meanings are evident.

Symbol	Meaning	Section
$A(v)$	Diffraction voltage	8.3.2
A_b	Surface area within beam	2.2.3
A_c	Area of resolution cell	2.2.3
A_m	Azimuth sector width	2.6.1
A_p	Flat-plate area	5.1.2
A_r	Effective receiving aperture	2.2.4
A_r	Amplitude of target ray to reflection point	H.2.1
A_t	Surface area within pulse	2.2.3
A_t	Amplitude of target ray to radar	H.2.1
a	Earth's radius	2.2.2
a	Radius of sphere	3.1.1
$a(t)$	Voltage waveform	G.1
a_t	Target acceleration	2.5.2
B	Spectrum bandwidth	1.1.4
B_f	Width of spectral line	1.1.4
B_j	Jammer bandwidth	2.2.4
B_n	Effective noise bandwidth of receiver	3.1.2
B_v	Video bandwidth	4.4.6
B_1	Bandwidth of single IF stage	11.3.1
C_B	Bandwidth factor for display	3.1.7
C_x	Detectability factor	4.3.1
C_0	Clutter spectral density	1.1.4
C_Δ	Clutter power in difference channel	11.2.2
c	Velocity of light	1.1.1
c_g	Velocity of propagation in waveguide	6.6.5

Symbol	Meaning	Section
D	Antenna diameter	1.1.1
D	Detectability factor	2.4.2
D	Divergence factor	8.2.2
D_c	Detectability factor for coherent system	4.4.1
D_e	Detectability factor with diversity	4.5.2
D_u	Duty factor	1.1.3
D_x	Effective detectability factor (including losses)	3.1.6
D_0	Detectability factor for steady target	3.2.1
D_1	Detectability factor for Case 1 target	4.5.1
d	Spot size on CRT	4.4.6
d	Element spacing in array	6.6.1
E	Voltage	1.1.1
E	Signal energy	1.1.4
E_n	Envelope detected noise voltage	4.1
E_1	Energy of single received pulse	3.1.3
E_Δ	Error voltage	10.4.1
e	Target voltage	11.2.2
e_c	Cross-polarized component of target voltage	11.2.2
F	Pattern-propagation factor	2.2.2
$F(f)$	Voltage spectrum	7.1
F_a	Fraction of active elements in thinned array	6.6.2
F_c	Pattern-propagation factor for clutter	2.2.3
F_d	Roughness factor	H.2.2
F_n	Receiver noise factor	3.1.2
F_j	Pattern-propagation factor for jammer	2.2.4
f	Frequency	1.1.1
$f(t)$	Received waveform	7.1
$f(\theta)$	Antenna voltage pattern	6.1
$f_a(\theta)$	Array voltage pattern	6.6.2
f_c	Target correlation frequency	4.5.4
f_d	Doppler shift	1.1.4
$f_e(\theta)$	Array element voltage pattern	6.6.2
f_r	Pulse repetition frequency	1.1.2
f_s	Received carrier frequency	1.1.4
f_t	Transmitted carrier frequency	
f_0	Carrier frequency	1.1.3
G	Antenna gain	1.1.5
G_d	Diversity gain	4.5.4

Symbol	Meaning	Section
$G_d(\theta)$	Derivative of gain pattern	6.6.3
G_e	Array element gain	6.6.2
G_i	Video integration gain	4.4.2
G_j	Jammer antenna gain	2.2.4
G_m	Gain on antenna axis	6.3
G_r	Receiving antenna gain	3.1.1
G_s	Sidelobe gain	6.3
G_t	Transmitting antenna gain	2.4.2
G_0	Gain of uniformly illuminated aperture	6.1
$g(x)$	Illumination function	F.1
$H(f)$	Filter frequency response	1.1.4
h	Target altitude	1.1.2
h	Antenna aperture height	2.4.5
$h(t)$	Impulse response of filter	7.1
h_a	Altitude of radar platform	2.6.1
h_r	Radar antenna height	2.2.2
h_r'	Radar antenna height above tangent plane	9.3.1
h_t	Target altitude	2.2.2
h_t'	Target antenna height above tangent plane	9.3.1
I	MTI improvement factor	7.2.3
I_d	Diffuse multipath power	H.2.2
I_r	Power density of received signal	3.1.1
I_s	Power density of transmitted signal	3.1.1
I_s	Specular multipath power	H.2.3
I_v	Diffracted multipath power	H.2.3
I_Δ	Interference power in difference channel	2.5.3
i	Integer	1.1.4
J_0	Power spectral density of jamming	5.6.1
K	Number of duo-degrees of freedom in chi-square distribution	4.5.3
K_a	Acceleration error constant	2.5.2
K_B	Beaufort wind scale number	2.2.3
K_v	Velocity error coefficient	11.2.2
K_θ	Beamwidth constant	1.1.5
K_θ'	Beamwidth constant for synthetic aperture radar	9.5
k	Earth's radius factor (usually 4/3)	2.2.2
k	Boltzmann's constant, 1.38×10^{-23} W-s/K	2.2.4

Symbol	Meaning	Section
k	Number of scans	2.4.1
k	$2\pi/\lambda$	5.1.2
k	Ratio of target amplitudes	5.4.1
k_m	Monopulse error slope	2.5.3
k_p	Error slope in sector scan	10.2.2
k_s	Conical scan error slope	10.2.1
k_{sh}	Wind shear coefficient	5.5.2
k_t	Ranging error slope	10.5.2
k_z	Error slope in z coordinate	2.5.3
k_α	Atmospheric attenuation coefficient, dB/km (two-way)	2.2.1
L	Loss factor	2.4.2
L	Target dimension	5.1.2
L_a	Antenna dissipative loss	3.1.2
L_c	Collapsing loss	4.4.6
L_{cd}	Clutter distribution loss	C.7.2
L_d	Scan distribution loss	3.2.4
L_{ea}	Straddling loss in angle scan	7.3.4
L_{ec}	Eclipsing loss	7.3.4
L_{ef}	Straddling loss in doppler filters	7.3.4
L_{eg}	Transient gating loss	7.3.4
L_{er}	Straddling loss in range gates	7.3.4
L_{ev}	Velocity response loss	7.3.4
L_f	Fluctuation loss	3.2.4
L_i	Video integration loss	3.2.4
L_k	Crossover loss in conical scan	10.2.1
L_m	Receiver matching loss	3.1.4
L_{mf}	Doppler filter matching loss	3.1.4
L_n	Antenna beam pattern constant	1.1.5
L_p	Beamshape loss	2.2.3
L_q	Quantizing loss	6.6.3
L_r	Receiving line loss	3.1.2
L_r	Target radial dimension	4.5.4
L_s	Search loss factor	2.6.1
L_t	Transmission line loss	3.1.5
L_x	Miscellaneous signal processing loss	3.1.7
L_x	Target cross-range dimension	4.5.4
L_α	Two-way atmospheric loss to target	3.1.5
$L_{\alpha j}$	One-way atmospheric loss from jammer to radar	2.2.4

Symbol	Meaning	Section
L_{ot}	Total two-way loss through atmosphere	8.1.4
M	Receiver matching factor	3.1.7
m	Synthetic aperture beamwidth factor	2.6.1
m	Binary integration threshold	4.4.3
m	Number of excess noise samples	4.4.6
m	Number of bits in A/D converter	6.6.3
m	Number of IF stages	11.3.1
N	Noise power	3.1.2
N	Atmospheric refractivity	8.4.1
N_0	Noise spectral density	1.1.4
n	Number of pulses integrated	
n	Number of elements in array	6.6.2
n	Atmospheric refractive index	8.4.1
n'	Number of independent outputs from coherent integrator	3.1.6
n_a	Number of array elements in row	6.6.2
n_c	Number of independent clutter samples	11.2.2
n_d	Number of detection decisions	2.3
n_e	Effective number of interference samples	2.5.3
n_e	Effective number of target samples	4.5.2
n_e	Number of array elements in column	6.6.2
n_f	Number of doppler channels	4.4.5
n_t	Number of range gates	4.4.5
P	Pressure of atmosphere	8.4.1
P_{av}	Transmitter average power	1.1.3
P_c	Cumulative probability of detection	2.4.1
P_d	Detection probability	2.3
P_e	Probability density function of target voltage	5.3.1
P_{fa}	False-alarm probability	2.3
P_i	Power incident on target	3.1.1
P_j	Jammer transmitting power	2.2.4
P_t	Transmitter peak power	1.1.3
P_v	Probability density of Gaussian variable	4.1
P_σ	Probability density function of radar cross section	5.3.1
p	Single-pulse detection probability in binary integration	4.4.3
p	Partial pressure of water vapor	8.4.1
R	Target range	1.1.2
R_h	Horizon range	2.2.2
R_j	Range of jammer	2.2.4

Symbol	Meaning	Section
R_m	Maximum detection range	2.2.2
R_u	Unambiguous range	1.1.3
R_0	Minimum range	2.6.1
r	Precipitation rate	5.5.4
r	Maximum radius of curvature of knife edge	8.2.3
S	Received signal power	2.4.2
$S(f)$	Power spectral density	1.1.4
$(S/N)_m$	Signal-to-noise ratio at peak of beam	10.2.1
$(S/N)_o$	Output signal-to-noise ratio	2.4.2
s	Sweep speed on CRT	4.4.6
s	Distance of slots along waveguide	6.6.5
$s(t)$	Waveform of linear FM signal	7.1.2
T	Number of elements in array	6.1
T	Temperature of atmosphere	8.4.1
T_a	Antenna noise temperature	3.1.2
T_a'	Sky noise temperature	3.1.2
T_d	Matched-filter time delay constant	7.1
T_e	Receiver noise temperature	3.1.2
T_j	Effective input noise temperature of jammer	2.2.4
T_r	Noise temperature of receiving line	3.1.2
T_s	Effective input noise temperature of radar	2.2.4
T_{tr}	Physical temperature of receiving line	3.1.2
t	Time	1.1.1
t_c	Target signal correlation time	4.5.4
t_d	Time delay	1.1.2
t_{dc}	Time delay of clutter	7.3.4
t_f	Coherent processing time	2.6.1
t_{fa}	False-alarm time	2.3
t_g	Transient gate duration	7.3.4
t_o	Observation time	1.1.4
t_r	Pulse repetition interval	1.1.2
t_s	Search frame time	2.4.1
t_v	Scan integration time	4.4.6
V_c	Volume of resolution cell	2.2.3
V_0	Visibility factor for display	3.1.7
V_{oc}	Visibility factor in clutter	7.2.3
v	Voltage	4.1
v	Knife-edge diffraction parameter	8.3.2

Symbol	Meaning	Section
v_a	Velocity of radar platform	2.6.1
v_b	Blind speed	1.1.4
v_p	Predicted estimate of velocity in tracker	11.2.2
v_r	Target radial velocity	1.1.4
v_s	Smoothed estimate of velocity in tracker	11.2.2
v_w	Wind velocity	5.5.1
W	Weight of chaff	2.2.3
w	Width of antenna	1.1.5
x,y,z	Rectangular coordinates in space	1.1.2
x,y	General variables	4.1
x_m	Peak error	11.1.1
x_o	Current observed data point in tracker	11.2.2
x_p	Predicted estimate in tracker	11.2.2
x_s	Smoothed estimate in tracker	11.2.2
x_0	Distance to specular reflection point	8.2.1
x_{50}	Probable error	11.1.1
Z	Low-grazing-angle correction factor	H.2.2
z	Any radar coordinate	2.5.3
z_3	Half-power width of resolution cell in z coordinate	10.1.2
α	Azimuth angle of wind	5.5.2
α	Phase angle of reflection	8.2.2
α	Diffraction angle	8.3.2
α	Parameter of sampled-data tracking filter	11.2.2
α_0	Rms time duration of pulse	10.6.1
β	Rms signal bandwidth	10.5.1
β	Parameter of sampled-data tracking filter	11.2.2
β_n	Servo bandwidth	2.5.2
β_0	Rms slope of surface	5.5.1
γ	Surface clutter reflectivity	2.2.3
Δ	Difference pattern voltage	10.1.2
Δ_c	Cross-polarized difference voltage response	11.2.2
ΔE	Elevation angle to horizon	2.2.2
ΔF	Bandwidth of swept signal	7.1
Δf	Agility or diversity bandwidth	4.5.4
Δ_r	Range resolution	2.6.1
Δ_t	Data delay to cancel range-doppler coupling error	11.3.2
Δ_v	Resolution in radial velocity	2.4.6
Δ_v	Velocity notch width	7.3.4

Symbol	Meaning	Section
Δ_{vf}	Resolution of coherent system in radial velocity	2.4.6
Δ_x	Cross-range resolution	2.6.1
δ	Frequency step	7.1.4
δ_t	Range-doppler coupling error	11.3.2
δ_0	Pathlength difference of reflected ray	8.2.1
ε_a	Acceleration lag error	2.5.2
ε_θ	Beam steering error from quantization	6.6.3
η_a	Aperture efficiency	6.3
η_b	Blockage efficiency	6.5
η_f	Feed efficiency	C.1.2
η_i	Illumination efficiency	6.5
η_m	Front-face mismatch efficiency	C.1.2
η_{ph}	Phase-shifter efficiency	C.1.2
η_q	Quantizing efficiency	C.1.2
η_s	Spillover efficiency	C.1.2
η_t	Surface tolerance efficiency	6.5
η_v	Volume clutter reflectivity	2.2.1
η_x	Illumination efficiency in azimuth	6.4
η_y	Illumination efficiency in elevation	6.4
θ	Beamwidth	1.1.1
θ_a	Azimuth beamwidth	1.1.5
θ_e	Elevation beamwidth	
θ_m	Maximum elevation angle	3.2.1
$\theta_m{}'$	Effective maximum elevation angle for time budget	3.2.2
θ_n	Separation of nulls in multipath reflection	2.2.2
θ_t	Target elevation angle	
θ_0	Minimum elevation angle	3.2.1
θ_1	Elevation angle to upper edge of mainlobe	3.2.2
θ_2	Elevation angle to upper edge of $\csc^2$ coverage	3.2.2
θ_3	Half-power beamwidth	1.1.5
λ	Wavelength	1.1.1
λ_c	Critical wavelength of waveguide	6.6.5
λ_g	Wavelength in waveguide	6.6.5
μ	Mean of normal distribution	4.1
μ	Linear FM slope constant	7.1.2
ν	Detection decision rate	4.4.5
ρ	Collapsing ratio	4.4.6
ρ	Reflection coefficient	8.2.2

Symbol	Meaning	Section
ρ_d	Diffuse reflection coefficient	
ρ_s	Specular scattering coefficient	8.2.3
ρ_v	Vegetation factor	H.2.1
ρ_0	Fresnel reflection coefficient	H.2.1
Σ	Sum pattern voltage	10.1.3
Σ_c	Cross-polarized sum pattern response	11.2.2
σ	Radar cross section of target	2.4.2
σ	Standard deviation of normal distribution	4.1
σ_a	Rms error in azimuth data	2.2.5
σ_c	Radar cross section of clutter	2.2.3
σ_c	Cross-polarized target cross section	11.2.2
σ_f	Rms error in frequency	10.6.1
σ_g	Rms glint error	5.4.1
σ_h	Rms surface height deviation	2.2.5
σ_s	Rms scintillation error	8.2.2
σ_s	Rms error in smoothed track	11.2.2
σ_t	Rms error in time delay	10.5.1
σ_v	Standard deviation in velocity	5.5.1
σ_{vv}	RCS of target, transmit and receive ant. pol. vertical	C.1.2
σ_{rr}	RCS of target, transmit and receive ant. rt. circ. pol.	C.1.2
σ_x	Rms error in x coordinate	11.1.1
σ_z	Rms error in measurement of z coordinate	2.5.3
σ^0	Surface clutter reflectivity	2.2.3
σ_1	Rms error in single data point	11.2.2
σ_θ	Rms error in angle data	2.2.5
σ_ϕ	Rms phase error	6.3
τ	Transmitted pulsewidth	1.1.3
τ_n	Processed pulsewidth	2.2.3
τ_p	Compressed pulsewidth	7.1
τ_o	Output pulse width	10.5.1
ϕ	Phase angle	5.4.1
ϕ	Angle coordinate	6.2
χ	Response function	6.2
χ	Ambiguity function	G.1
ψ	Grazing angle	2.2.3
$\psi(t)$	Complex waveform	G.1
ψ_B	Brewster angle	8.2.1
ψ_b	Solid angle of beam	2.2.3

Symbol	Meaning	Section
ψ_c	Critical grazing angle	5.5.1
ψ_s	Solid angle of search or mapping coverage	2.6.1
Ω	Phase excursion	1.1.4
ω	Target rotation rate	2.6.2
ω_a	Azimuth scan rate	4.4.6
ω_a	Rate of change of target aspect angle	4.5.4
ω_d	Radian doppler frequency, $2\pi f_d$	1.1.4
ω_e	Elevation scan rate	4.4.6
ω_e	Target elevation rate	4.4.6
ω_m	Maximum target azimuth rate	11.2.2

Appendix C

IDEAL ASSUMPTIONS AND LOSS FACTORS

Analysis of radar performance is often based on idealized models of the radar and its environment, adjusted to the real situation by inclusion of several "practical loss factors." In this appendix, the idealized models are summarized, and the needed loss factors are listed and described. Test and evaluation procedures for validation of the models and loss factors are given.

The loss factors represent departures from the idealized models, and hence are dependent on the starting assumptions of the selected model. If the detection performance is based on a prediction of single-pulse signal-to-noise power ratio, S/N, given by (3.1.17) and (3.1.18), the losses must account for the actual received signal power S, the actual noise power N, and the ability of the signal processor to convert the ratio S/N into a detection with probability P_d. If the search radar equation (3.2.3) is used to predict range on the basis of signal energy E received during the search frame time t_s, in the presence of noise with spectral density N_0, the losses must include those in the process of integration or accumulation of signal information over t_s, and its conversion to the desired detection probability P_d.

Similar relationships apply to prediction of tracking accuracy and radar imaging performance. In each case, the effects of clutter and jamming must also be included. Some of these effects occur only when the clutter or jamming is present, adding to the interference residue at the processor output. Others will affect the performance even in the absence of clutter or jamming, as a result of circuits and processes which have been included in anticipation of such interference (e.g., CFAR detection processing). These issues, and means of evaluating the several losses, have been discussed in the body of this handbook, and hence will be listed and discussed only briefly here, with references to the chapters in which more complete discussions appear.

A loss may be described as factor $L > 1$, placed in the denominator of the radar equation, or equivalently as an efficiency factor $1/L = \eta < 1$, multiplying the gains in the numerator of the equation.

C.1 Radar Equation for Received Signal Power

Received signal power S delivered by the antenna, at the peak of the pulse, from a target in free space on the beam axis, is given by:

$$S = \frac{P_t G_t G_r \lambda^2 \sigma}{(4\pi)^3 R^4}$$
(3.1.7)

C.1.1 RF Losses

Ideal Assumptions

No internal or external loss at RF.

Losses to be Considered

To be included in (3.1.20) for actual snr:

(a) *Transmission line loss* L_t, from transmitter port at which P_t is measured to antenna port at which G_t is measured;

(b) *Atmospheric attenuation* L_α, through atmosphere to target and back to radar.

Other factors to look for:

(c) Losses within the transmitter, if P_t is based on the output of a tube or solid-state device, should be evaluated and included in L_t.

(d) Components of P_t in spectral regions not intended for transmission, and not considered in calculation of receiver matching loss L_m, may be measured by an output power meter. The power P_t should be adjusted to exclude these components.

C.1.2 Antenna Losses

Ideal Assumptions

(a) Uniform illumination of aperture area A, ideal gain $G_0 = 4\pi A/\lambda^2$.
(b) Directive gain (no dissipative losses), $G_a = G_0 \eta_i$
(c) Power gain with matched polarizations, G_t and G_r

Losses to be Considered

Starting with assumption (a), uniform illumination, the directive gain is derived by applying the following efficiency factors:

Illumination (taper) efficiency $\eta_i = \eta_x \eta_y$, where x and y are the two aperture coordinates;

Cosecant-squared loss. For a cosecant-squared elevation pattern, η_y should include the effects of this extension of the elevation beam;

Surface tolerance efficiency η_t for reflector systems;

Effects of phase and amplitude errors η_t for array systems;

Blockage efficiency η_b in reflectors and space-fed reflectarrays;
Spillover efficiency η_s in reflectors and space-fed reflectarrays;
Phase-shifter quantizing efficiency η_q in phased arrays;
Front face mismatch η_m in phased arrays.

The directive gain is $G_a = G_0\eta_i\eta_t\eta_b\eta_s\eta_q$.

Starting with assumption (b), directive gain, the power gain is derived by applying the following further efficiency factors:

Feed horn or network efficiency η_f;
Phase-shifter efficiency η_{ph} in phased arrays.

Mismatch efficiency from rear face, in space-fed lens arrays, should be included in η_f.

Mismatch efficiency in a feed network, in constrained-fed arrays, should be included in η_f.

In a phased array, the projected aperture area A, and hence the directivity, varies as the cosine of the scan angle θ. Mismatch η_m usually varies as $\cos^{1/2}\theta$, leading to G_t and G_r proportional to $\cos^{3/2}\theta$.

The power gain G_t or $G_r = G_a/L_a = Ga\eta$

Starting with assumption (c), power gains G_t and G_r for matched polarizations, the following loss must be considered:

Polarization loss. It is assumed in the derivation of the radar equation that the antenna gains G_t and G_r apply to particular polarizations, to which σ is also applicable. Thus, if the antenna polarizations are both vertical, the value σ_{vv} is used, corresponding to vertical incident and scattered polarizations. In the case of circular polarizations, if both antennas are right-handed, the target cross section σ_{rr} should be used, and this is usually less than σ_{vv} by several dB. If only a single value for σ is available, it may be necessary to include a polarization mismatch loss as a separate term in the radar equation, or as a component of the pattern-propagation factor.

C.1.3 Pattern-Propagation Factor

Ideal Assumption

Free-space propagation to target on beam axis.

Losses to be Considered

Departure from free-space, beam axis operation are included in the pattern-propagation factor F, with different values for the transmitting and receiving paths if antenna characteristics differ.

(a) For the case of multipath lobing, F is defined by (8.2.5), including both the deviation of the target from the beam axis and the effects of surface reflections.

(b) For low-elevation paths, the factor F is dominated by diffraction (over the smooth sphere or over a knife edge), as discussed in Section 8.3.

(c) At low angles, an atmospheric lens loss must also be included, as a reduction in F (see Section 8.4).

(d) At low angles, ducting may occur, radically changing the value of F either upwards or downwards (see Section 8.4.4).

(e) When the beam scans across the target (e.g., in azimuth, for a 2D search radar), the effect of azimuth deviation from the axis is expressed as a one-coordinate ***beamshape loss***, L_p (see Section 3.1.5). The factor F then expresses the propagation effects and the effect of deviation in the nonscanning coordinate.

(f) When the beam performs a raster scan, or when performance is averaged over the elevation beamwidth, the effect of deviations is expressed as a two-coordinate beamshape loss $L_p^2 \approx 2.5$ dB (see Section 3.2.4). The factor F then expresses only the propagation effects.

C.2 Receiver Noise Level

C.2.1 Input Noise Temperature

Ideal Assumption

Noise-free receiver viewing space through lossless receiving line and antenna.

Losses to be Considered

The system input temperature T_s (3.1.11) is controlled by three components:

(a) Antenna temperature (3.1.12) is in turn controlled by three components:

Sky temperature T_a' (Figure 3.1.5). If the beamwidth covers an elevation sector, this temperature must be averaged over beam.

Earth-directed sidelobes. Blake assumes a temperature component of 36 K, included in antenna temperature, (3.1.12).

Antenna ohmic losses. The loss $L_a = 1/\eta_j\eta_{ph}$ contributes a temperature component as described by (3.1.12).

(b) Receiving line temperature T_r depends on ***receiving line loss*** L_r (3.1.13).

(c) Receiver noise temperature T_e is determined by ***noise factor*** (also called noise figure) F_n (3.1.14), and is multiplied by line loss to refer it to the antenna terminal in (3.1.11).

C.2.2 Equivalent Input Noise Level and Output S/N

Ideal Assumption

(a) A filter matched to the entire transmission is used, in which the noise bandwidth of the filter is equal to the reciprocal of the observation time (or time on target), $B_f = 1/t_o$, and the output snr is thus equal to the ratio of total received energy to noise spectral density: $E/N_0 = nS\tau/N_0 = S_{av}t_o/N_0 = S_{av}t_o/kT_s$. There are n equal pulses received during t_o with the full gain from the antenna beam axis.

(b) A filter matched to the single pulse is used, in which the noise bandwidth of the receiver is equal to the reciprocal of the transmitted pulsewidth, $B_n = 1/\tau$, and the output snr is thus equal to the ratio of received pulse energy to noise spectral density: $E_1/N_0 = S\tau/N_0 = S\tau/kT_s$.

Losses to be Considered

(a) With respect to total energy ratio, there are two separate matching losses, and the effect of beamshape loss, which must be considered:

Matching loss to single pulse, L_m, depends on pulsewidth τ, receiving filter noise bandwidth B_n, and the shape of the filter [C.1, p. 84].

Matching loss to the pulse train, L_{mf}, depends on the observation time t_o, doppler filter bandwidth B_f, and shape of the doppler filter [C.1, pp. 251-257 and 339-348].

Beamshape loss is $L_p = 1.3$ dB for one-coordinate scan, $L_p^2 = 2.5$ dB for two-coordinate scan.

A portion of L_m may be recovered, in the case of $B_n > 1/\tau$, by video filtering, at the expense of a *collapsing loss* (see Section 4.4.6).

A portion of L_{mf} may be recovered, in the case of $B_f > 1/t_o$, by low-pass filtering (noncoherent integration) at the output of the doppler filter, at the expense of noncoherent integration loss.

(b) With respect to single-pulse energy, L_m and L_p are calculated as in (a) preceding. Losses in using the information in multiple pulses are included in the detectability factor calculation.

C.2.3 Effect of STC

Ideal Assumption

The receiver operates at full gain and sensitivity during the portion of the pulse repetition interval not occupied by the transmission.

Losses to be Considered

The receiver noise factor is increased in the time-delay region where gain is reduced by sensitivity time control (STC).

C.3 Signal Processing Losses

Radar detection range is calculated by setting the actual received energy ratio (for a single pulse or for an extended observation time) equal to the detectability factor, or required energy ratio.

C.3.1 Detectability Factor

Ideal Assumptions

(a) Detection is based on total signal energy received during the observation time t_o, integrated coherently in a matched filter and converted to baseband by a synchronous detector with the reference phase matched to signal phase. The target returns a steady signal and competing interference is Gaussian noise of known power.

(b) Detection is based on total signal energy as above, but an envelope detector is used at the matched filter output.

(c) Detection is based on coherent integration over a fraction of the observation time, followed by envelope detection and postdetection integration.

(d) Detection is based on an IF matched filter for the individual pulse, followed by envelope detection and optimum postdetection integration over the observation time.

Losses to be Considered

Assumption (a) is the ideal reference, never achieved in practical radar systems because the signal phase cannot be known prior to target detection. The detectability factor for this ideal case, calculated from the integral of the normal distribution, is given in [C.2, pp. 63-66]. The following losses represent increases in signal energy required to achieve a given detection performance, relative to this ideal case.

Envelope detector loss. Under assumption (b), the basic detectability factor (Figure 4.3.2) is calculated from the Rician distribution (Figure 4.3.1). The increase, compared to assumption (a) is the envelope detector loss (Figure 4.3.3 and Eqs. 4.3.1 and 4.3.2). When calculation of detectability factor starts with Figure 4.3.3, envelope detector loss for the single-pulse case is already included.

Integration loss. Under assumptions (c) and (d), the increase in envelope detector loss, when signal power is reduced to take advantage of postdetection integration, is described as integration loss (Figure 4.4.1).

Binary integration loss. When the postdetection integration in (c) or (d) is performed in an optimum binary integrator, a loss of 1.5 dB results (Figure 4.4.2).

Cumulative detection loss. When no postdetection integration is used, but detection probabilities are combined in a cumulative process, a larger loss is incurred (Figure 4.4.2). When this process extends over several scans, the loss is described as *scan distribution loss.*

Collapsing loss. This loss (Table 4.1, Eq. 4.4.13) results when extra noise samples are included in the postdetection integration.

Integrator weighting loss. This loss (Section 4.4.6) results from failure to match the weighting function of the postdetection integrator to the envelope of the pulse train.

Fluctuation loss. This loss L_f (Figure 4.5.2) describes the increase in average signal power necessary to preserve a given detection probability when the target fluctuates. The use of diversity may reduce this loss (Section 4.5.4).

CFAR loss. When an adaptive threshold (CFAR) is used to control false alarms, there is a CFAR loss [C.2, p. 90].

Eclipsing loss. This loss results when the radar operates in medium- or high-PRF modes, in which case targets beyond the unambiguous range may be eclipsed by the transmission. In a tracking radar, it may be eliminated by proper choice of PRF.

Straddling losses. These losses are described as (a) range straddling loss, when range gates or sampling strobes at regular intervals may not be centered on the signal delay; (b) filter straddling loss, when doppler filters may not be centered on the spectral line; and azimuth straddling loss, when pulse bursts not centered on the envelope of the pulse train are batch processed.

MTI and doppler losses. These losses [C.2, pp. 250, 268] are the result of filter responses to targets and to noise, and possible deficiencies in phase detector implementation.

Operator loss. This is the loss describing the inability of the operator to equal the performance of the ideal threshold detector process.

The product of the several losses listed above, along with matching loss and beamshape loss, multiply the basic single-pulse, steady-target detectability factor to give the effective detectability factor $D_x(n)$, $D_x(1)$ or $D_x(n')$ in radar equations (3.1.22), (3.1.23), or (3.1.24), respectively.

C.4 Losses in Search Radar Equation

C.4.1 Ideal Assumptions

The search radar scans uniformly over a solid angle ψ_s in frame time t_s, using a rectangular transmitting beam of solid angle ψ_b, with receiving beam(s) matched to the transmitting beam, and integrates echoes from steady targets in matched filters. Propagation is in free space, and there are no RF or signal processing losses.

C.4.2 Losses to be Considered

When the search radar equation (3.2.3) is used, the following losses must be added to those listed in Sections C.1 - C.3 above:

Beam pattern constant, L_n. This factor in (3.2.1) adjusts the relationship between gain and beam solid angle. Its minimum value is $1.1 = 0.4$ dB for an ideal antenna with any illumination function, and it must be increased to include the effects of other efficiency and loss terms listed in Section C.1.2.

The pattern-propagation factor F^4 must be included as a loss component $1/F^4$ in this equation and scan distribution loss L_d must be included when the search radar performs more than one scan of the volume within the search frame time t_s.

C.5 Tracking Loss Factors

C.5.1 Acquisition by a Tracking Radar

The acquisition range of a tracking radar is calculated using the same equations as for search radar, taking into account the reduced search solid angle ψ_s and resulting increased observation time t_o for a given beamwidth. Losses are as given in C.1 - C.4.

C.5.2 Losses Reduced or Eliminated during Tracking

> Beamshape loss, if tracking is on the beam axis;
> Binary integration loss;
> Collapsing loss;
> Integrator weighting loss;
> Fluctuation loss;
> CFAR loss;
> Eclipsing loss, if PRF is adaptively changed;
> Straddling losses;
> Operator loss.

C.5.3 Specialized Tracking Losses

Ideal Assumptions

The target is tracked on the beam axis using a monopulse process with sum and difference patterns optimized for minimum error in a thermal noise environment [C.2, pp. 381-382, 400]. Range and doppler tracking is performed with optimized gates and filters.

Losses to be Considered

In addition to the losses applicable to the normal radar equation, reduced as noted in Section C.5.2, the measurement slope constants are reduced by using illumination and weighting functions with lower sidelobes.

Error slope. Difference channel illuminations for monopulse tracking are normally chosen to give voltage slopes 0.7 to 0.5 times that of the idealized pattern [C.2, p. 403]. The increase in *S/N* required to regain a given accuracy in thermal noise is 3 to 6 dB. This requirement is already included in the error equation (10.4.4). Similar compromises apply to range and doppler tracking, Sections 10.5 and 10.6. For conical scanning, the slope is described by Figure 10.6, for use in (10.2.1). For sector scanning, the slope is described in (10.2.2).

Fluctuation loss. In a monopulse tracker using instantaneous normalization on fluctuating targets, increase in *S/N* is required to maintain the error level of the steady target of equal average RCS [C.2, pp. 411-412].

Multiplexing loss. When two-channel monopulse is used (Figure 10.18), half the error signal energy is discarded in multiplexing, leading to a 3-dB loss in effective *S/N* for accurate tracking.

Crossover loss. In conical-scan trackers, this loss (Section 10.2.1) accounts for the squinting of the beam axis from the tracking axis.

C.6 Imaging Loss Factors

C.6.1 Ideal Assumptions

An imaging synthetic aperture radar (SAR) performs unweighted coherent processing over the entire time t_s during which the real beam passes over a given surface region. Free-space propagation applies, with no RF losses and no signal processing losses.

C.6.2 Losses to be Considered

System temperature T_s. Since the radar beam is directed toward the surface, the antenna temperature approaches 290 K, rather than a lower sky temperature.

Pattern-propagation factor F. Free-space propagation, $F = 1$, is normal, except at the lowest usable grazing angles. Beamshape losses are included in the search loss factor to account for antenna gain variations over the mapped region.

Synthetic aperture beamwidth factor, *m*. Use of weighting in the coherent processor results in $m \approx 0.5$ in most systems, broadening the cross-range resolution. The increased size of the resolution cell cancels the matching loss, so no change in output *S/N* ratio results.

Search loss factor L_s. The remaining components of the normal search radar equation apply, except for omission of fluctuation loss, CFAR loss, and eclipsing loss.

C.7 Effects of Clutter and Jamming

C.7.1 Ideal Assumptions

(a) Only thermal noise is present to compete with target echoes.

(b) Jamming or clutter is present, characterized by Rayleigh amplitude distribution and uncorrelated from pulse to pulse.

C.7.2 Losses to be Considered

Under assumption (a), and jamming or clutter encountered by the system increases the background level, causing false alarms unless the detection threshold is raised proportionately. The loss in performance is given by the interference-to-noise ratio $I/N = (N+J+C)/N$, modified by effects of non-Rayleigh amplitude distribution and clutter correlation as described below.

Under assumption (b), a CFAR system is employed to maintain an acceptable false-alarm rate while retaining sensitivity in regions of low interference. The following losses are encountered:

CFAR loss (see Section C.3.1).

Clutter distribution loss. When clutter has a broad amplitude distribution, the CFAR threshold must be raised by a larger amount than for Rayleigh clutter of equal average power. The increased threshold leads to the clutter distribution loss [C.2, p. 93].

Clutter detectability factor D_{xc}. The S/C ratio required to achieve a given detection performance will be greater than the corresponding S/N ratio by the product of clutter distribution loss L_{cd} and the effect of clutter correlation time [C.2, p. 95]. Clutter correlation time larger than the PRI leads to a reduction in the number of independent samples integrated, with corresponding reduction in integration gain and increase in D_{xc}.

C.8 References for Appendix C

[C.1]   D. K. Barton and H. R. Ward, *Handbook of Radar Measurement*, Artech House 1984.

[C.2]   D. K. Barton, *Modern Radar System Analysis*, Artech House 1988.

Appendix D

Tables of Typical Target Cross Sections

The following tables of RCS values have been derived from the sources indicated. All values have been converted to dBm^2, rounded to the nearest dB. The range of values reported is indicative of the wide variation to be expected in targets of similar size but different construction, observed at slightly different aspect angles, with averages taken in a variety of ways.

Table D.1 RCS of Selected Targets

Type of Target	Aspect	RCS (dBm^2)
Small jet fighter or small commercial jet	Nose, tail	-5 to +10
	Broadside	+7 to +25
Medium bomber or midsize airline jet such as 727, DC-9	Nose, tail	+6 to +20
	Broadside	+23 to +29
Large bomber or large airline jet such as 707, DC-8	Nose, tail	+10 to +27
	Broadside	+25 to +27
Wooden minesweeper, 144 ft long, from airborne radar (5 - 10 GHz)		-10 to +25
Small bird, 450 MHz	Average overall	-56
Large bird, 9 GHz	Broadside	-20
Insect (bee), 9 GHz	Average	-28
Large insect (5 cm), 9 GHz	Average	-18

[D.1, p. 129]

Table D.2 Median Aircraft RCS for Various Classes
and Aspect Regions

Class	Aspect	RCS in dBm^2					
		VHF 0.03-0.3	UHF 0.3-1.0 GHz	L 1-2	S 2-4	C 4-8	X 8-12
Large, heavy bomber or 707, DC-8 size jet	Nose			+15	+16	+10, +16	+14
	Tail			+12, +18	+24		+14, +27
	Side			+27	+27		+25
	Ave.		+10	+14, +16	+14, + 16		+18
Medium attack bomber, 727, DC-9 size jet	Nose	+18, +20	+1, +15	+8, +13			+6
	Tail	100	6	20	20		
	Side	+25	+23	+25	+24		+29
	Ave.			+11	+11		+10
Small fighter or four-passenger jet	Nose	+10	+8	-5,+9	-5,+10	-1,+3	+1
	Tail	+10	+5	+3	-2,+12		
	Side			+7,+20	+15, +25		+13, +18
	Ave.			-1,+1	+1,+3		+1,+7

[D.2, p. 15]

Table D.3 Median RCS of Birds and Insects

		RCS in dBm^2		
	Aspect	UHF 0.3-1.0	S 2-4	X 8-12
Sparrow	Head			-46
	Side			-32
	Tail			-47
	Ave.	-56	-28	-38
Pigeon	Head			-40
	Side			-20
	Tail			-40
	Ave.	-30	-21	-28
Duck	Head	-12		
Grackle		-43	-26	-28
Hawkmoth, 5 cm		-54	-30	-18
Worker bee, 1.5 cm		-52	-37	-28
Dragonfly		-52	-44	-30

[D.2,p. 16]

Table D.4 RCS of World War II Aircraft

Aircraft Type	RCS (dBm^2)
OS2U	+12
Curtis-Wright 15D	+16
J2F Grumman amph.	+16
B-18	+18
B-17	+19
SNB	+14
AT-11	+13
PBY	+17
Taylorcraft	+12

[D.3,p. 78]

Table D.5 RCS of World War II Ships

Ship Type	RCS (dBm^2)	RCS (dBm^2)
	S-band	X-band
Tanker	+24	+24
Cruiser	+42	+42
Small freighter	+22	+22
Medium freighter	+39	+39
Large greighter	+42	+42
Small submarine (sur-faced)	+16	+22

[D.3 p. 80]

Table D.6 RCS of Aircraft

Target designation	Target name	RCS (dBm^2)
Aero 500	Aero Commander	+8
B-29	Super Fortress	+20
B-47	Stratojet	+12
B-52	Strato Fortress	+21
B-57B	Canberra	+10
707	Boeing 707	+15
720 and 727	Boeing	+14
Britannia	Bristol Britannia	+14
C-121	Constellation	+20
C-130	Hercules	+19
	Caravelle	+12
	Comet	+16
Cessna 180		+2
Cessna 310B		+6
Convair 240, 340, 440	Metropolitan	+16
DC-3	Dakota	+13
DC-8		+15
DC-9	+14	
F-1/FJ	Sabre	+7
F-4	Phantom	+10
F-27	Fokker Friendship	+14

Table D.6 (Continued) RCS of Aircraft

Target designation	Target name	RCS (dBm^2)
F-86	Sabre	+7
F-104	Starfighter	+7
IL-28	Beagle	+9
	Javelin	+9
	Lamps	+14
MIG-21	Fishbed	+6
P-3A	Orion	+19
P-3B	Orion	+20
TU-16	Badger	+14
TU-20	Bison	+16
TU-95	Bear	+21
	Viscount	+12

[D.4,p. 183]

Table D.7 RCS of Ships

Target designation		RCS (dBm^2)
Small cargo ship		+22
Large cargo ship		+42
Submarine surfaced		+27
Submarine schnorkel		+7
Aircraft carrier		+35
Destroyer		+31
Patrol boat		+19
Trawler		+12

[D.4,p. 183]

Table D.8 Example RCS at Microwave Frequencies

Target	RCS (dBm^2)
Conventional, unmanned winged missile	-3
Small, single-engine aircraft	0
Small fighter, or 4-passenger jet	+3
Large fighter	+8
Medium bomber or medium jet airliner	+13
Large bomber or large jet airliner	+16
Jumbo jet	+20
Small open boat	-17
Small pleasure boat	+3
Cabin cruiser	+10
Ship at zero grazing angle	See equation[1]
Ship at higher grazing angles	See equation[2]
Pickup truck	+23
Automobile	+20
Bicycle	+3
Man	0
Bird	-20
Insect	-50

[1] Equation for ship RCS at zero grazing angle:
$\sigma = 52f^{1/2}D^{3/2}$, where f is in MHz and D is full-load displacement in kilotons.
[2] Ship RCS at higher grazing angle: σ in m^2 = D in kilotons.

[D.5, p. 44]

References for Appendix D

[D.1]   L. V. Blake, *Radar Range Performance Analysis*, Artech House, 1986.
[D.2]   F. E. Nathanson, *Radar Design Principles*, McGraw-Hill, 1969.
[D.3]   L. N. Ridenour, *Radar System Engineering*, McGraw-Hill, 1947.
[D.4]   P. Rohan, *Surveillance Radar Performance Prediction*, Peter Peregrinus, 1983.
[D.5]   M.I. Skolnik, *Introduction to Radar Systems*, McGraw-Hill 1980.

Appendix E

CLUTTER AND JAMMING CALCULATIONS

E.1 Radar Range with Clutter

The calculation of target detection range in a background of clutter is one of the most difficult calculations to perform with accuracy. This is so for several reasons. The first reason is that the reflectivity of clutter sources, and the corresponding propagation factors, are only approximately known. The second is that the variation of clutter power with range is extremely uncertain, especially in the case of land clutter. The third is that the amplitude statistics of clutter may depart significantly from the Rayleigh distribution that characterizes random noise, and for which the detectability factor $D_x(n)$ has been determined. The fourth reason is that attenuation of clutter in MTI or doppler processors is difficult to predict with good accuracy. As a result of all these factors, range equations that purport to predict a range at which targets can be detected must be used with caution. Nonetheless, we will present here some equations which are useful when the output of the receiver-processor is dominated by clutter, with suggestions as to when these equations can be used. Equations that predict the signal-to-clutter ratio will also be used in the next section, when we consider multiple simultaneous sources of interference.

E.1.1 Volume Clutter: Rain and Chaff

These clutter sources are considered first because they are more homogeneous in spatial distribution and more nearly noise-like (being essentially Rayleigh-distributed in amplitude) than other clutter types. The radar resolution cell, at a given clutter range R_c, is characterized by its volume (Figure E.1).

$$V_c = \left(\frac{R_c \theta_a}{L_p} \right) \left(\frac{R_c \theta_e}{L_p} \right) \left(\frac{\tau_n c}{2} \right) \tag{E.1.1}$$

where θ_a and θ_e are the azimuth and elevation beamwidths in radians, $L_p = 1.33$ is the beamshape loss τ_n is the width of the processed pulse (e.g., after pulse compression, if that technique is used), and c is the velocity of light. The radar section σ_c of clutter filling this volume is

$$\sigma_c = V_c \eta_v \tag{E.1.2}$$

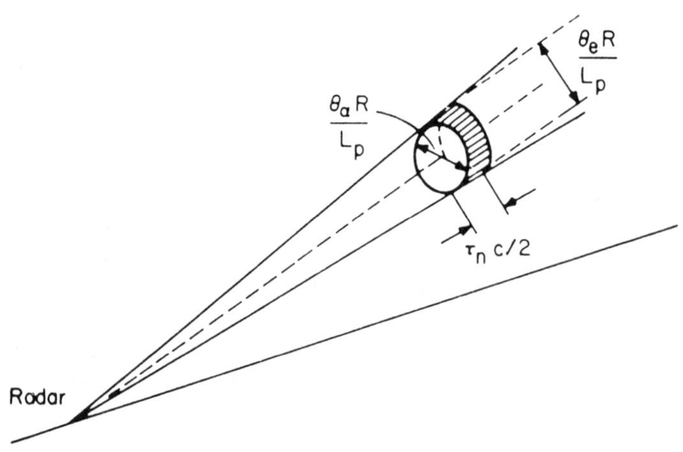

Figure E.1 Geometry of volume clutter.

where η_v is the volume clutter reflectivity (in units of m^2/m^3). If η_v and radar parameters are known, the received clutter power from this cell may be calculated:

$$C = \frac{P_t G_t G_r \lambda^2 \sigma_c F_c^4}{(4\pi)^3 R_c^4 L_t L_{\alpha c}} = \frac{P_t G_t G_r \lambda^2 \theta_a \theta_e (\tau_n c/2) \eta_v F_c^4}{(4\pi)^3 R_c^2 L_t L_{\alpha c} L_p^2} \tag{E.1.3}$$

where F_c^4 and $L_{\alpha c}$ are the pattern-propagation factor and atmospheric attenuation for the radar-to-clutter path. Both of these factors should be averaged over the resolution cell, in cases where they would change significantly with elevation angle. For an upward-looking narrow beam filled with clutter, the factor $F_c^4 \approx 1$, since the average effect of antenna pattern shape has been included as L_p2. However, for horizontal beams which illuminate an underlying reflecting surface, the factor may take the form:

$$F_c(\theta) = 2\sin(2\pi h_r \theta/\lambda) \tag{E.1.4}$$

and the average of F_c^4 is then equal to six. Intermediate values between one and six will apply to beams which are tilted slightly above the horizontal, or for operation over surfaces which are not totally reflecting.

If the clutter competes with a target at range R, the input C/S ratio will be

$$C/S = \frac{R^4\theta_a\theta_e(\tau_n c/2)L_\alpha\eta_v F_c^4}{\sigma F^4 L_p^2 R_c^2 L_{\alpha c}} \tag{E.1.5}$$

When the radar has no range ambiguities in which clutter is present, $R_c = R$, $L_{\alpha c} = L_\alpha$, and we obtain

$$C/S = \frac{R^2\theta_a\theta_e(\tau_n c/2)\eta_v F_c^4}{\sigma F^4 L_p^2} \tag{E.1.6}$$

When clutter is present in multiple range ambiguities,

$$C/S = \frac{R^4\theta_a\theta_e(\tau_n c/2)L_\alpha}{\sigma F^4 L_p^2} \sum_{i=1}^{j} \frac{\eta_{vi}F_{ci}^4}{R_{ci}^2 L_{\alpha ci}} \tag{E.1.7}$$

where each of the factors in the summation can have a separate value in each of the j ambiguities. In many cases, the $1/R_{ci}^2$ variation of terms in the summation permits us to omit all but the first term, and (E.1.6) can be used with the values for $i = 1$.

E.1.2 Detection Range in Volume Clutter

If we assume that the clutter, after possible reduction by MTI or doppler processing, remains the dominant source of background interference at the radar output, we may write

$$(S/C)_{out} = \frac{I_m}{C/S} = D_{xc}(n), \quad \text{or} \quad C/S = \frac{I_m}{D_{xc}(n)} \tag{E.1.8}$$

where I_m is the MTI or doppler improvement factor and $D_{xc}(n)$ is the clutter detectability factor, defined as the value of $(S/C)_{out}$ required to achieve the desired detection performance (means of calculating this factor are given in Chapter 4). For example, with no range ambiguities.

$$R^2 = \frac{I_m\sigma F^4 L_p^2}{D_{xc}(n)\theta_a\theta_e(\tau_n c/2)\eta_v F_c^4} \tag{E.1.9}$$

When the first range ambiguity is the dominant clutter source,

$$R^4 = \frac{\sigma F^4 L_p^2}{D_{xc}(n)\theta_a\theta_e(\tau_n c/2)L_\alpha}\left[\frac{R_{c1}^2 L_{\alpha 1}I_m 1}{\eta_{v1}F_{c1}^4}\right] \tag{E.1.10}$$

When subsequent ambiguities are significant sources, the term in brackets is replaced by

$$\left[\sum \frac{\eta_{vi} F_{ci}^4}{R_{ci}^2 L_{\alpha ci} I_{mi}} \right]^{-1} \tag{E.1.11}$$

Note that the improvement factor may be dependent on the range or spectral spread of the clutter in each ambiguity, and hence must be included in the summation.

E.1.3 Example of Detection Range in Rain

Assume a low-PRF radar (with no visible clutter beyond the unambiguous range $R_u = c/2f_r$), with rain clutter ($\eta_v = 2 \times 10^{-8} \, m^2/m^3$) filling the beam within R_u, the radar characteristics being those used previously in Figure 3.1.7. No effective MTI improvement will be assumed ($I_m = 1$) because of the wind velocity of the rain cloud. However, the clutter spectrum will be assumed narrow enough to reduce the integration gain to that for a system integrating only 4.5 pulses: $D_{xc}(24) = D_x$ $(4.5) = +21.1$ dB. From previous calculations, $\theta_a = 1.3° = 0.023$ r, $\theta_e = 2.0° = 0.035$ r, $\tau_n = \tau = 1$ μs. An upward tilted beam will be assumed, over a poorly reflecting surface, to give $F_c^4 = 1$, and the target will be assumed at the beam elevation, $F = 1$. Thus,

$$R^2 = \frac{1 \times 1 \times 1 \times 1.33^2}{129 \times 0.023 \times 0.035 \times 150 \times 2 \times 10^{-8} \times 1}$$

$$= 5.7 \times 10^6 \text{m}^2$$

$$R = 2380 \text{ m}$$

In order to obtain any significant target detection range, the radar would have to use a circularly polarized antenna, achieving a net improvement in S/C of about 18 dB. Then,

$$R^2 = 5.7 \times 10^6 \times 63 = 3.6 \times 10^8 \text{m}^2$$

$$R = 18.96 \text{ km}$$

Clearly, since clutter reduced the radar range by a factor greater than four, some type of doppler signal processing would be essential in this radar to provide useful surveillance range in rain.

E.2 Surface Clutter: Land and Sea

We may approach range calculation with a simple, homogeneous clutter model, in which the clutter fills the azimuth beamwidth out to a horizon range given by

$$R_h = \sqrt{2kah_r} \tag{E.2.1}$$

where *ka* is the effective earth radius (8.5×10^6 m) and h_r is the height of the radar antenna above the clutter surface. A more refined model, which takes into account the propagation factor for land and sea clutter, is described in Chapter 8.

The area of surface within the radar resolution cell (Figure 2.2.6) is

$$A_c = \left(\frac{R\theta_a}{L_p}\right)\left(\frac{\tau_n c}{2}\right) \sec\psi \tag{E.2.2}$$

where for surface-based radar $\sec\psi \approx 1$. The clutter cross section is

$$\sigma_c = A_c\sigma^0 \tag{E.2.3}$$

where σ^0 is the surface clutter reflectivity (a dimensionless quantity). The best simple model for reflectivity, at a grazing angle ψ, is

$$\sigma^0 = \gamma\sin\psi = \gamma h_r/R_c \tag{E.2.4}$$

where $\gamma \approx 0.1$ for land (values from 0.03 to 0.15 characterize different terrain types). For the sea, γ is dependent on wind conditions and radar wavelength (see Chapter 5). Using this model, σ_c is constant with range:

$$\sigma_c = \frac{h_r\theta_a}{L_p}\frac{\tau_n c}{2}\gamma \tag{E.2.5}$$

The clutter-to-signal ratio is

$$C/S = \frac{R^4 h_r\theta_a(\tau_n c/2)L_x\gamma F_c^4}{\sigma F^4 L_p R_c^4 L_{\alpha c}} \tag{E.2.6}$$

Then, for a system with no range ambiguities containing clutter, the C/S ratio becomes

$$C/S = \frac{h_r\theta_a(\tau_n c/2)\gamma F_c^4}{\sigma L_p F^4} \tag{E.2.7}$$

In the region well within the horizon, where $F = F_c = 1$, the output S/C ratio will be

$$(S/C)_{out} = I_m S/C = \frac{\sigma L_p I_m}{h_r\theta_a(\tau_n c/2)\gamma} \tag{E.2.8}$$

and if this exceeds the required $D_{xc}(n)$, the target will be detectable throughout this region. If the target is not detectable in the region where $F_c = 1$, the target may become detectable at longer range, where the F_c is reduced more rapidly than F. Models that predict this effect are given in Chapter 8.

When range ambiguities are occupied by clutter, (E.2.6) becomes

$$C/S = \frac{R^4 h_r\theta_a(\tau_n c/2)L_\alpha\gamma}{\sigma F^4 L_p}\sum_{i=1}^{j}\frac{F_{ci}^4}{R_{ci}^4 L_{\alpha ci}} \tag{E.2.9}$$

As with volume clutter, it is usually the first ambiguity, $i = 1$, which dominates the result, and can then be used with clutter values for that ambiguity.

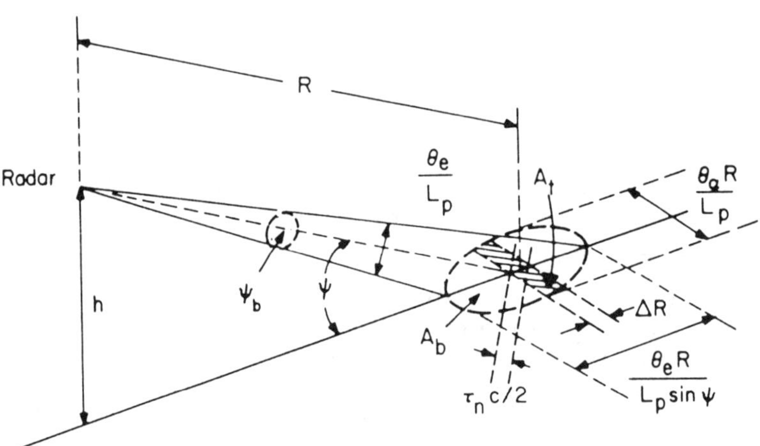

Figure E.2 Surface clutter geometry.

E.2.1 Example of Detection Range over Land

Consider first the case of unambiguous range. For example, with $I_m = 30$ dB, $h_r = 10$ m, $\gamma = 0.1$, we have

$$(S/C)_{\text{out}} = \frac{1 \times 1.33 \times 10^3}{10 \times 0.023 \times 150 \times 0.1} = 386, \quad \#2 \text{ or } +25.9 \text{ dB}$$

Targets of 1 m^2 would be detectable above clutter at all ranges in this case. With $\sigma = 0.3$ m^2 or $I_m = 25$ dB, and $D_{xc}(n) = 21.1$ dB, the detection requirement is almost met at all ranges within the clutter horizon, and would be exceeded beyond this horizon. With yet lower σ or I_m, the target is below the detection threshold at short ranges, but may become detectable when F^4/F_c^4 increases, near and beyond the horizon.

Now consider a high-PRF land-based radar, operating at $f_r = 100$ kHz, for which $R_u = 1500$ m. For a long-range target appearing in the center range gate (e.g., $R = 750$ m $+ iR_u$), and assuming $F_c = L_{xc} = 1$ at 750 m, (E.1.16) becomes

$$C/S = \frac{R^4 h_r \theta_a (\tau_n c/2) L_\alpha \gamma}{\sigma F^4 L_p (750)^4} = I_m/D_{xc}(n)$$

$$R^4 = \frac{I_m \sigma F^4 L_p (750)^4}{D_{xc}(n) h_r \theta_a (\tau_n c/2) L_\alpha \gamma}$$

Assuming $I_m = 80$ dB for a high-PRF doppler processor, and $F = L_x = 1$, with $D_{xc}(n) = D_x(n) = +16.1$ dB, we have

$$R^4 = \frac{10^8 \times 1 \times 1.33 \times 750^4}{40.7 \times 10 \times 0.023 \times 150 \times 0.1} = 3.0 \times 10^{17}$$

$$R = 23.4 \text{ km}$$

This shows the extreme requirements for doppler improvement factor in a range-ambiguous system.

E.2.2 Land Clutter Detectability Factor

Most land clutter exhibits a broad amplitude distribution, requiring that the threshold be set higher relative to the average clutter output than for random noise, if false alarms are to be controlled. In addition, the pulse-to-pulse correlation of land clutter, if not destroyed by effective MTI or doppler processing, can lead to less effective integration. Both of these effects lead to an increase in the clutter detectability factor: $D_{xc}(n) > D_x(n)$. This must be taken into account in calculating target detection performance over land.

E.3 Noise Jammer Examples

E.3.1 The Stand-Off Jammer

Consider the case of a stand-off barrage noise jammer (SOJ), directed at the S-band surveillance radar used in the example range calculation of Figure 3.1.7 in Section 3.1. We will assume the following jammer parameters:

$P_j G_j = 100$ kW$\qquad\qquad L_{xj} = 1.0$ dB
$B_j = 300$ MHz$\qquad\qquad P_j G_j / B_j = 3.33 \times 10^{-4}$ W/Hz
$R_j = 200$ km

With this jammer in the main lobe of the radar, $F_j = 1$, we find, from (5.6.2), the effective jamming temperature:

$$T_j = \frac{10^5 \times 10^4 \times 10^{-2} \times 1}{(4\pi)^2 \times 1.38 \times 10^{-23} \times 3 \times 10^8 \times (2 \times 10^5)^2 \times 1.25} = 3 \times 10^8 \text{K}$$

The resulting receiver noise level is 3.2×10^5 or $+ 55$ dB above receiver noise, reducing range by a factor of 23.7 to only 3.9 km.

However, if the jammer is in a region where average receiving antenna sidelobe levels are 50 dB below the main lobe gain ($F_j^2 = 10^{-5}$), the jammer will be $T_j = 3 \times 10^3$K, leading to $T_s + T_j = 3957$ K. This is only 6.1 dB above the normal receiver noise level. The detection range will be reduced by a factor of 1.42 to

61.2 km, providing significant surveillance performance in all sectors where the low sidelobe levels are directed toward the jammer. In other words, the jammer would be effective in preventing detection of a 1 m^2 target that lay between the radar and the SOJ only if the target were at the same azimuth as the jammer.

E.3.2 Self-Screening and Escort Jamming

The *self-screening jammer* is by definition, located on the target itself, such that $R_j = R$ and $F_j = F$. In calculating the range at which a target echo becomes detectable above the jamming (e.g., for ranging purposes), we find that our example radar must obtain an energy ratio:

$$\frac{E_1}{J_0} = \frac{P_t \tau G_t \sigma F^2 B_j}{4\pi P_j G_j R_{bt}^2 L_t L_{\alpha j}} = D_x(n) = 40.7 = +16.1 \text{dB}$$

Solving for burn-through range R_{bt}, we obtain:

$$R_{bt}^2 = \frac{P_t \tau G_t \sigma F^2 B_j}{4\pi P_j G_j L_t L_{\alpha j} D_x(n)}$$

This computation rarely yields a useful target detection range. In our previous radar example, if the self-screening noise jammer has an ERP of 10 KW and $B_j = 300$ MHz we find

$$R_{bt}^2 = \frac{10^5 \times 10^{-6} \times 10^4 \times 1 \times 1 \times 3 \times 10^8}{4\pi \times 10^4 \times 1.26 \times 1.26 \times 40.7} = 3.69 \times 10^4 \text{m}^2$$

$$R_{bt} = 192 \text{ m}$$

The escort jammer closes in range with the target, but may lie outside the main lobe in which the target is to be detected. If the sidelobes directed toward the escort jammer are at a level F_j^2 relative to the mainlobe, the burn-through equation will have F^4 instead of F^2 in the numerator, and F_j^2 in the denominator, greatly increasing the burn-through range on the target. In our example, if $F_j^2 = 10^{-4}$ or -40 dB (for an escort jammer outside the first sidelobes), we find

$$R_{bt}^2 = 3.39 \times 10^8 \text{m}^2$$

$$R_{bt} = 18.4 \text{ km}$$

If the escort jammer is in the skirt of the main lobe or in one of the principal sidelobes, even this range may not be available. It should be noted that self-screening noise jammers play a dangerous game, for although the jammer can, if it jams the radar mainlobe, deny range information to that radar, it simultaneously provides a strong beacon signal that actually enhances detection and angle track by the victim radar. Thus, in the case of penetrating jammers, the radar system response is not to await burn-through, but to identify the angles of the jammers and to attempt triangulation with data from adjacent radar sites. The jammers are then tracked passively in angle, and missile guidance is accomplished by using track-on-jam or home-on-jam seeker, when the target is within the missile range.

Appendix F

TYPICAL ANTENNA PATTERNS
AND ILLUMINATION FUNCTIONS

In Chapter 6, it was shown that antenna gain is maximized when the antenna aperture is uniformly illuminated. That is, with uniform illumination, each antenna element (or equivalently, for a continuous reflector antenna illuminated by a horn feed, an antenna "patch" of dimensions ($\lambda/2$ x $\lambda/2$), will see the same value of electric field intensity at the same distance y (from each element and perpendicular to the antenna face) if the elements are perfectly in phase. The process of designing the *antenna illumination function*, or the *aperture distribution* so that the electric field intensity is not uniform over the aperture, but biased so that it is greater near the center called *tapering*. The object of tapering is to reduce the antenna sidelobe levels, and this occurs because with tapering more of the total energy is directed to the mainlobe. Tapering also causes a broadening of the mainlobe, and a reduction in mainlobe gain relative to that achievable with uniform illumination.

F.1 Illumination Function

We define the illumination function $g(x,y)$ such that it describes the illumination amplitude (normalized to unity at the center) in terms of distances x and y from the center, then for uniform illumination:

$$g(x,y) = 1, \ |x| < w/2, \ |y| < h/2$$

where w and h are the antenna planar dimensions. The two-dimensional antenna pattern, in terms of azimuth angle θ and elevation angle φ, can be found by taking the Fourier transform of the illumination function:

$$f(\theta, \psi) = \frac{1}{A'} \int_{-h/2}^{h/2} \int_{-h/2}^{w/2} g(x, y) \exp[j(2\pi/\lambda)(x\sin\theta\cos\phi + y\sin\theta\sin\phi)]dx\,dy$$

where $A' = \iint g(x, y)dx$ is an effective area which normalizes $f(0,0) = 1$.

F.2 Aperture Gain

The gain of the tapered aperture can be expressed as

$$G_m = \frac{4\pi \mid \int\int g(x,y)dxdy \mid^2}{\lambda^2 \int\int g^2(x,y)dxdy} = G_0\eta_a$$

For patterns not extending too far from the z axis, $\sin\theta \approx \theta$, and we may express the azimuth and elevation patterns as

$$f(A,E) \leftrightarrow g(x,y)$$

where $\leftrightarrow$ represents the Fourier transform.

F.3 Antenna Beam Pattern

In a rectangular array, the illumination and pattern functions are usually separable into functions in the two principal planes:

$$g(x,y) = g(x)g(y), f(A,E) = f(A)f(E)$$

Setting $\varphi = 0$ so that θ represents azimuth angle A, we may write

$$f(\theta) = \frac{2}{A_x}\int_{-w/2}^{w/2} g(x)\exp[j(2\pi/\lambda)\times\sin\theta]dx$$

where $A_x = \int g(x)dx$ normalizes the pattern. Similar expressions apply to elevation with $\varphi = \pi/2$, y replacing g and h replacing w. Figure F.1 illustrates the Fourier transform relationship in one plane. Substitution of $g(x) = g(y) = 1$ yields the familiar $(\sin x)/x$ pattern.

$$f(\theta,\phi) = \frac{\sin[(\pi w/\lambda)\sin\theta\cos\phi]}{(\pi w/\lambda)\sin\theta\cos\phi}\frac{\sin[(\pi h/\lambda)\sin\theta\sin\phi]}{(\pi h/\lambda)\sin\theta\sin\phi}$$

For tapered illumination the beamwidth and efficiency will vary with level of the first sidelobe. In a rectangular aperture with separable illuminations, the aperture efficiency η_a will be

$$\eta_a = G_m/G_0 = G_m\lambda^2/4\pi A = \eta_x\eta_y$$

F.4 Results for Typical Illumination Functions

Table F.1 lists the first sidelobe ration, the beamwidth constant, and the efficiency for some common taper functions used in linear and rectangular arrays. More distant sidelobes will have amplitudes varying reciprocally with the square of the angle θ, or more accurately, with the quantity $[2.4\theta/\theta a(2-1)]$. Only the first sidelobe will exceed -30 dB, for cosine illumination.

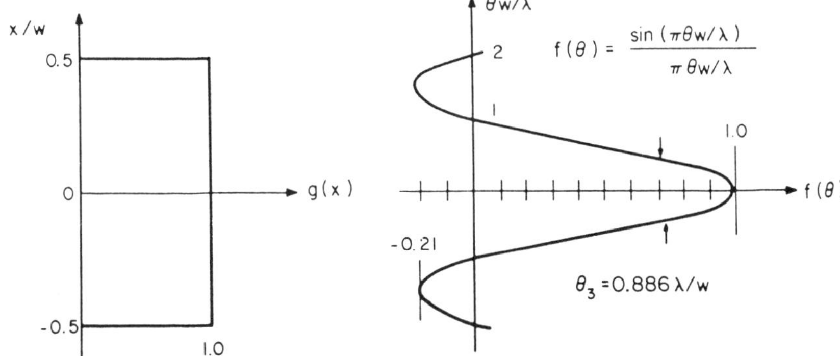

Figure F.1 Fourier transform relationship between aperture illumination and beam pattern.

The relationship between gain and beam solid angle remains almost constant, for all practical illuminations:

$$G_m = 11.3/\theta_a\theta_e \text{ (in radians)} = 37,100/\theta_a\theta_e \text{ (in degrees)}$$

This corresponds to $L_n = 1.1 = 0.5$ dB in Eq. (3.2.1). Note that the adjustment L_n is less than for uniform illumination, indicating that the use of taper increases the beamwidth more rapidly than it decreases the gain. This is the result of transferring power from the sidelobes into the mainlobe. The constant 37,100 may be compared with the frequently used relationship $G_m = 25,000/\theta_a\theta_e$ for reflector antennas, corresponding to $L_n = 1.65$ or 2.2 dB, which is indicative of the greater loss of illumination power in spillover and blockage lobes. The area required in the aperture, for a given gain or beamwidth product, varies as $1/\eta_a = 1/\eta_x\eta_y$, as does the number of $\lambda/2$ spaced elements covering this area.

Table F.1 Antenna Parameters for Tapered Rectangular Apertures

Illumination function, $g(x)$	Beamwidth $\theta_3 w/\lambda$	Sidelobe G_s/G_m (dB)	Efficiency η_x (dB)	(ratio)	Efficiency η_x^2 (dB)	(ratio)
$\cos(\pi x/w)$	1.19	-23	-0.9	0.80	-1.8	0.64
$\cos^2(\pi x/w)$	1.44	-32	-1.8	0.66	-3.6	0.44
$1 - (2x/w)^2$	1.18	-21	-0.8	0.83	-1.6	0.69
$[1 - (2x/w)]^2$	1.37	-27.5	-1.6	0.69	-3.2	0.48
$1 - |2x/w|$	1.27	-27.1	-1.3	0.74	-2.6	0.55
Taylor, $n = 3$	1.05	-25	-0.5	0.90	-1.0	0.81
$n = 4$	1.12	-30	-0.7	0.85	-1.4	0.72
$n = 5$	1.18	-35	-0.9	0.80	-1.9	0.64
$n = 6$	1.25	-40	-1.2	0.76	-2.4	0.58

F.5 Elliptical and Circular Apertures

When low-sidelobe tapers are used for $g(x)$ and $g(y)$, the combined $g(x,y)$ near the corners of the aperture is very small, and those elements or portions of the aperture can be omitted with little effect on performance. The result is an elliptical aperture with gain and pattern that are essentially the same as those of the rectangle having the same width w and height h. The area of the ellipse and the number of elements are $\pi/4$ times those of the rectangle, and hence its aperture efficiency η_a will be greater by about $4/\pi = 1.05$ dB.

Consider a circular aperture of diameter D formed by omitting the corners from a square, $w = h = D$, and designed with 40 dB Taylor weighting in both coordinates (Figure F.2). The original square aperture gave

$$\theta_3 = 1.25\lambda/w, \eta_x = \eta_y = 0.763, \eta_a = 0.582$$

The circle, with illumination adjusted slightly for -40 dB circular sidelobes, has

$$\theta_3 = 1.29\lambda/D, \eta_a = 0.706, \eta_a\pi/4 = 0.555$$

Considering the reduced aperture area of the circle, the circle is 0.84 dB more efficient than the square, with a beamwidth only three percent greater. The solid angle occupied by the principal sidelobes will, on the otherhand, be greater than for the rectangular aperture, but in many cases the actual levels outside the main beam will be set by random errors in the illumination rather than by the intended $g(x,y)$. The relationship between gain and beamwidth is essentially the same as for the rectangular aperture. Table F.2 compares parameters for common circular taper functions with those of rectangular arrays.

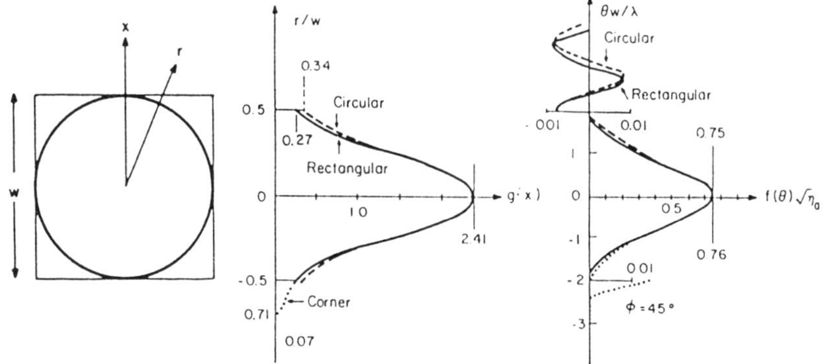

Figure F.2 Circular aperture derived from square.

Table F.2 Typical Parameters for Rectangular and Elliptical Antennas
(Taylor Illumination Functions)

Sidelobe Level	Parameter	Rectangular Aperture	Elliptical Aperture
-25	K_θ	1.05	1.12
	η_a	0.81	0.91
-35	K_θ	1.18	1.23
	η_a	0.65	0.77
-45	K_θ	1.30	1.33
	η_a	0.53	0.65

NOTE: $K_\theta = \theta_3 w/\lambda$ or $\theta_3 D/\lambda$. In all cases, $L_n = 4\pi/\theta_3^2 = 1.11$.

F.6 Typical Antenna Patterns

Figures F.3 - F.6 show antenna patterns, plotted on linear voltage scales and logarithmic scales, for four typical illumination functions:

(a) Uniform illumination;

(b) Parabolic illumination;

(c) Cosine illumination; and

(d) The difference pattern formed as the first derivative of the pattern for cosine illumination.

Many other patterns are plotted in [F.1].

F.7 Reference for Appendix F

[F.1] D. K. Barton and H. R. Ward, *Handbook of Radar Measurement,* Artech House, 1984.

x := -5,-4.99 ..5   x = θ/(θ sub 3)

$$f(x) := \frac{\sin(2.7831 \cdot x)}{2.7831 \cdot x}$$

$$f'(x) := 20 \cdot \log(|f(x)|)$$

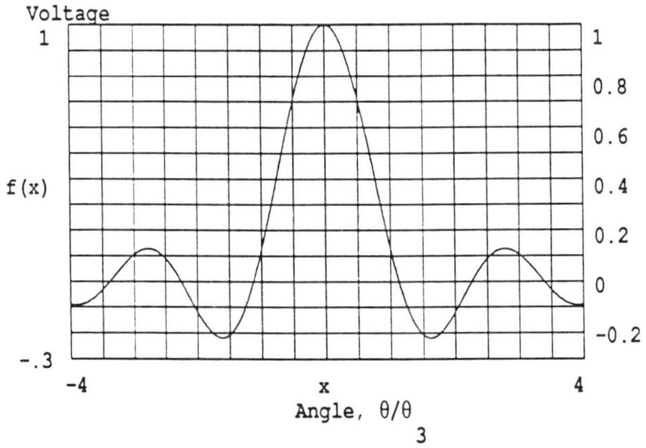

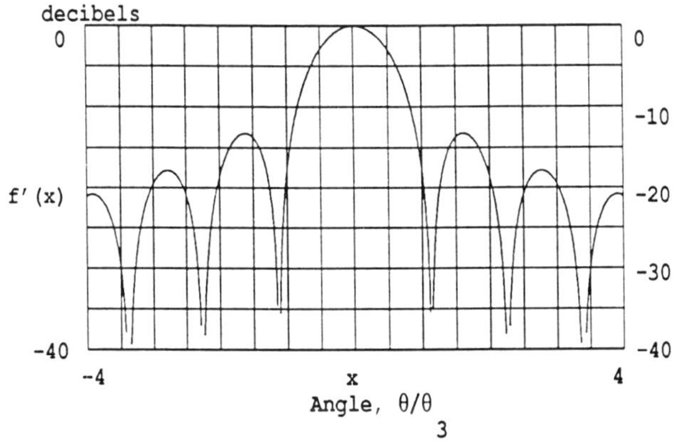

Figure F.3 Antenna pattern for uniform illumination.

$$x := -15, -14.8 \,..15 \qquad\qquad x = \theta/(1.15\theta_3)$$

$$f(x) := \frac{\sin(x)}{x} \cdot 1.5 + \frac{d}{dx}\left[\frac{d}{dx}\left[\frac{\sin(x)}{x}\right]\right] \cdot 1.5$$

$$f'(x) := 20 \cdot \log(|f(x)|) \qquad\qquad f(0.001) = 1$$

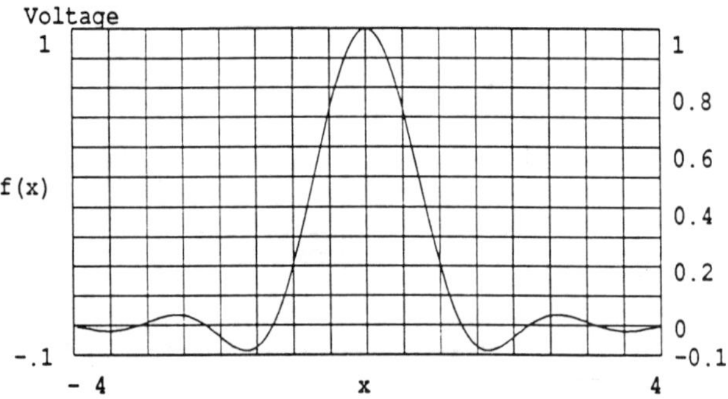

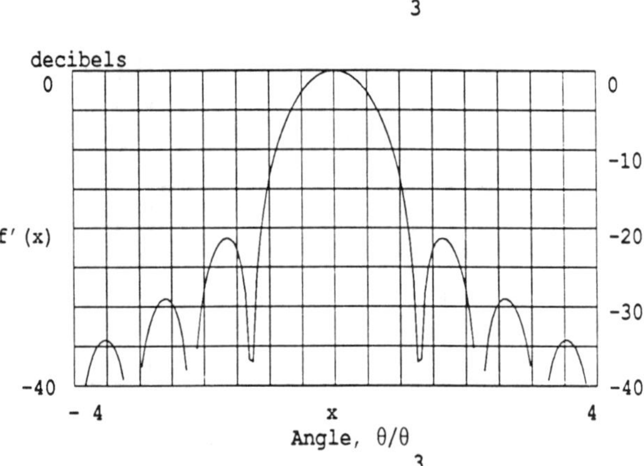

Figure F.4 Antenna pattern for parabolic illumination.

$$x := -5, -4.99 \dots 5 \quad x = \theta/\theta_3$$

$$f(x) := \frac{\cos(3.7352 \cdot x)}{1 - 5.6544 \cdot x^2}$$

$$f'(x) := 20 \cdot \log(|f(x)|)$$

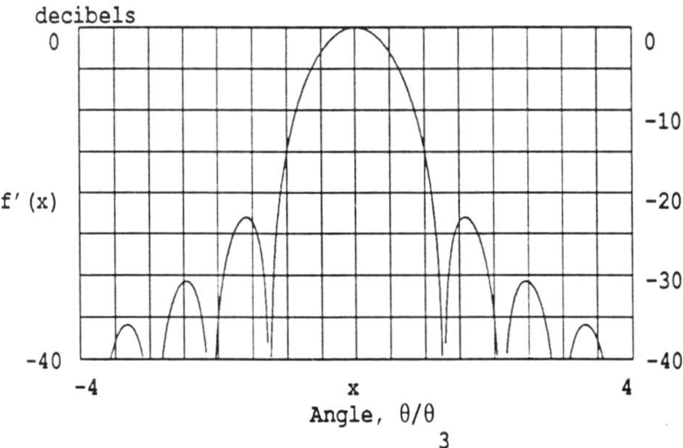

Figure F.5 Antenna pattern for cosine illumination.

$$f(x) := \frac{6 \cdot \left[\left[\frac{\pi^2}{4} - u(x)^2\right] \cdot \sin(u(x)) - 2 \cdot u(x) \cdot \cos(u(x))\right]}{\left[\frac{\pi^2}{4} - u(x)^2\right]^2}$$

$$f'(x) := 20 \cdot \log(|f(x)| + 0.001)$$

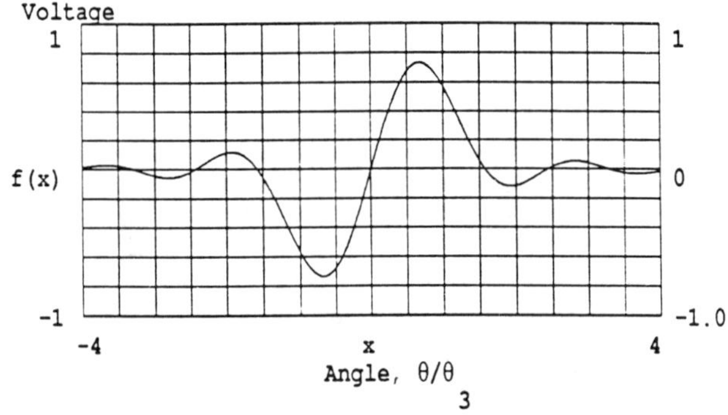

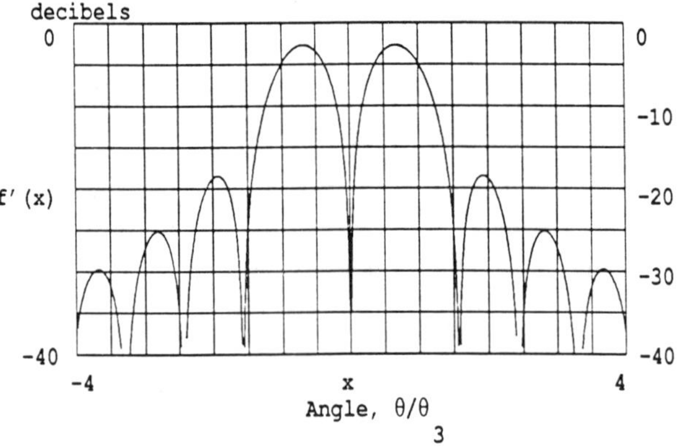

Figure F.6 Difference pattern formed as first derivative of pattern for cosine illumination.

Appendix G

THE RADAR AMBIGUITY FUNCTION

G.1 The Radar Ambiguity Function

The radar ambiguity function is a mathematical concept that provides a measure of the "ambiguity" inherent in a radar waveform with respect to the time intervals (target ranges) and doppler shifts (target velocities) of interest. Criteria for defining the "goodness" of a radar waveform, in the context of the specific target and interference (including clutter) environment, are generally based on the radar's ability to distinguish the differences superimposed by two or more targets on their respective reflected radar signals. One such criterion is the "mean square departure" of the waveform from its shifted self in time (t_a) and frequency (f_d). Mathematically, this mean square difference criterion is expressed as (G.1)

$$\int_{-\infty}^{\infty} |\psi(t) - \psi(t - t_d, f_0 - f_d)|^2 \, dt \tag{G.1}$$

where $\psi(t)$ is the complex representation of the waveform $f(t)$ given by

$$\psi(t) \;=\; a(t)\exp[j2\pi f_0 t + \theta(t)] = u(t)\exp[j2\pi f_0 t]$$

and

$$a(t) \;=\; \text{signal envelope}$$

$$\theta(t) \;=\; \text{phase modulation added to the carrier frequency } f_0$$

The expression $u(t) = a(t)\exp j\theta(t)$ is referred to as the complex signal envelope and can be substituted in (G.1) to perform the integration. This results in the following expression as the only result of the integration that depends on the time difference t_d and the frequency difference f_d

$$\chi(t_d, f_d) \;=\; \int_{-\infty}^{\infty} u(t)u^*(t - t_d)\exp[j2\pi f_d t]\,dt \tag{G.2}$$

Squaring (G.2) yields

$$|\chi(t_d, f_d)|^2 \;=\; \left| \int u(t)u^*(t - t_d)\exp[j2\pi f_d t]\,dt \right|^2 \tag{G.3}$$

This last relationship is designated as the radar ambiguity function, which can be plotted as a surface above the t_d - f_d plane. The height of this surface is a measure of the ambiguity (or interference) generated by the waveform at a point displaced from a target's true position ($t_d = 0$) and true doppler shift ($f_d = 0$) by an amount (t_d, f_d). Conversely, the ambiguity function predicts the interference created at the location of a desired signal ($t_d = 0$, $f_d = 0$) by an undesired target at range displacement t_d and doppler offset f_d.

G.2 Physical Interpretation of the Ambiguity Function

Examination of (G.1) for the special case of $f_d = 0$ leads to the conclusion that $\chi(t_d, 0)$ is the waveform observed at the output of a matched filter under ideal matched conditions. By extension, (G.2) describes all of the possible matched filter outputs when the doppler shift varies from the maximum negative value (target receding) to the maximum positive value (target approaching). Figure G.1 illustrates this physical interpretation for the rectangular pulse whose carrier is linearly swept over a frequency interval Δf and the doppler shift varies from Δf to $-\Delta f$. Here it can be seen that the effect of a doppler shift is to cause the signal to shift in range. For small doppler shifts the range shift is approximately linear, being given by $\Delta t_d = f_d \tau / \Delta f$. Under these conditions there is very little change in the shape of the compressed pulse. This creates the well-known range-doppler ambiguity associated with linear FM pulse compression. Thus, in absence of prior knowledge of target location or doppler shift, one cannot tell from observing the matched filter output if the compressed pulse indicates the true range of a zero-doppler signal or the offset range (by Δt_d) of a signal with doppler shift f_d.

G.3 Volume Constraint for Ambiguity Functions

The three-dimensional ambiguity function shown in Figure G.1 provides a wealth of detail that may not be particularly useful to a first-order understanding of different types of radar waveform applications. In making preliminary judgments on the suitability of various waveforms it is often helpful to visualize an ambiguity function contour for some level below (say -3 dB or -6 dB) its peak value at $t_d = 0$, $f_d = 0$.

Figure G.2 illustrates examples of such contours for a wide pulse, narrow pulse, and a wide pulse that has had linear FM added to its carrier frequency. Assuming the same energy content for each pulsed signal, the contours shown, designated as uncertainty ellipses, provide an indication of the potential utility of each waveform. For example, the narrow pulse uncertainty ellipse in Figure G.2(a) would be effective in an environment of relatively closely spaced targets that have very little doppler spread among them. The wide pulse uncertainty ellipse, on the other hand, would not be effective in this situation, but would be useful against targets that are widely spread in doppler shift. In this case, a bank of parallel matched filters, tuned to frequencies offset from the transmitted frequency by multiples of $1/\tau$, would be able to detect individual targets at different doppler shifts that overlapped in range.

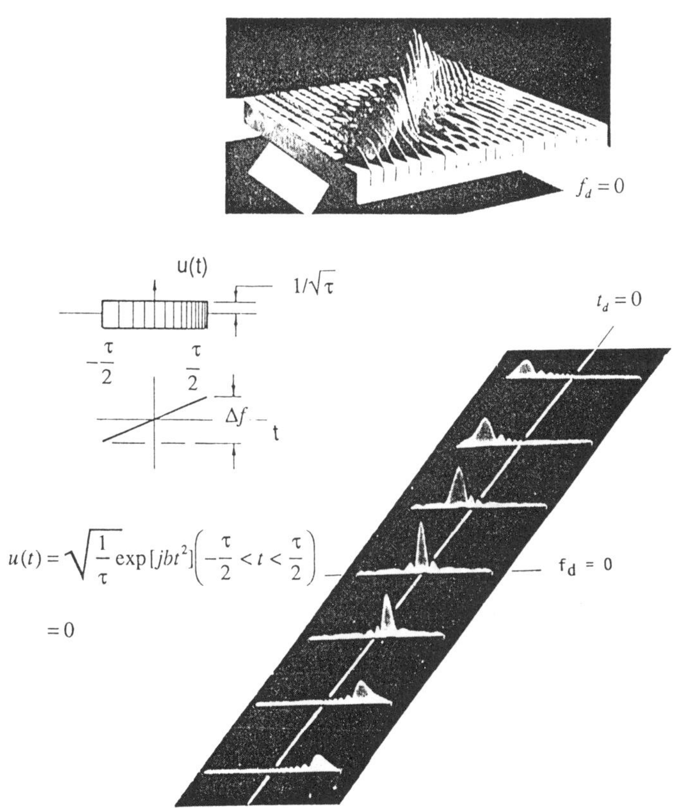

Figure G.1 Linear FM pulse waveform and response function properties for $\tau\Delta f = 10$. (a) Waveform. (b) Calculated surface.
(c) Experimental composite surface.

The uncertainly ellipse for the linear FM pulse, Figure G.2(c), shows that the addition of linear FM to the carrier causes the uncertainty ellipse to rotate in the time/frequency plane, thus achieving greater resolution along the time axis. However, the tilted aspect of the linear FM uncertainty ellipse shows that signals with doppler offsets from the matched condition will cause apparent range, or time, shifts at the matched filter output. This creates ambiguity as to the actual target location. This may not be of practical significance if the doppler shifts are small. However, if the doppler shifts are relatively large, then additional processing would be required to sort out the linear FM range/doppler ambiguity.

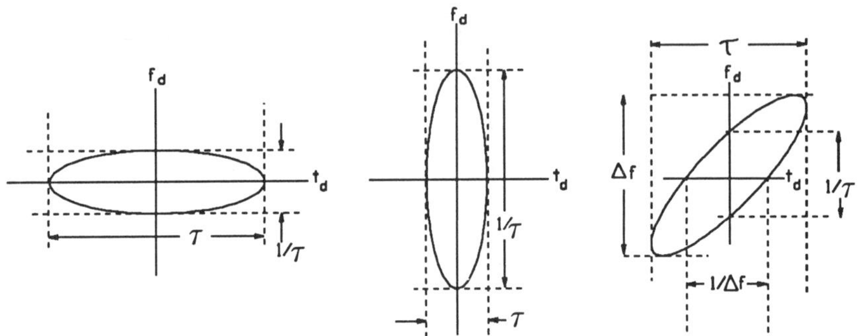

Figure G.2. Ambiguity function contours. (a) Long pulse. (b) Short pulse. (c) Linear FM pulse.

If the areas contained within the contours of Figure G.2 are considered to be a measure of the total ambiguity associated with each waveform, then one can make the intuitive observation that the choice of any one of the three waveform examples does not change the total amount of waveform ambiguity, but only results in its redistribution within the time/frequency plane [G.2]. This intuitive conclusion is actually confirmed by strict mathematical analysis that results in the most important result of the extensive studies of ambiguity function properties. This result is known as the *volume invariance property* of ambiguity functions and is expressed mathematically as

$$\int\int |\chi(t_d, f)|\, dt\, df \;=\; 2E^2 \tag{G.4}$$

where E is the energy contained in the waveform of interest. If, for convenience, $2E = 1$, then the integral representation of (G.4) is also equal to unity. Thus, with this normalization, both the peak of the ambiguity function ($t_d = 0, f_d = 0$) and the total volume under its surface are equal to one.

The implication of equation (G.4) is that no one waveform has any less total ambiguity than any other. Their only differences are how the ambiguity is distributed in the time/frequency plane as shown by the simple examples of Figure G.2. Thus, without prior knowledge of a radar's mission, it is not possible to tell which are the "best" waveforms to use. The analogy used by Siebert [G.2] likens the total waveform ambiguity to the sand in a sandbox. It is theoretically possible to shove it around and pile it up in different locations of the sandbox, but it is not possible to add to or take away from the total amount of sand in the box. The sandbox is analogous to the time/frequency plane, and the best that the radar designer can hope to do is to choose a waveform that results in the least amount of ambiguity in those portions of this plane where targets are expected to exist. This is, more often than not, difficult to achieve and the ultimate choice of a radar waveform, or set of waveforms, is based on practical compromises.

G.4 The Ideal Ambiguity Function

Since the object of choosing a radar waveform is to achieve the best possible measurement capability in both time and frequency, a number of speculations have been made as to the possibility of designing an "ideal" waveform to accomplish this. Since the theoretical time resolution of a waveform is proportional to 1/(bandwidth), it was thought that some very large time-bandwidth product waveform might exist having the spike-like ambiguity function shown in Figure G.3. Since the volume contained under this spike becomes vanishingly small as both Δf and τ increase, it was realized that this type of ambiguity function was an impossibility since it would violate the volume constraint given by equation (G.4). An approximation to the ideal ambiguity function was postulated as a spike sitting on a low-level, uniform-height platform of width 2τ in the time dimension and $2\Delta f$ in the frequency dimension, as shown in Figure G.4. This form of ambiguity function, designated as the "thumbtack," is of interest since it can be approximated by certain types of phase-coded pulse compression signals.

In order to meet the volume constraint of equation (G.4) for $|\chi|^2$ the thumbtack platform, with area $4\tau\Delta f$ would have to have a height of $1/4\tau\Delta f$. In addition, a single resolution cell would have an area of $\tau\Delta f$, so that there would be a total of $(2\tau\Delta f)^2$ resolution cells in the bounded time/frequency plane of Figure G.4. The question then follows; Is this a "good" ambiguity function? The answer, unfortunately, is ambiguous. Assuming that $|\chi(0,0)| = 1$ and there are only a few competing targets at other time/frequency locations in the ambiguity plane, then the signal-to-interference ratio caused by any one of these other targets at the desired target location $(t_d = 0, f_d = 0)$ is proportional to $4\tau\Delta f$. Thus, the thumbtack ambiguity function would provide excellent discrimination against these few additional targets if the $\tau\Delta f$ product is large. However, if there are very many distributed scatterers, such that there is at least one in each ambiguity plane resolution cell, then the resulting ambiguity-function generated interference results in a signal-to-interference ratio at the desired target location that is proportional to $1/\tau\Delta f$. In this latter target situation, which is representative of very many moving targets having a large extent in both range and doppler shift, the signal-to-interference ratio becomes worse as the signal time-bandwidth product increases. It should be noted when this latter type of target environment prevails, that no waveform has an ambiguity function that performs any better than the thumbtack [G.4, 7.1].

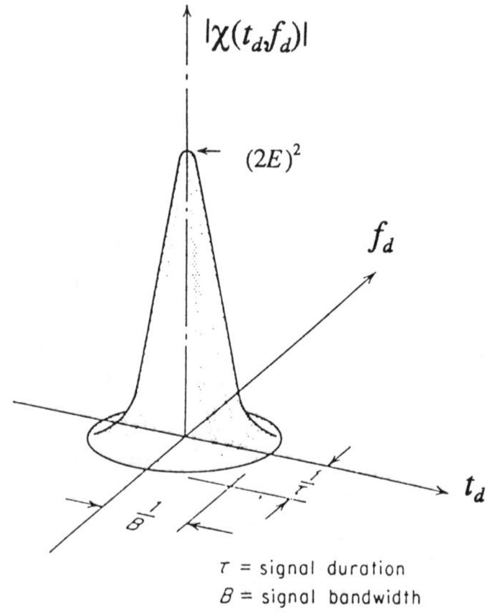

Figure G.3 An approximation to the ideal ambiguity function, taking account of the restrictions imposed by the fixed value 2E. (After Skolnik, [G.3].)

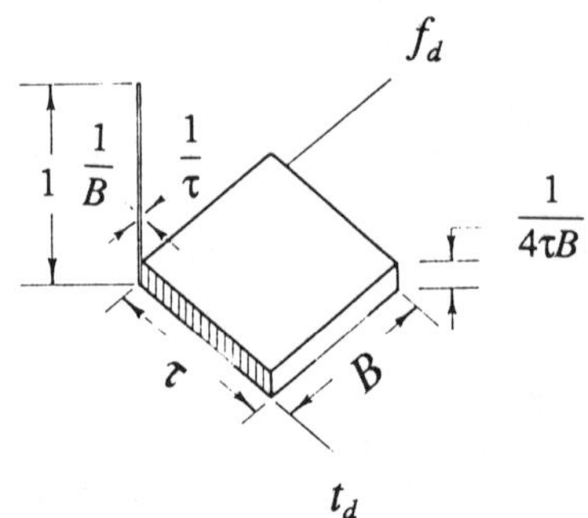

Figure G.4 A thumbtack ambiguity function.

In general, increasing the time-bandwidth product in an over-dense target environment only serves to degrade the signal-to-interference ratio, since this creates more resolution cells that can reflect ambiguity responses into the desired cell under observation. As pointed out in Chapter 2, one possible method of combating the effects of this kind of environment is to increase the carrier frequency of the transmitted signal. This has the effect of increasing the doppler shifts of the received signals and thus diluting their concentration in the ambiguity plane. Whether this is ultimately effective in practice depends on the actual number of interfering scatterers and their relative time/frequency locations.

G.5   Role of the Ambiguity Function in System Design and Evaluation

It has been shown that the ambiguity function corresponds to the time/frequency response function observed at the output of the filter matched to the transmitted waveform. As such, the ambiguity function is a measure of the ability of the radar system to distinguish among similar pulse compression or coded waveforms at the receiver that differ in time of arrival and/or frequency shift. Thus, the ambiguity function can be taken also as a measure of (1) how accurately and unambiguously the range and velocity of a single target can be estimated, and (2) how reliably two targets can be resolved in time and frequency. In the most severe radar environments, targets of interest compete with other discrete targets at different ranges and with a wide range of received signal strengths. In addition, the desired targets may also compete with signal returns that occupy extended regions of range and/or doppler shift, such as land, sea, or weather clutter. Thus, during the system design phase, and later when the prototype or production system is being evaluated, the specified system parameters that affect the observed matched filter outputs, or ambiguity responses, must be thoroughly reviewed in the context of the expected target environment. These include not only the details of the waveform and its ideal matched filter response, but also active and passive phase and amplitude response errors in both the transmitter and receiver that can distort the matched filter output. These are discussed extensively in [7.1].

In many design studies there is no clear cut answer to the question: What should the radar waveform be? In attempting to design and evaluate the waveform (or set of waveforms) used by the radar, primary emphasis will be placed on the kinds of target distributions and regions of the range-Doppler ambiguity plane that are of principal concern. These may include:

(1) The entire (for all practical purposes) ambiguity plane.
(2) Limited regions in range and Doppler shift.
(3) Limited Doppler shift region or regions.
(4) Limited range interval or intervals.
(5) Single, multiple, or dense target distributions for each of the above.

Thus, during system evaluation each waveform, and its resulting effect on the receiver signal processor output, must be measured and compared to the relevant system specifications to determine if the radar will meet its operating requirements for target detection and acquisition. It is incumbent on the system manufacturer to provide the theoretical details of the waveforms being used, as well as data on system errors and predictions of system performance under the specified target environments. Chapter 7 provides summary descriptions of the major features of the most common types of waveforms that are found in radar systems.

G.6 References for Appendix G

[G.1] P. M. Woodward, *Probability and Information Theory, with Applications to Radar,* Pergamon Press, Oxford, 1955.
[G.2] W. M. Siebert, "A Radar Detection Philosophy," IRE Trans., **IT-2**, pp. 204-219, 1956.
[G.3] M. I. Skolnik, *Introduction to Radar Systems*, McGraw-Hill, 1980.
[G.4] E. C. Westerfield, R. H. Prager, and J. L. Stewart, "Processing Gains Against Reverberation (Clutter) Using Matched Filters," IRE Trans., **IT-6**, pp. 342-352, 1960.

Appendix H

MULTIPATH ERROR IN RADAR MEASUREMENT

Multipath error was discussed briefly in Section 2.2.5, in connection with the fundamental system design process. In Section 8.2, the physics and geometry of reflection from the earth's surface was discussed. In this appendix we will describe the methods of evaluating the error in angle tracking caused by multipath reflection and diffraction.

H.1 Tracking Error Expression

The rms tracking error in any coordinate z caused by an interference component I_Δ entering the difference channel of the radar is given by

$$\sigma_z = \frac{z_3}{k_z \sqrt{2(S/I_\Delta)n_e}} \tag{10.1.3}$$

where z_3 is the width of the resolution cell, k_z is the error slope, and n_e is the number of independent samples of the error averaged in the observation time t_o of the measurement system. It is assumed that S/I_Δ is sufficiently large that linear operation of the error detection circuits is possible.

In the case of multipath error, the interference component is the power of the multipath reflection, weighted by the difference-channel response of the tracker in the z coordinate. The most serious error usually appears in the elevation coordinate, for which z_3 is represented by the elevation beamwidth θ_e, and $k_z = k_m$ (for a monopulse system). The difference-channel response is Δ_θ, the elevation difference pattern of the antenna.

H.2 Multipath Power

The power of the multipath reflection is given by the sum of a specular reflection component, a diffuse reflection component, and a possible diffraction component arising from knife-edge obstacles beneath the path to the target.

H.2.1 Specular Multipath

The amplitude of the specular component, relative to the direct ray, is given by the reflection coefficient $\rho = \rho_0\rho_s\rho_v$, where

ρ_0, the Fresnel reflection coefficient, is plotted in Figure 8.2.2;

ρ_s, the specular scattering coefficient for a rough surface, is plotted in Figure 8.2.3 and expressed by (8.2.8)

$$\rho_s^2 = \exp\left[-\left(\frac{4\pi\sigma_h}{\lambda}\right)^2\right] \tag{8.2.8}$$

The vegetation factor, ρ_v, accounts for the absorption of energy in the layer of vegetation covering the surface (see Section 8.2.4).

The angle from which the specular component arrives is below the horizontal, and equal to the grazing angle ψ_0 from (8.2.2). Figure H.1 shows the geometry for specular reflection. Because the target may scatter differently toward the radar and toward the reflection point, voltage factors A_t and A_r are introduced to multiply the direct and reflected signals, respectively. For a beam at elevation angle θ_b (which may be equal to the target elevation θ_t for null tracking on the target), the specular multipath is at an angle $\theta_b + \psi$ from the beam axis. The voltage gain of the difference channel at this angle, Δ_s will be applied as a weight to the specular multipath voltage, resulting in an interference component

$$I_s = (A_r^2/A_t^2)\Delta_s^2\rho_0^2\rho_s^2\rho_v^2 \tag{H.1}$$

The coefficients ρ_0, ρ_s and ρ_v in this expression are evaluated at the specular point, with grazing angle ψ_0 given by (8.2.2).

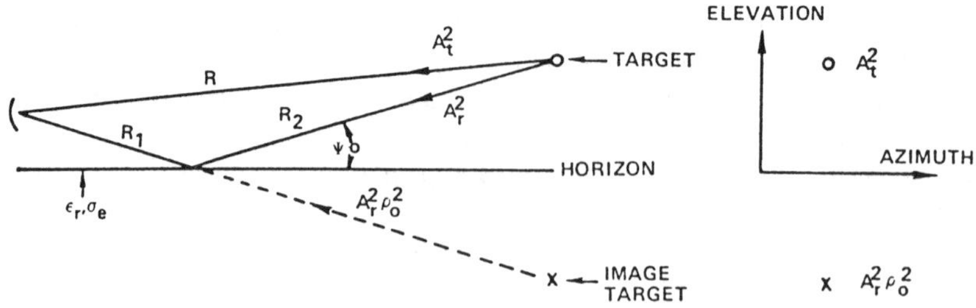

Figure H.1 Specular reflection from a flat surface.

H.2.2 Diffuse Multipath

The rms amplitude of the diffuse component is given by $\rho = \rho_0\rho_d\rho_v$, where ρ_d is the diffuse scattering factor. The diffuse reflections arrive from a broad region surrounding the point of specular reflection, with varying intensity. This region is called the glistening surface [H.1], and is illustrated in Figure H.2. The intensity of reflections from different portions of this surface were derived in [H.2] and [H.3], where it was found that two correction factors had to be applied to the theory of [H.1] to account for conservation of energy over surfaces that could support some specular reflection and the special low-elevation angle geometry commonly encountered in radar tracking. The combination of specular and diffuse reflection is shown in Figure H.3.

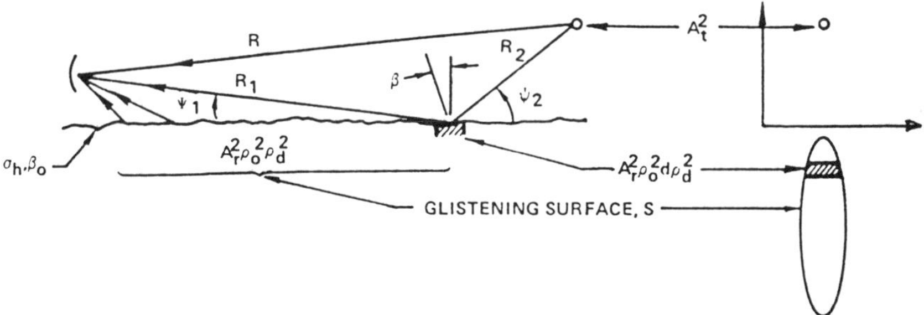

Figure H.2 Diffuse reflection from a completely rough surface.

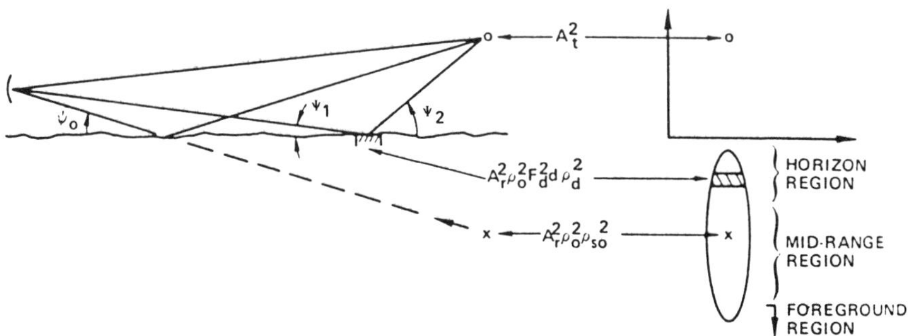

Figure H.3 Reflection from a partially rough surface.

The resulting diffuse reflection effect can be calculated by summing the powers I_i reflected and received in the difference pattern from elements of surface area subtending angle increments $\Delta\theta$ at the radar and having powers given by

$$I_i = (A_r^2/A_t^2)\Delta_i^2\rho_{0i}^2\rho_v\eta_{d_i}\Delta\theta \qquad (H.2)$$

where Δ_i is the difference pattern voltage gain to the surface element, ρ_{si} is the Fresnel coefficient at the element, and η_{di} is the power density of diffuse reflection. The Fresnel coefficient for the element is calculated using a grazing angle equal to the mean of ψ_1 and ψ_2.

For a surface element at range x_i from the radar, the diffuse reflection density is [H.3]

$$\eta_d = \frac{R}{R - x_i} \frac{\psi_1 + \psi_2}{4\sqrt{\pi}\,\beta_0 \psi_1} \exp\left[-\left(\frac{\psi_1 - \psi_2}{2\beta_0}\right)^2\right] F_d Z \tag{H.3}$$

where R is the range from radar to target, ψ_1 and ψ_2 are the grazing angles of the radar-to-surface and surface-to-target paths, respectively, β_0 is $\sqrt{2}$ times the rms slope of the surface, and F_d and Z are the roughness factor [H.2] and low-angle correction factor [H.3].

The roughness factor depends of the specular scattering factors calculated for the two grazing angles at the surface element:

$$F_d^2 = \sqrt{1 - \rho_{s1}^2}\,\sqrt{1 - \rho_{s2}^2} \tag{H.4}$$

The factor Z depends of the smaller of the two grazing angles. Normalized angles are calculated as

$$a = \frac{\psi_1}{2\beta_0}, c = \frac{\psi_2}{2\beta_0} \tag{H.5}$$

$$a' = \min(a,c)$$

$$b = \frac{\psi_c}{2\beta_0} \tag{H.6}$$

where $\psi_c = \lambda/4\pi\sigma_h'$ is the critical angle below which the surface appears essentially smooth.

The effective surface roughness is corrected for shadowing of the lower surface regions by the higher:

$$\sigma_h' = \sigma_h \sqrt[5]{4a'}, \quad 4a' \le 1 \tag{H.7}$$

$$\sigma_h' = \sigma_h, \quad 4a' > 1 \tag{H.8}$$

The low-angle correction factor is

$$Z_1 = \frac{8\sqrt{1 - \rho_{sa}^2}}{(1/a') + 4 + 3a' - (b^2/3a') - b}, \quad \text{for } b \le 1 + a' \tag{H.9}$$

$$Z_2 = \frac{24}{(2/a'^2) + (9/a') + 12 + 5a'}, \quad \text{for } b \ge 1 + a' \tag{H.10}$$

The coefficient $\rho_{sa} = \exp[-(a'/b)^2]$.

Using these relationships, the contribution I_i from surface element i may be calculated, and the values summed over angles from the horizon to the lower edge of the significant portion of the difference channel pattern, to arrive at the total diffuse interference component I_d.

H.2.3 Diffraction

The magnitude $A(v)$ of the diffraction component arriving at the antenna is calculated using the method described in Section 8.3.2. The geometry for diffraction is shown in Figure H.4. This is weighted by the difference-channel gain Δ_v at the angle of the top of the obstacle, and the diffraction interference component is then

$$I_v = (A_r^2/A_t^2)A^2(v)\Delta_v^2 \tag{H.11}$$

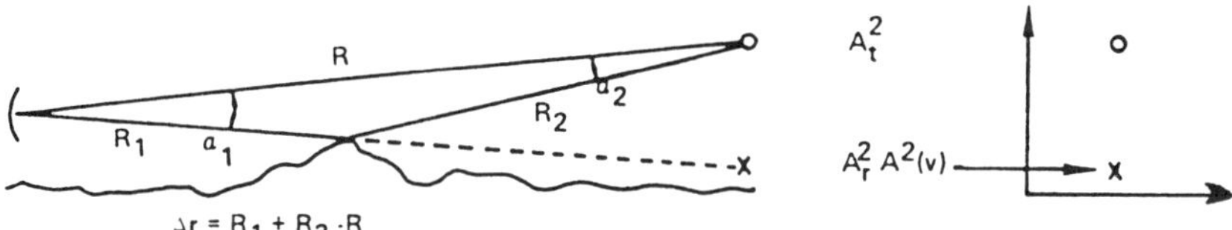

Figure H.4 Diffraction over an obstacle.

H.2.4 Total Multipath Interference

The total multipath interference used in (H.1) is then the sum $I_\Delta/S = I_s + I_d + I_v$. The number of independent samples n_e is often near unity. From these values, the error may be calculated for given beamwidth θ_e and error slope k_m. Typical curves for error as a function of target elevation are shown in Figure H.5, where both the target elevation and the error are normalized to beamwidth.

H.3 Multipath Error in Other Coordinates

The errors in azimuth, range, and doppler caused by multipath reflections are not as serious as those in elevation. Data on these other errors may be found in the references.

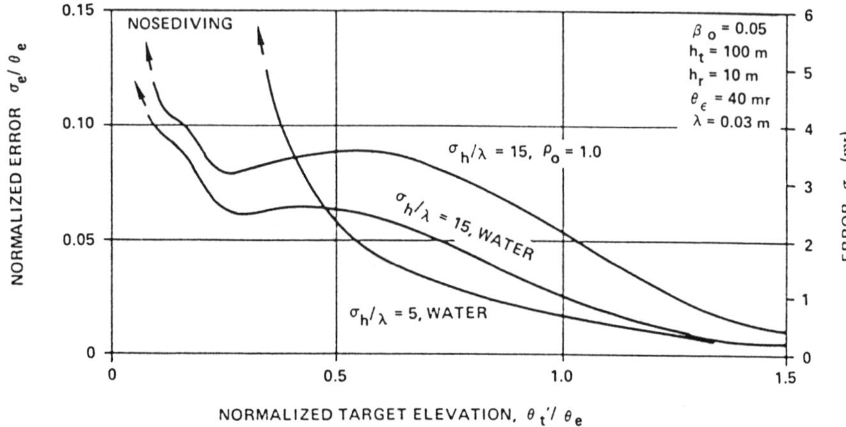

Figure H.5 Typical multipath error from surface producing diffuse reflection.

H.4 Multipath on Surface Targets

When the target is above the surface, as shown in Figures H.1 - H.4, the glistening surface is extended in elevation, leading to relatively large errors in that coordinate. Targets on the surface give a much smaller glistening surface, often little more that twice the height of the target. The diffuse elevation errors are reduced proportionately. However, if viewed from surface-based radars, the grazing angles are often small enough that specular reflection and nosediving occur.

H.5 Multipath in Different Radar Bands

H.5.1 VHF and UHF Bands

At UHF and below, many surfaces appear smooth, as a result of the λ dependence of ρ_s in (8.2.8). Extreme errors may result from this condition, as the interference can approach or even exceed the direct signal. The radar may experience a phenomenon known as ***nosediving*** [H.4, p. 524], in which the antenna oscillates between an angle about 0.7 beamwidths above the horizon and the horizon itself. Tracking in all coordinates may be lost under these circumstances.

H.5.2 Microwave Bands

From L-band to K_u-band, the surface usually has both specular and diffuse components, and the analysis of both given here is applicable.

H.5.3 Millimeter-Wave Bands

In K_a-band and higher frequencies, most surfaces appear rough and give essentially diffuse reflections, as analyzed in Section H.2.2. Nosediving is unlikely, but diffuse errors appear proportional to the elevation beamwidth of the radar.

H.6 References for Appendix H

[H.1]   P. Beckmann and A. Spizzichino, *The Scattering of Electromagnetic Waves from Rough Surfaces*, Artech House, 1987.
[H.2]   D. K. Barton, "Low-Angle Radar Tracking," Proc. IEEE **62**, No. 6, June 1974, pp. 687-704.
[H.3]   D. K. Barton, "Low-Altitude Tracking over Rough Surfaces I: Theoretical Predictions," IEEE Eascon Record, 1979, pp. 224-234.
[H.4]   D. K. Barton, *Modern Radar System Analysis*, Artech House, 1988.

Appendix I

SPECIFICATION FOR RADAR ECCM PERFORMANCE

Specifying the required performance of a radar designed to operate in a military environment, it is essential that ECCM requirements be called out in detail. ECCM features are integral to the radar and should not be considered as fixes to be incorporated after the basic design is established. In fact, ECCM features may affect basic radar performance measures such as detection range and SCV in a benign (no ECM) environment. Hence, these performance trade-offs should be considered in specifying and designing the radar.

I.1 Basis for ECCM Performance Specification

This specification should be based on the best estimates of the ECM threat environment in which the subject radar will have to operate. Such estimates would include known ECM capabilities and projected new capabilities to be introduced over the deployment lifetime of the radar. Some of the latter capabilities, if in doubt with regard to feasibility or affordability, may be included in the threat model as growth options, and the resulting specification may require provision for appropriate ECCM features as growth capabilities. The implications of the latter requirements may be, for example, that basic design approaches be avoided if they would preclude the growth capabilities from being added after initial production has begun.

The ECCM performance specification should include both the ECM threat model and the radar performance measures under ECM conditions.

I.1.1 Characterizing the ECM Threat

The ECM threat model is based upon representative scenarios, including assumed deployment of offensive and defensive forces and, more specifically, the deployment (Chapter 12, Section 12.2) of ECM against the subject radars. The technical characteristics of the deployed ECM equipment would, for example, be described in the manner of Chapter 12, Section 12.3 and Section 12.4. It is important that older types of jammers (e.g., swept noise jammers) be included in the threat model, even though a sophisticated enemy is postulated. If these somewhat outmoded techniques are ignored in the radar design, the door would be open to exploitation of the radar by reintroduction of older equipment or by a less sophisticated adversary.

I.1.2 Radar Performance Degradation in an ECM Environment.

The part of the specification that deals with performance in an ECM environment can be treated either as a companion specification to the basic (benign environment) specification or on the basis of degradation from basic performance when the radar is in the ECM threat environment. The latter approach is chosen for the example specification of Section I.2.

Realistic degradation should be allowed for the various types and levels of ECM, since most types will have some negative effects upon even the best designed radar. In fact, the radar minimum acceptable performance measures should be those which are achievable in the expected ECM environment, and the benign environment performance will typically be better. Thus, the performance specification of a radar that is to perform its function exclusively in an ECM environment should be based primarily on the ECCM performance specification.

I.1.3 Radar Design Implications of the ECCM Specification

ECCM design techniques are discussed in Chapter 12, Section 12.5, and their application relative to the various ECM types is summarized in Chapter 12, Section 12.6. A review of these design techniques clearly demonstrates their integral relationship to the overall radar design. The transmitter design is impacted by waveform requirements and frequency agility. The antenna subsystem has special requirements for sidelobe performance, adaptive nulling or sidelobe cancellation, auxiliary antenna beams, etc. The receiver and signal processor designs are heavily influenced by ECCM processing and logic requirements.

I.2 Example ECCM Performance Specification

A representative radar performance specification for operation in an ECM environment is described here. The radar is assumed to be a ground-based multi-function electronically scanned array system, having both search and tracking modes, and designed to support a medium range air defense missile system. The associated threat model is typical for a battlefield scenario, with attacking aircraft and ECM originating from one side of the FLOT (to be referred to here as behind the FLOT) and the air defense system located 20 to 30 km from the FLOT on the opposite side.

The radar is assumed to operate at C-band, with a rapid tuning capability over a 10% bandwidth. An instantaneous bandwidth of 10 MHz is representative, and various PRF values would be associated with the different radar functions (search, target track, missile track, and missile control or target illuminate) and the different operating modes that may be required to accommodate siting and environmental conditions (e.g., different ground clutter conditions, weather variations, different attack scenarios, and ECM deployment). The waveform duty cycles would usually be different for the different modes and functions, with the largest value assumed to be 10%. The radar average duty cycle would depend upon the sequence of waveforms that are interleaved to meet the specific radar performance requirements for a given battlefield engagement scenario. Waveform dwell times would also be different for different radar modes and would fall within the region of 1 to 10 msec.

The nominal radar detection range is assumed to be 100 km against a 1 m^2 target cross section (Swerling Case 1 fluctuation model). It is also assumed that the radar receiver and detection circuits will have sufficient dynamic range to permit the detection and tracking of much smaller (low cross section) targets at the expected reduced free space ranges. Monopulse tracking is assumed.

For the purposes of this appendix, the radar is given the designation AN/MPQ-XX.

I.2.1 Typical ECM Threat Model

The AN/MPQ-XX radar shall be designed to perform in accordance with the performance specification for ECM environments of Section I.1. The elements of the ECM environments which form the basis for the design are described in this section, together with representative deployment considerations.

Stand-off Jammers

Stand-off jammers are deployed behind the FLOT and direct their jamming signals toward the FLOT for the purposes of degrading performance of the subject radars and other friendly battlefield and associated airborne radars and communications equipment. The SOJ threat model to be used for the AN/MPQ-XX is given in Tables I.1 and I.2, which present SOJ transmission characteristics and a deployment model, respectively. The SOJ transmission power levels shown are for a single SOJ carrying aircraft. The narrowband spot noise ERP level is based on the assumption that a single SOJ would divide the available (narrow band noise) power among ten victim radars.

Table I.1 SOJ Transmissions

Characteristics	Broadband Barrage Noise	Narrowband Spot Noise	Narrowband Swept Noise
ERP	+60 dBW	+50 dBW	+ 60 dBW
Duty cycle	100%	90%	100%
Frequency	Center of AN/MPQ-XX operating band	Rapid set-on to AN/MPQ-XX transmission frequency	Swept over AN/MPQ-XX operating band
Noise bandwidth	600 MHz	20 MHz	20 MHz
Sweep Rate	N/A	N/A	10^3 MHz/sec to 10^6 MHz/sec

Table I.2 SOJ Deployment

Aircraft Location:	
Distance behind FLOT	300 km
Altitude	10 km
Number of SOJs and Flight path	10 Aircraft in elongated racetrack configuration - parallel to FLOT. Racetrack approximately 500 km long and 2 km wide.
Aircraft spacing	100 km along racetrack.

Swept noise jammers are included in the SOJ threat model, even though they are an older technique and are not very effective against a properly designed modern radar. The reason for this is that they could be used (e.g., by a third world country), and may be effective, especially if the radar is not "properly designed" from an ECCM point of view.

Escort Jammers

The escort jammer model assumes a single aircraft equipped with ECM capabilities flying in close proximity to a number of attacking aircraft which it is protecting. The number of protected aircraft is between 2 and 5 per escort jamming aircraft. The attack flight path is assumed to originate behind the FLOT and to proceed toward the victim radar in an attack mode or to pass near the victim radar (within its normal detection range) toward assets protected by the radar (typically, at distances from the FLOT greater than that of the radar). The protected aircraft are assumed to be located within a cylindrical volume with its axis lying in the range dimension of the victim radar and with the cylinder being no greater than 500 m in diameter and 5 km long.

Table I.3 describes the transmission characteristics of a single escort jammer aircraft. In the case of narrowband spot noise jamming, the ERP level is based on the assumption that the escort aircraft can simultaneously spot jam three AN/MPQ-XX radars operating at different frequencies. The escort jamming aircraft is also assumed to carry jammers which are capable of jamming the other radars and radar missiles that would normally be deployed in the same battlefield scenario.

Table I.3 Escort Jammer Transmissions

Technique	ERP	Duty Cycle	Frequency	Transmission Characteristics
Broadband Barrage Noise	+40 dBW	100%	Center of AN/MPQ-XX operating band	600 MHz noise bandwidth
Narrowband Spot Noise	+35 dBW	90%	Rapid set-on to AN/MPQ-XX transmission frequency	20 MHz noise bandwidth
Narrowband Swept Noise	+40 dBW	100%	Swept over AN/MPQ-XX operating band	20 MHz noise bandwidth sweep rates from 10^3 to 10^6 MHz/sec
Repeaters	+30 dBW	Same as AN/MPQ-XX	Same as AN/MPQ-XX	Pulses identical to AN/MPQ-XX Range gate pull-off (RGPO) Velocity gate pull-off (VGPO)
Decoys (towed, dropped, or rocket launched)	+ 10 dBW	Same as AN/MPQ-XX	Same as AN/MPQ-XX	Pulses identical to AN/MPQ-XX
Special Jammer	+30 to +40 dBW depending on (technique - as above)	Up to 100% (depending on technique - as above)	Skirt frequency jammer operates at edge of instantaneous signal band at 100% duty cycle. Two frequency jammer operates within radar tuning band, with two signals separated by radar first IF. Image frequency jammer operates at frequency separated from radar transmission frequency by twice radar first IF.

Table I.4 SSJ Transmissions

Technique	ERP	Duty Cycle	Frequency	Transmission Characteristics
Broadband Barrage Noise	+30 dBW	100%	Center of AN/MPQ-XX operating band	600 MHz noise bandwidth. In formation, could be blinking at 50% duty cycle at rates from 1 to 100 Hz.
Narrowband Spot Noise	+35 dBW	90%	Rapid set-on to AN/MPQ-XX transmission frequency	20 MHz noise bandwidth. In formation, could be blinking at 50% duty cycle at rates from 1 to 100 Hz.
Narrowband Swept Noise	+30 dBW	100%	Swept over AN/MPQ-XX operating band	20 MHz noise bandwidth. Sweep rates from 10^3 to 10^6 MHz/sec
Repeaters	+30 dBW	Same as AN/MPQ-XX	Same as AN/MPQ-XX	Pulses identical to AN/MPQ-XX. Range gate pull-off (RGPO). Velocity gate pull-off (VGPO)
Decoys (towed, dropped, or rocket launched)	+ 10 dBW	Same as AN/MPQ-XX	Same as AN/MPQ-XX	Pulses identical to AN/MPQ-XX
Special Jammer	+30 to +40 dBW (depending on technique - as above)	up to 100% (depending on technique - as above)	Skirt frequency jammer operates at edge of instantaneous signal band at 100% duty cycle. Two-frequency jammer operates within radar tuning band, with two signals separated by radar first IF. Image frequency jammer operates at frequency separated from radar frequency by twice radar first IF. In formation, could be blinking at 50% duty cycle at rates from 1 to 100 Hz.	

Self-Screening Jammers

Self-screening jammers (SSJs) are on board attack aircraft operating without an escort, either singly or in a formation of three to five aircraft. In the latter case, blinking noise jammers may be used, with approximately 50% duty cycle for each aircraft on broadband (barrage) noise. Narrowband noise may be used to deny range, but deceptive repeater techniques are more likely to be used against radar tracking modes. Table I.4 describes the transmission characteristic of a single SSJ.

For the purposes of this threat model, the attack profiles shall be the same as the flight paths described above for the escort jammer case. The growth aspects of this threat model include cross polarization and cross eye repeater jammers.

SSJs can be expected to turn on jammers and select jamming techniques in accordance with the specific tactical situation; e.g., jamming signals may be turned on only when the SSJ's radar warning receiver (RWR) detects certain radar transmissions (frequencies, waveforms, etc.) and specific signal power levels. The selected jamming technique will depend upon the attack geometry and the operating mode of the victim radar. In addition, an SSJ can be expected to execute evasive maneuvers while activating jammers, up to the maximum g-force capabilities defined for the specified threat aircraft.

Chaff

Chaff clouds will be produced by both friendly and enemy forces. This threat model shall include a chaff corridor, sown from 10 km altitude along a path roughly paralleling the FLOT, produced by the enemy behind the FLOT at a distance between 100 km and 200 km. This corridor will extend through the full azimuth angle coverage of the AN/MPQ-XX. Another similar chaff corridor sown by friendly forces, will extend for the same length, paralleling the FLOT at a distance from the FLOT (behind the radar) between 100 km and 200 km.

The rate of chaff sowing will be 1 kg per kilometer, and the chaff radar cross section model (Chapter 5, Section 5.5) will be

$$\sigma = 22 \times 10^3 \lambda W \quad \text{m}^2 \tag{I.1}$$

where λ is the radar wavelength and W is the chaff weight in kg.

The velocity spectrum for chaff will be based on the weather model of the primary specification, with the mean velocity being that of the wind radial velocity at the center of the radar beam and standard deviation of its (Rayleigh) velocity spectrum being as given in Chapter 5, Section 5.5, as

$$\sigma_v = 0.3 k_{sh} R \theta_e \cos \alpha \tag{C.2}$$

where, R is in km, α is the azimuth angle between the beam and the wind direction, and the elevation (one-way 3 dB) beamwidth θ_e is in radians. The constant k_{sh} shall be taken as 3 m/s per km.

Chaff may also be dropped or forward launched from escort and self screening aircraft, with individual bundle sizes being 0.1 kg in weight. Ejection or launch rates will be of the order of one bundle per second, and a given defensive action by the SSJ will involve up to 10 chaff bundles.

SSJs can be expected to execute evasive maneuvers while ejecting or forward launching chaff. The model for evasive maneuvers shall conform to the maximum g-force capabilities for the specified threat aircraft.

I.2.2 ECCM Performance Specification for AN/MPQ-XX Radar

The AN/MPQ-XX radar shall perform in accordance with the following ECCM specification when operating in the ECM environment represented by the threat model of Section I.2.1. The total environment shall include the natural environment (ground clutter and weather) models of the primary specification. Specified ECCM performance shall be achieved with the combined natural environment and ECM environment. The ECM environment shall consist of the SOJ model plus either an escort with protected aircraft or three SSJs, together with chaff corridors and SSJ chaff, as described in Section I.2.1. Quantitative performance levels are presented as degradations from AN/MPQ-XX radar performance in the specified natural environment (ground clutter plus weather).

Broadband Noise and Narrowband Noise Jammers

Broadband noise and narrowband noise jammers operating in accordance with the threat model of Section 2.1 shall effect no more than a 10% reduction in detection range at all mainbeam look angles except within ± 1 beamwidth (azimuth and elevation) of the direction toward any SOJ, escort jammer, or SSJ. The radar shall provide strobe designations with an accuracy of ± 0.1 radar beamwidth (azimuth and elevation) of each SOJ, escort jammer, or SSJ. The radar computer shall include the capability of combining strobe data from one or more (up to 3) additional AN/MPQ-XX radars to determine jammer positions consistent with strobe angle measurement accuracy.

Target position (range and angle) tracking accuracy, as well as doppler frequency measurement accuracy, shall not be degraded by more than 10% at all mainbeam pointing angles except within ± 1 beamwidth (azimuth and elevation) of the direction toward any SOJ, escort jammer, or SSJ. Hence, any parameter having a mean deviation of measurement of σ_x shall have a mean deviation no greater than $1.1\,\sigma_x$ under these conditions.

Under no conditions shall the radar fail to detect and generate strobe data on a radiating SOJ, escort jammer, or an SSJ.

Narrowband Swept Noise Jammers

Narrowband swept noise jammers operating in accordance with the threat model of Section I.2.1 shall effect no more than a 5% reduction in the single scan probability of detection as averaged over the coverage volume. The probability of detection measured within ± 1 beamwidth of the direction to any SOJ, escort jammer, or SSJ shall not be degraded by these jammers by more than 10% for 2 out of 3 scans, unless the frequency sweep rate is sufficiently slow to vary by less than one instantaneous bandwidth of the radar during two radar scans. In the latter case, the performance degradation limits shall be as specified alone for the broadband and narrowband continuous noise jammers.

Target tracking accuracy shall be degraded no more than 10% within ± 1 beamwidth of an SOJ, escort jammer, or SSJ, except when the sweep rate of the jammer is sufficiently slow to put it in the category of a continuous noise jammer, as above. The radar shall provide strobe designations to an accuracy of 0.1 radar beamwidth (azimuth and elevation) of each SOJ, escort jammer, and SSJ.

The AN/MPQ-XX shall employ techniques to cope with repeater jammers as follows:

(1) Waveforms (e.g., multiple PRFs) shall be employed to prevent classification as real threats any false targets that are made to appear at ranges less than the range (from the radar) to any repeater jammer with up to three repeaters operating simultaneously.

(2) When multiple (threat classified) targets are detected within one radar antenna beamwidth, tracking priority shall be given to approaching targets. The highest priority shall be given to the target with the shortest calculated time to reach the nearest point of approach to the radar, and descending priorities shall be assigned in accordance with the increasing approach times calculated for the additional targets. This prioritization shall hold for up to 5 targets within a single radar beam.

(3) Range-gate pull off and velocity-gate pull off repeater signals shall be classified as non-threat targets after no more than 5 target tracking dwell times following their initial detection.

Decoys

Decoys shall be classified as non-threat targets after no more than 5 target tracking dwell times following their initial detection.

Special Jammers

Special jammers, including skirt frequency jammers, two-frequency jammers, and image frequency jammers shall be no more effective against the AN/MPQ-XX, when employing any of the above jamming techniques, than these techniques would be when operating within the radar operating band. Hence, the radar performance degradation shall be no more than specified alone for the particular jamming techniques. Furthermore, the radar shall not fail to report special techniques jamming strobes and it shall achieve the above strobe measurement accuracies.

Chaff

Chaff, when deployed in accordance with the model of Section 2.1, shall not degrade the probability of target detection within the chaff cloud by more than 15%. Target tracking within a chaff cloud shall not be degraded in accuracy by more than 10% in any measurement parameters (angle, range, and velocity).

Cross-Polarization and Cross-Eye Jammers

Cross-polarization and cross-eye jammers shall be considered as potential future threats. Upgrading the radar to cope with these threats shall be feasible without a new design approach for any major subsystem. In particular, the antenna subsystem design shall anticipate these requirements and shall, with minimum modification, provide the capability of achieving the radar tracking accuracies specified above.

I.3 Evaluation of Radar Performance in an ECM Environment

Methods for evaluating radar performance in an ECM environment are described in Chapter 13. The use of simulation facilities (simulating targets and jammers) is usually the most efficient way of conducting experimental evaluations of radar response to active ECM signals. Evaluation of radar responses to passive ECM (chaff and decoys) may require full radar operation in conjunction with real ECM.

Appendix J

TYPICAL RADAR SYSTEMS

Representative modern radars are described in this appendix. The Raytheon-produced RAMP (RAdar Modernization Project) primary surveillance radar, which will replace all of Canada's older primary and secondary air traffic control radars, is presented as an example of a modern search radar design. The Hughes Aircraft-produced AN/TPQ-36 and AN/TPQ-37 Firefinder radar systems represent electronically scanned acquisition and tracking radars, and the Texas Instruments-produced AN/APS-134 (V) maritime surveillance radar represents a modern airborne radar.

This appendix is largely adapted from papers by H. R. Ward [J.1], D. A. Ethington [J.8], and J. M. Smith of Texas Instruments, Inc. [J.10].

J.1 The RAMP Surveillance Radar

This new primary radar is entirely solid-state, including the transmitter, and uses the latest digital technology for signal processing, control, and monitoring. Seven independent transmitters combine to generate the RF signal for fail-safe operation. A 12-bit digital moving target detector processes both target and weather returns. The system is capable of unattended operation through a remote control and monitoring subsystem [J.1] - [J.5].

Air traffic control radars have evolved steadily since World War II [J.6]. The antennas have improved with the addition of circular polarization, a second high elevation receive beam to reduce land clutter, and vertical patterns shaped to compensate for receiver attenuation at short range. The receivers and signal processors have gone from tube to solid-state technology, while many magnetron transmitters have been replaced by klystrons. Signal processing has progressed from analog to digital MTI. As these changes occurred, the fundamental parameters such as range, data rate, resolution, and accuracy have remained essentially unchanged. The RAMP radar represents a recent step in this evolution, with major improvements to the transmitter and signal processor.

Table J.1 RAMP Radar Parameters

Frequency	L-Band
Peak Power	20.8 kW
Transmitted Pulsewidths	1 μs (CW)
	100 μs (Chirp)
Repetition Frequency (Avg.)	676 Hz
Antenna Gain	33.5 dB
Beamwidth (Az)	1.5 Degrees
Rotation Rate	12 rpm
Noise Figure	1.5 dB
Pulse Compression Ratio	100:1
A/D	12 Bits
Doppler Filters	4

J.1.1 System Architecture

The block diagram in Figure J.1 shows complete redundancy of all electronic components. This is a traditional architecture for an air traffic control radar, where very high system availability is required, except for the transmitter, which consists of seven independent channels. Table J.1 lists the RAMP radar parameters.

The receiver/exciter generates the high duty waveform required to efficiently utilize the solid-state transmitter. This waveform consists of a pulse pair transmitted in groups of four at a fixed repetition frequency, which is called a coherent processing interval or CPI. From CPI to CPI the repetition frequency and the RF frequency change. The pulse pair is made up of a short (1 microsecond) CW pulse used to provide coverage to 8 nmi. After the short range returns have been received, a 100 microsecond "nonlinear chirp" pulse is transmitted to provide target coverage from 8 to 80 nmi and weather coverage from 8 to 100 nmi. As the antenna beam scans past a target there is time for 15 pulses (4 CPIs) to be radiated. The target returns from each CPI are processed as a batch by the four-pulse moving target detector (MTD).

Clutter filters in the moving target detector (MTD) attenuate the zero doppler returns to give a subclutter visibility of 41 dB. False alarms from land returns, second time targets, and stationary targets are removed by the CFAR, zero doppler clutter map, binary integration, range-azimuth gating, and low velocity editing.

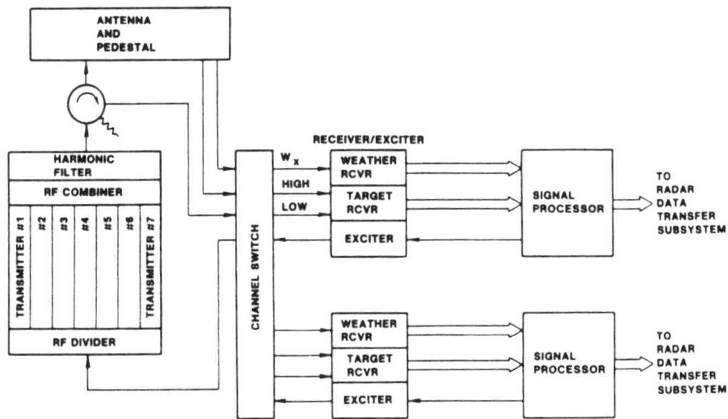

Figure J.1 RAMP system block diagram.

In addition to the target channel receiver and processor, the radar contains a parallel weather channel receiver and processor (see Figure J.1). This channel is connected to the orthogonally polarized antenna port, which does not reject the weather return. Weather returns are integrated to derive contours that bound regions of intense weather.

The free space coverage of the radar, illustrated in Figure J.2, shows the region in which a 2 square meter fluctuating target will be detected with a probability exceeding 80% using circular polarization and six of seven transmitters. Also shown is the system coverage for the same conditions using the high receive beam. This beam is used to prevent receiver saturation by reducing the amplitude of strong surface clutter returns. Figure J.3 shows the RAMP antenna without its radome.

J.1.2 Solid-State Transmitter

The RAMP radar solid-state transmitter represents a major improvement over a magnetron or klystron transmitter by not requiring high voltage, magnetics, oil immersion, or physically large components. It is also a transmitter that is more reliable and easier to maintain. While its predecessors used dual transmitters to achieve the necessary availability, the RAMP PSR uses seven small solid-state transmitters. Since a tube transmitter must meet the systems coverage requirements with a single tube operating, the second transmitter, needed only for redundancy, results in a 100% increase in transmitter hardware. In the solid-state architecture

(Figure J.1), seven transmitters are used, where six are sufficient to meet the system requirement. The necessary reliability is provided by the seventh transmitter for an increase of only 17% in transmitter hardware.

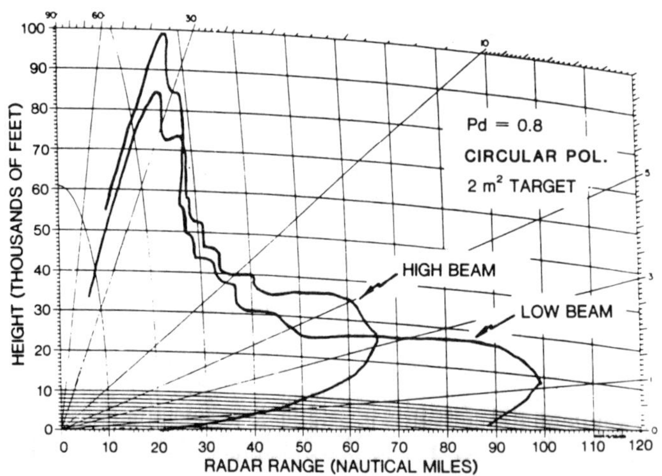

Figure J.2 RAMP vertical coverage diagram.

Figure J.3 RAMP antenna.

Broad bandwidth is another fundamental advantage the solid-state transmitters have over the magnetron and klystron transmitters. This allows the system to utilize single-channel frequency diversity, which requires 2.6 dB less radiated power to reliably detect and track aircraft. Klystron or magnetron transmitters may only realize the benefits of frequency diversity when both channels are operating together, but this mode of operation does not provide high availability. The RAMP radar, because of the wideband transmitters, is able to utilize frequency diversity while allowing a standby transmitter to be available for maintenance.

By structuring the transmitter as a number of parallel power amplifiers, multiple failures can be tolerated without shutting down the system, thus improving the system's survivability. The modules are small in size and operate at low voltage, so that they may be replaced during operation.

As illustrated in Figure J.1, the transmitter units consist of seven transmitters operating in parallel, with their inputs divided and outputs summed with passive combiners. Each transmitter is a self-contained unit consisting of two RF amplifier modules with their own power supply, air cooling, BITE, and control interface.

The peak power of 308 watts from a driver is divided among 14 solid-state modules. Each module receives 50 W, which it amplifies to 2 kW at the module output by 42 solid-state devices. Each module is air cooled and operates at 35 volts with an efficiency of 25% at a maximum duty cycle of 8.2%. When the outputs of all 14 modules are combined, the peak output power exceeds 25 kW over the operating band from 1250 to 1350 MHz.

The high power combiner sums the power of the seven transmitters (14 modules) using microwave slab line techniques. It is a symmetric network composed of hybrids and resistive loads. The structure assumes that the module ports are isolated and that any combination of failed modules will not cause the power rating of any load to be exceeded.

As the radar operates in the frequency diversity mode, the transmit spectrum resembles that of a conventional ATC radar operating in a dual diversity mode. Four frequencies spaced in two close-spaced pairs serve to decorrelate rain return.

J.1.3 Pulse Compression

Available high power transistors are most cost effective when operated at a duty ratio between 5% and 10%. This consideration has led to the use of pulse compression in the RAMP PSR. By transmitting a 100 microsecond pulse which is compressed to 1 microsecond in the receiver, high duty operation is achieved without compromising range resolution.

The particular modulation used by the RAMP pulse compression system is a nonlinear chirp. In the past, linear chirp was more often used, but this required mismatching the receiver to achieve low range sidelobes, and the mismatch loss was the order to 1.5 dB. With nonlinear chirp, mismatch loss is reduced to 0.3 without compromising range sidelobe levels.

The nonlinear chirp pulse is generated by phase modulating a stable oscillator signal. Precise waveform samples stored in PROMS are used to generate a constant amplitude pulse that is matched to the dispersive surface wave delay line in the receiver for pulse compression.

J.1.4 Signal Processing

The signal processor consists of the four functional elements shown in Figure J.4. Target returns from the main receiver pass through the moving target detector (MTD) where they are coherently integrated and separated from clutter and the background noise. The correlation and interpolation processor does the necessary parameter estimation while the surveillance processor tracks targets to edit plots from stationary objects and attach target tags that provide data continuity from scan to scan. Weather returns are processed and the weather contour data are merged with the target data stream in the correlation and interpolation processor.

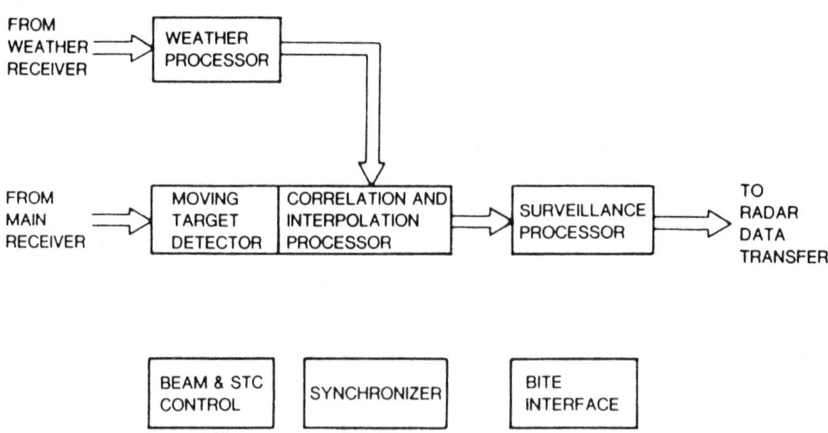

Figure J.4 RAMP signal processor block diagram.

The MTD is a pulsed doppler processor operating on four pulse batches of received data. Each batch or CPI passes through four finite impulse response (FIR) filters with a log-magnitude detector at each filter output. A separate constant false alarm rate (CFAR) threshold at each filter output is adjusted to follow the background return in that filter. Land return, integrated for many scans in a zero doppler map, is also scaled and added to the threshold to prevent land return from being detected [J.6].

The clutter filters are of hard wired digital design, each capable of implementing an arbitrary FIR filter, with the necessary 15 bit filter coefficients stored in ROM. The particular filter responses used by the RAMP PSR have been optimized for detection sensitivity; however, the design allows for a variety of other responses to be implemented.

The CFAR and clutter map circuitry are also hard wired digital designs with the flexibility of a variable CFAR range window and a variable integration time in the clutter map. A combination of the background averages taken before and after the cell of interest in range along with a scaled value from the clutter map output produce a different threshold at each range resolution cell.

Returns that exceed this threshold are called alarms, and these are output to the 2-of-4 binary integrator to combine the returns of four CPIs of data as the beam scans past a target.

A beam switch that connects the receiver to either the high or low antenna beams and receiver attenuation are adaptively controlled to prevent receiver saturation from strong stationary clutter returns. Receiver gain is varied as a function of range and azimuth according to the clutter level in the clutter map. Controlling saturation in this way avoids coverage loss due to blanking and provides the maximum subclutter visibility over the strong clutter cells where it is most needed.

The C&I processor performs two functions, correlation of multiple detections and target position estimation. Since large targets often produce more than one detection, a programmable processor is used to correlate these detections and interpolate to find the best estimate of the target's range and azimuth position. This position along with the target's amplitude is called a plot and is formatted and output to the primary surveillance processor.

The primary surveillance processor consists of a set of programmable single board computers, which track all of the radar plots that repeat from scan to scan. Track history information is used to edit plots from slowly moving objects, to tag each plot so that it may be more easily combined with the secondary radar plot from the same target, and to attach an estimate of target heading and speed to the target report. The surveillance processor is sized to handle up to 400 targets within the radar's coverage. However, it has a modular architecture that may be adjusted to match the needs of a particular site.

Weather data are processed in parallel with the target data by the weather processor. This processor, which also operates on a four-pulse CPI, consists of a land return filter and a weather map integrator. The land return filter has a rejection notch at zero doppler, which is varied as a function of range and azimuth, depending on the amplitude of the land return as determined by the land clutter map in the MTD. After the land return filtering, target returns are edited from the weather data. Then the weather data are collapsed into 1 nmi × 1.4° cells and integrated for five scans. The map data are then compared with the two amplitude levels to define weather contours. These contours are encoded to form the weather data, which are merged with the target data by the C&I processor.

J.1.5 False Alarm Control

Every primary surveillance radar must reject many targets other than aircraft which would be false alarms if detected. This problem is solved by identifying all of the known sources of false alarms and providing a means for suppressing each of these sources.

Alarms from distributed targets such as weather and land, as well as background noise, are controlled by the cell average CFAR process at each doppler filter output. Not all land return is distributed, however. Much land return consists of discrete clutter points, and these will be prevented from causing alarms by the adaptive clutter map which maintains an estimate of their individual amplitudes.

Moving point clutter, such as birds and cars, is a more difficult class of return because these returns resemble the aircraft returns that the system is designed to detect. Bird returns, however, are much smaller in amplitude than the smallest targets of interest and, therefore, the STC is an effective means of minimizing bird detection. While cars are comparable with small aircraft in cross section, they are confined to roadways which are stored and blanked by the C&I processor. The low velocity edit function in the surveillance processor is also an effective means of eliminating bird and car returns.

Pulsed interference, which is most likely caused by other L-band radars, will be dispersed by the pulse compression network in the receiver in such a manner that the CFAR operation will prevent it from being detected. The 2-of-4 binary integrator also prevents interference detection as well as second-time target and clutter return detection. Because the repetition frequency is changed between processing intervals, a second-time return will not fall in the same range gate on successive intervals. Distributed second-time returns will be rejected by the CFAR process.

Systems which do not use frequency diversity have difficulty controlling the false alarms from rain return. This is because the alarm rate set up for noise background increases when correlated rain return is received. To avoid this source of alarms, the system incorporates frequency diversity that decorrelates rain alarms entering the 2-of-4 binary integrator. This causes the output false alarm rate to remain the same for a rain background as it is for a noise background.

J.2 The Firefinder AN/TPQ-36 and AN/TPQ-37

The Firefinder system provides automatic, first-round location of hostile artillery positions. The system employs two different radars (AN/TPQ-37 and AN/TPQ-36) which have similar architecture [J.8]. Their antenna transceiver groups of equipment are different but their operations shelter with signal processing, computer displays, and communications are identical.

The radars use electronic scanning antennas, transmit coherent pulse trains, and use digital range-gated pulse doppler signal processing to resolve projectiles from various types of clutter. Radar operation is fully automatic, under control of a general-purpose computer. Software algorithms further reject birds, aircraft, and clutter and then estimate the projectile trajectory. This trajectory is extrapolated backwards to the weapon firing point and forward to the impact point.

J.2.1 System Architecture

The Firefinder system was developed by the U.S. Army to provide automatic, first-round location of hostile artillery (cannon, guns, rockets, and mortars) positions. The system is designed to locate simultaneous fire from numerous weapons on the battlefield. The high data rate requirements for search and tracking multiple simultaneous targets have resulted in the use of electronic scanning antennas, each with the inherent ability to rapidly reposition the antenna beam anywhere over a 90-degree azimuth sector. The antennas remain stationary during normal operation.

The radars are designed to operate in worldwide environmental extremes of temperature (-46°C to +52 °C), humidity, clutter, and interference. Two different microwave frequency bands are employed in order to optimize each for its technical performance and transportability. The AN/TPQ-36 operates at X-band and the AN/TPQ-37 operates at S-band [J.9, pp. 191-193]. The radars are transported by surface vehicles or by helicopter lift. They may be rapidly emplaced at a new tactical site (30 minutes set-up time for AN/TPQ-37 and 15 minutes to set up the AN/TPQ-36). Likewise, they can quickly be displaced from the site (on the move in 15 minutes for the AN/TPQ-37 and 5 minutes for the AN/TPQ-36).

J.2.2 AN/TPQ-37 Characteristics

The AN/TPQ-37 is capable of locating artillery at ranges beyond 30 kilometers. Figure J.5 shows the system with antenna erected. The transmitter and receiver are also located on this trailer. The system employs phase scanning to steer the antenna beam in azimuth and elevation. The antenna beam dimensions are approximately 0.6° in elevation and 0.9° in azimuth [J.9, p. 193]. This phase scan is implemented with diode phase shifters. The beamsteering network provides a 90-degree azimuth sector scan and an elevation scan of a few degrees. The weapon locating mission is accomplished with this limited elevation scan and therefore the number of phase shifters in the vertical dimension are greatly reduced by having each phase shifter feed a vertical subarray of six elements. The phase shifter devices are integrated into subarray modules that include a stripline power divider and dipole radiating elements. These modules can be inserted/removed from the array similar to circuit card assemblies. The antenna feed provides monopulse data (sum and two difference beams) for tracking in azimuth and elevation.

The transmitter final power amplifier is traveling wave tube (TWT) controlled by a shadow grid. All prior stages in the amplifier chain are solid state. In each beam position, the transmitter is pulsed at a fixed PRF, forming a train of coherent pulses. The pulse train includes extra pulses to ensure cancellation of ambiguous range clutter. The transmitter final amplifier, the transmit/receive duplexer, and a dummy load are liquid cooled.

The radar receiver uses a gallium arsenide field effect transistor (GaAs FET) preamplifier for each of the monopulse channels. The preamp is protected with a cascaded network of pre-TR tube, ferrite limiter, and diode limiter. A sensitivity time control (STC) is integrated with the diode limiter. After frequency down-conversion, the monopulse channels are time multiplexed into a single channel by using surface acoustic wave (SAW) delay lines. The individual pulses of the coherent train are coded with a non-linear frequency modulation (FM). This code is generated and decoded using SAW delay lines. Two different time bandwidth products are generated by two lines. The decoded IF (intermediate frequency) data are transmitted to the operations shelter over coaxial cable. Automatic gain control (AGC) under computer control is used to keep strong targets within the required dynamic range while tracking. It is also used to maintain the receiver gain within required tolerances over the temperature extremes. A pilot pulse is periodically processed through the entire receiver chain (all three monopulse channels) to keep the monopulse channels phase and gain calibrated with respect to each other.

Figure J.5 Firefinder AN/TPQ-37 weapon locating radar.

For operation, the antenna is oriented to grid north after the antenna trailer is placed over a known survey point. An operator then views a distant survey point with known azimuth from the local reference by boresighting through a telescope mounted on the array. The known azimuth offset is entered into the radar computer. A 14 bit shaft angle encoder is used to measure the instantaneous azimuth of the antenna. In addition, there is a tilt sensor (14 bits) mounted on the antenna that instantaneously measures the array tilt and cross-tilt attitude using force-balanced accelerometers. These data from the azimuth and tilt sensors are automatically, continuously provided to the system computer for use in stabilization calculations.

J.2.3 AN/TPQ-36 Characteristics

The AN/TPQ-36 provides comparable weapon locating performance (artillery and mortars) to a range that is approximately half of the AN/TPQ-37. It also is configured with antenna, transmitter, and receiver on a trailer (see Figure J.6). This trailer also carries a 10 kW turbine generator for system prime power. Total weight of the loaded trailer is 3,000 pounds. Its antenna electronically scans a 90-degree azimuth sector and a limited elevation sector similar to AN/TPQ-37. It uses a series end-feed to distribute the RF power in the horizontal plane. It uses ferrite phase shifters (reciprocal type) to electronically scan in azimuth. These phase shifters are latching, Faraday-rotator flux-driven devices which handle the necessary levels of peak and average power while requiring no bias to maintain the phase setting. RF power is distributed in elevation through individual waveguides, with slots cut in the narrow wall to form the radiating elements. The frequency-dispersive characteristic of the waveguide is used to electronically scan the beam in elevation by changing the radiated frequency. Sequential lobing a pair of beams in azimuth and another pair of beams in elevation is used for precision angle measurements in track.

Figure J.6 The Firefinder system includes the smaller, lightweight, highly mobile AN/TPQ-36 radar.

The transmitter has two stages of traveling wave tube amplifiers. The final amplifier is shadow grid controlled and uses an inverter power supply. The transmitter (and all other electronics in the AN/TPQ-36) are air-cooled. The receiver uses a GaAs FET preamplifier , and the IF is transmitted on coaxial cable to the operations shelter as in the AN/TPQ-37. All the electronics on the trailer are located in an enclosure above the azimuth bearing and thus rotate with the antenna when the antenna is repositioned to new sector. The system (including the operations shelter) uses less than 10 kW of 400 Hertz prime power.

Just as in the AN/TPQ-37, a boresight scope attached to the array is used for orienting the system. A shaft angle encoder measures the azimuth of the array and a tilt sensor mounted on the back of the array measures the attitude of the array in two orthogonal planes.

J.2.4 AN/TPQ-36 and AN/TPQ-37 Operation Shelter Electronics

The signal processing/data processing electronics are housed in a S250 shelter. Figure J.7 is a block diagram of the signal processor. The received IF data are synchronously detected with quadrature detectors and converted to 11 bit digital words. The quadrature channels are processed through a two-stage moving target indicator (MTI) canceller that serves as a pre-filter for the subsequent doppler filtering. This pre-filtering of the large amplitude clutter minimizes the truncation effects of using only 6 bit multiplying coefficients in the doppler filtering process. The digital MTI cascaded with digital doppler filters is essentially identical to the technology identified as Moving Target Detector (MTD). The doppler filters are formed in conjugate pairs with both phase and amplitude weighting to achieve the desired sidelobe suppression throughout the PRF interval. The low sidelobe design is based on optimizing the coefficients for each filter in order to maximize the signal to clutter plus noise ratio for that filter in the presence of an assumed spectral distribution of ground, rain, and bird clutter. The number of filters and their specific coefficients are different for the AN/TPQ-36 and AN/TPQ-37. This selection is under computer control as are several other signal processor parameters in order to realize a common set of hardware that can serve either mission.

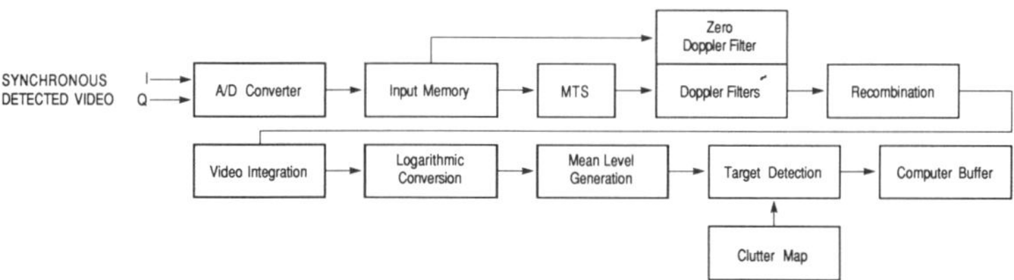

Figure J.7 Block diagram of Firefinder signal processor.

The doppler filter quadrature outputs are recombined into a single channel and can be multiple hit non-coherently integrated (again under computer control). The digital video from each filter is then converted to a base 2 logarithm scale so that subsequent thresholding processing can be carried out in decibel scale with additions and subtractions. A mean level detector with a computer-selected threshold offset is used to detect targets. A hardware clutter map is used to remember the spatial and doppler extent of slow-moving clutter. Note that the projectile targets of interest have radar cross sections that are extremely small, comparable to the radar cross section of birds. This map helps minimize the demand for radar energy to track birds and other slow-moving targets.

A general-purpose computer with 32K of 16 bit memory is used to control the radar. An additional 32K of memory is provided for automatic height correction of trajectories to the local terrain height. A 33-column line printer provides a permanent record of radar initialization parameters and weapon locations. A range-versus-azimuth B-display is provided for the operator to monitor radar search and tracking activity. However, the entire weapon location process is all automatic. A specially designed weapons location unit (WLU) is used to present to the operator the individual weapon location positions. This unit plots locations on a 1:50,000 scale universal transverse Mercator (UTM) projection map which is rolled onto a rotating transparent drum. Under computer control, the drum is rotated to the proper northing coordinate and a small light spot is projected from inside the drum through the map to indicate the easting coordinate. Light-emitting diode displays of weapon coordinates and other amplifying information are simultaneously provided to the operator. Communications to the system computer by the operator are through a keyboard and special switches, also located on this unit.

J.2.5 Firefinder Operation

The radar antenna is pointed to the sector of interest under remote control by the operator in the operations shelter. While the antenna operates stationary in that sector, the sector can be changed in a few seconds by demand of the operator. Each radar electronically scans the horizon of the sector with a single row of beams searching the sector a few times each second. The search rate is commensurate with detecting the projectile before it penetrates through the search fence. A new target is detected in one of the beams and is verified with a similar waveform whose pulse repetition frequency (PRF) has been slightly changed to permit resolution of any range or doppler ambiguities. During the search process, major changes in PRF are used to prevent blind zones in range and doppler. Both systems are basically operating range unambiguous for the vast majority of targets. However they do permit the acquisition, tracking, and location of weapons at ambiguous ranges.

Targets are tracked by updating the measurements (range, azimuth, elevation, doppler, and amplitude) several times each second. The tracking of multiple targets is time interlaced with the search process. Each target is tracked for a few seconds, thus measuring a small portion of the projectile trajectory. During the tracking process, a series of discriminates (tests on amplitude and time differentiated data) are supplied to the track in order to further reject birds, aircraft, and other clutter. The measurements along the trajectory are smoothed with a Kalman-type filter that estimates position and velocity in three orthogonal radar centered coordinates along with an estimation of a drag coefficient. This smooth estimate is then used for extrapolation down to the weapon firing altitude. The terrain height of the weapon position can be a trial height selected by the system computer and subsequently manually corrected by the operator or it can be automatically accessed from a digitized terrain data base stored permanently on magnetic tape cartridges, a portion of which can be located in the computer memory. The weapon location is transmitted to a fire direction center by either voice or automatic data link. A single operator can operate the radar although typically a team of two will be stationed in the shelter.

The radar's automatic weapon locating process (detect, verify track, discriminate extrapolate, display) is so rapid that the position of the hostile weapon is usually determined before the projectile impacts. The radar system computer can also extrapolate forward along the trajectory and indicate to the operator the expected impact point along with the estimated weapon firing position. The radars also have modes to register friendly weapons (cannon and mortars) by precision tracking of the friendly projectile trajectory. They also can be used to adjust fire predicting the impact point of the friendly weapon trajectory.

The AN/TPQ-36 also provides a 270-degree sectoring mode (extended azimuth) in which it will automatically search one 90-degree sector for a short period, track and locate targets in it, and then automatically turn to the next sector.

J.2.6 Firefinder System Maintainability

Maintainability and reliability were emphasized throughout the design of the system. More than 90% of all repairs can be performed in the field by the maintenance man normally assigned to the crew. The mean-time-to-repair (MTTR) is less than 30 minutes. There is built-in-test-equipment (BITE) throughout the radars. The BITE provides test stimuli under computer control and also reports to the computer the data from the test for computer analysis. There is an extensive series of tests that can be automatically executed to determine the system status prior to a mission. While operating during a mission, another set of on-line tests is automatically conducted on a continuous basis to detect the occurrence of any faults. To aid in the isolation of faults detected during the status test or during the on-line mission, a set of diagnostic software may be called in off-line. This diagnostic software is used to find the fault location and report that location to the maintenance man. In the vast majority of cases, this automatic isolation identifies the location down to the replaceable module.

The component packaging is modularized to be compatible with automatic fault isolation and rapid replacement. The modules are intended to be repaired at a higher echelon maintenance facility where special or general-purpose test equipment is available.

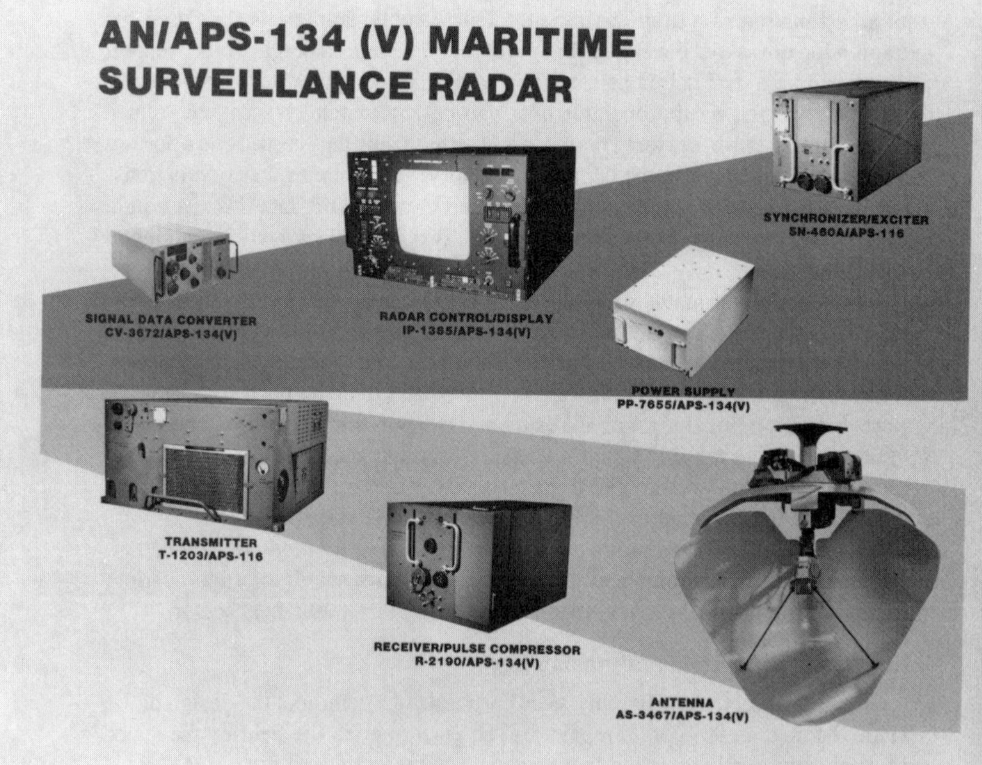

Figure J.8 AN/APS-134(V) maritime surveillance radar.

J.3 AN/APS-134(V) Maritime Surveillance Radar

The AN/APS-134(V) Radar System is the international successor to the U.S. Navy's AN/APS-116 Periscope Detecting Radar. The heart of the system is the fast-scan antenna and associated digital signal processing, which provide the means for eliminating sea clutter. The three operational modes are periscope detection, maritime surveillance, and navigation. Both the periscope and maritime surveillance modes employ rapid scan and scan-to-scan processing. The $512 \times 512 \times 8$ bit digital scan converter allows implementation of a multilevel (or shades-of-gray) processing scheme that provides for presentation of a background clutter map with overriding target information. System detection performance has been demonstrated in excess of 22 nmi on a 1 m^2 target in sea state 3. The maritime surveillance mode provides detection on larger targets out to 150 nmi in sea state 5. Figure J.8 identifies the units comprising the radar system.

J.3.1 AN/APS-134 (V) Design Considerations

The inherent problem of a maritime surveillance radar is detecting small targets in the sea-clutter environment. The combination of extremely high resolution coupled with scan-to-scan (S/S) processing provides effective means of overcoming this limitation. The AN/APS-134 incorporates both features to provide optimum antisubmarine and surface detection capability. The transmitted waveform is frequency modulated ("chirped") in a linear manner over a 500 MHz bandwidth. This permits the high average power necessary for long-range periscope detection in the noise environment. The received signal is pulse-compressed, using state-of-the-art surface acoustic wave devices, from a 0.5 microsecond (μs) to a 2.5 nanosecond (ns) pulse. This effectively reduces the sea surface area illuminated by the radar beam. Since the surface sea clutter return is a function of this area, the clutter is reduced by the pulse compression ratio (23 dB). Average clutter return however, is only part of the problem. Clutter spikes of 3 or more seconds compete with target returns to confuse the radar operator. To combat the spiky nature of the sea clutter return and further reduce the average value, fast scan processing is employed to time-decorrelate the clutter. The radar returns are processed on an S/S basis over a period of several seconds. The digital processing circuitry integrates the correlated target signals while rejecting the uncorrelated clutter.

J.3.2 AN/APS-134 (V) System Description

The basic radar set is comprised of modified AN/APS-116 units. The signal data converter (SDC) is the interface between the control commands and the radar set, and the input/output channels are tailored for the particular requirement. The SDC supplies composite radar video with 512 active lines for use by external multipurpose displays. Radar control is effected through a MIL-STD-1553B serial data bus.

In operation, the SDC, with the RC/D, provides processed radar video. It controls radar operation by performing aircraft and radar interface functions. Hence, the SDC is divided into two functional sections (plus a power supply). These are the digital scan converter (DSC) and the radar set control (RSC) sections. The DSC section accepts radar video and processes it by pulse-to-pulse (P/P) and S/S integration in the fast and intermediate scan speed modes. The processed data are stored in memory locations (write addresses) selected by the address generator. Read address from the generator recalls data from the memory for display. The data are converted to an analog signal and combined with synchronization pulses to provide a composite video signal for transmission to the display. The RSC section contains aircraft and radar interface and performs processing and timing functions. The processing and timing function provides the synchronization and processing needed to operate the radar and DSC. The timing portion provides clocks and triggers to the DSC for synchronization with radar timing. The processor stores and distributes the radar and DSC commands and performs arithmetic operations on the data from the aircraft systems, display control panel, and the antenna azimuth encoder. The results are used by the DSC.

J.3.3 AN/APS-134 (V) Operational Characteristics

Three operational modes are provided and they are discussed in the following paragraphs:

Mode I is optimized for the low-altitude detection of periscope-size targets of limited exposure time. The compressed pulse equates to a range resolution of less than 1.5 feet, which closely matches the physical dimension of periscope and snorkel classes of targets. The rapid scan antenna rotates at 150 rpm to provide the necessary S/S processing gain in a short time. Maximum range in this mode is 32 nmi.

Mode II is the navigation and weather avoidance mode and is designed for maximum sensitivity against land and other distributed targets. The techniques associated with the fast-scan modes emphasize point target detection while rejecting distributed or area targets (i.e., sea clutter). Therefore, this mode employs a slow antenna rotational rate and an uncompressed waveform. Maximum range is 150 nmi.

Mode III provides high altitude maritime surveillance to ranges of 150 nmi. Again, the use of pulse compression and fast-scan processing allows detection of patrol/fishing boat-size targets and larger under all sea conditions. These targets are continuously exposed, and longer integration times (i.e., S/S processing intervals) can be employed. Thus, the antenna rotational rate can be slowed to 40 rpm to minimize processing losses.

Table C-2 is a listing of the system parameters of the AN/APS-134.

Table C.2　AN/APS-134(V) System Parameters

<u>Antenna</u>

Radar:

Gain	35 dB (nominal)
Azimuth beamwidth	2.4 degrees
Elevation beamwidth	4.0 degrees
Sidelobes	Down at least 20 dB
Polarization	Vertical

Scan speed

Mode I	150 rpm
Mode II	6 rpm
Mode III	40 rpm
IFF (integral):	Standard IFF interrogator sets monopulse compatible

<u>Transmitter</u>

Peak power	500 kW
Average power	500W

Frequency

Mode I, III (linear FM sweep)	9.5 to 10.0 GHz
Mode II (random frequency agility)	9.6 to 9.9 GHz

PRF (four-pulse stagger)

Mode I	2,000 pps
Mode II, III	500 pps
Pulsewidth	0.5µs

<u>Receiver</u>

Noise figure (system)	4.5 dB

Intermediate frequency

Mode I, III	1300 MHz
Mode II (dual conversion)	1300 MHz and 100 MHz

IF bandwidth

Mode I, III	500 MHz
Mode II	2.4 MHz
Pulse compression gain (Mode I, III)	23.0 dB
Compression pulsewidth (Mode I, III)	2.5 ns
AGC	Three loops, CFAR

J.3.4 AN/APS-134 (V) Signal Processing

The choices of modes/parameters and post detection processing centered on the use of either P/P or S/S integration, or a combination of both. In the slow-scan antenna mode, P/P integration results in reduced signal-to-noise (S/N) ratio requirements at the input to the detector for a given detection probability and false alarm probability. Pulse-to-pulse integration does not improve performance in the sea clutter case because clutter returns are highly correlated for short time intervals (e.g., milliseconds) between pulses. Scan-to-scan integration over longer time periods (e.g., seconds) can improve both S/N and signal-to-clutter (S/C) ratio. The processing gain or degradation in S/N or S/C depends primarily on the time interval and number of samples for integration.

Figure J.9 (A) shows a generic block diagram of the signal processing chain, although not all elements of this chain are used in each mode of operation.

The digitizer converts the analog video to digital data either by two-level digitization (thresholding) or multiple-level digitization (quantization). Range stretching is a method of reducing the number of range resolution elements to an acceptable value for subsequent processing and display. The stretching consists of taking the peak value in a number of successive range cells as representative of that group, which is then treated as one stretched range cell. The total range stretch is determined by the range resolution inherent in the radar signal structure and the fact that the digital scan converter can accommodate only 512 stretched range cells. Figure J.9 indicates that the total range stretch can be apportioned in the processing chain to optimize trade-offs between memory requirements and/or processing rate and performance.

An additional signal processing feature of the AN/APS-134 is multilevel processing, which allows preservation of a background clutter map. This enhances the ability of the radar operator to discriminate between debris (or other nonsignificant returns) and small targets of interest. The clutter (or ocean) map is generated by allowing the large area returns to integrate within the digital scan converter memory to a preset amplitude (Level 80), which represents approximately one-third of the display dynamic range. Point targets are allowed to integrate to the full memory depth (Level 256) and thus be displayed at the maximum intensity. The presentation is an underlay representing the ocean surface with target returns appearing as intensified spots. Both surface and subsurface currents can be distinguished along with predominant wave structures. The relationship of the point targets to the clutter map provides the basis for operator discrimination. An additional benefit of the clutter-mapping technique is the case of detecting calm or smooth areas within the ocean structure. These are readily apparent and can be investigated for possible contamination from oil spills or leakage.

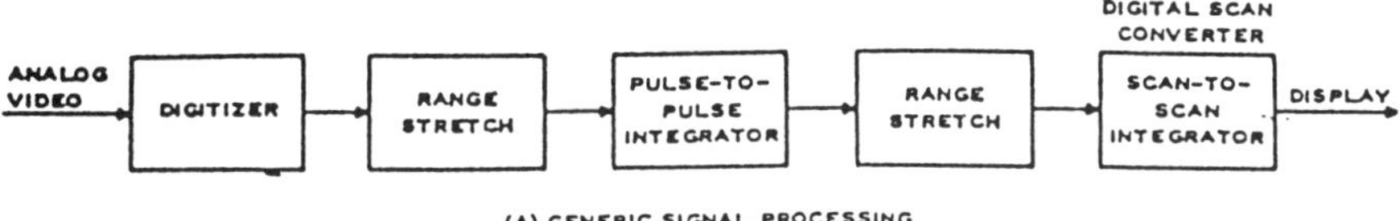

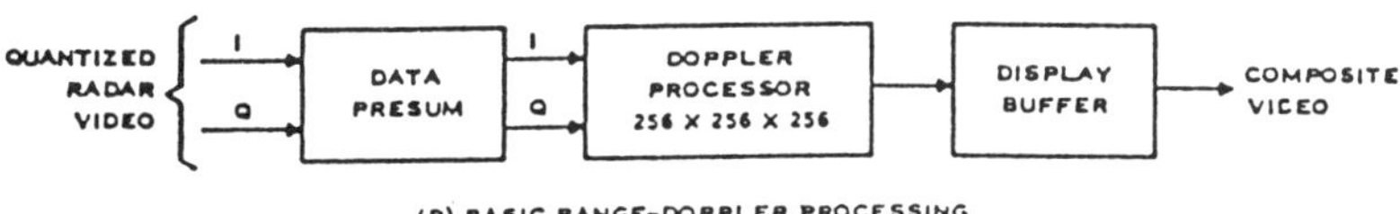

Figure J.9 Signal processing block diagram.

J.3.5 AN/APS-134 (V) Performance

Considerable analysis has been done in calculating and predicting detection ranges. Optimization routines with the basic AN/APS-116 parameters were used to develop new digital processing algorithms and antenna rotation rates. The system loss budget contains all installation losses including radome. Final results of the analyses are shown in Table J.3. These performance data have been confirmed by field testing, flight qualification, and evaluation testing.

Table J.3 AN-APS-134(V) Detection Performance

TARGET TYPE	RADAR CROSS SECTION (m^2)	DETECTION RANGE (nmi)	SEA STATE	ALTITUDE (FEET)
Periscope	1	22	3	2,500
Snorkel	10	38	4	8,000
Patrol/fishing boat	100	65	5	20,000
Small transport or destroyer	1,000	105	5	20,000
Cruiser	3,300	140	5	20,000
Aircraft carrier or weather	>5,000	>150	5	25,000

(Probability of detection (P_d) = 0.5, probability of false alarm (P_{FA}) = 10^{-6})

J.3.6 AN/APS-134 (V) Radar Imaging Growth

The AN/APS-134 is compatible with the inverse synthetic aperture radar (ISAR) imaging technique. This processing generates an image of a recognizable nature on surface ships for use in long-range ship classification. The U.S. Naval Research Laboratory developed the ISAR concept, which actually replies on ship motion.

The AN/APS-134 pulse compression chain and its attendant coherent operation provide the two key parameters of image generation.

The orthogonal axes used for the display of ISAR images represent directions parallel to the line of sight (range) and perpendicular to the line of sight (cross range). High resolution in the range dimension is obtained from the pulse compression. Radial velocity, and its effect on the returned signal phase, is used for resolution in the cross-range direction. Ship motion is the sole source of the radial velocities.

ISAR images are essentially maps indicating the location and reflexivity of the component scattering elements comprising a target. The positions of these elements are displayed in Cartesian coordinates proportional to range and cross range. Range locations are determined by the round trip time of the signals. Cross-range measurement is dependent on target motion, with the radial velocity of the target being proportional to its cross-range location. Amplitude of the returned signals is plotted as intensity and indicates the radar cross section (reflexivity) at that location.

Generating the cross-range location requires selecting a reference point on the target and measuring target rotational motion relative to that point. A fundamental prerequisite for imaging is that any relative motion between the radar and the reference point must be compensated so that only the effects of target rotation remain. This motion compensation is achieved through highly precise range tracking and doppler tracking circuit functions to stabilize the reference point in both dimensions.

The type of image generated is determined by the nature of the ship motion and the location of the radar with respect to that motion. Figure J.10 depicts yaw motion with the radar as shown. In this case, a plan view is produced. Figure J.11 represents pitch motion and generates the profile view. In the true environment, ship motion is a combination of yaw, pitch, and roll (also producing a profile view). These combined motions generate images that are isometric in appearance, which enhances the ability of the interpreter to recognize and classify the ship.

The ISAR image is basically a range-doppler plot generated from a time aperture of coherent radar video. Figure J.9 (B) is a block diagram illustrating the three basic functions that comprise ISAR range-doppler processing: data presum, doppler processing, and display buffering.

The presum provides coherent integration and data reduction, providing the bandwidth of the target data is much less than the radar pulse repetition frequency (PRF). The presum, therefore, reduces the effective PRF and performs integration that otherwise would be done by the doppler processor. This effective PRF reduction is important because it allows additional time to generate the required frequency data.

Spectral analysis is performed in the doppler processor. For each sample range, the doppler processor converts time domain data cells to the frequency domain. This function can be performed by various techniques, the most common being the fast Fourier transform (FFT). The FFT is an efficient algorithm when the required image update rate is on the order of the time aperture. As the image update rate increases, the FFT becomes less attractive. Each time the image is updated, a complete time aperture of data must be processed. This means that the data are processed over and over. An FFT system that would provide an output for every time input would require $(N/2)\log_2 N$ complex multiplications to produce each updated range cell output, where N is the number of samples comprising an aperture. This means that each range cell would require $N_2 \log_2 \sqrt{N}$ complex multiplications per time aperture.

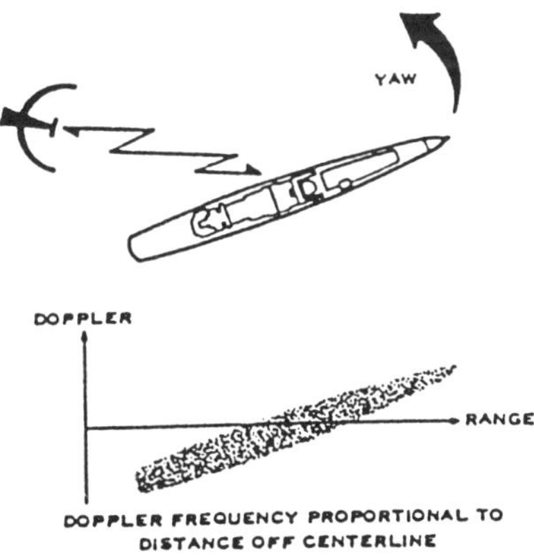

Figure J.10 Yaw motion imaging.

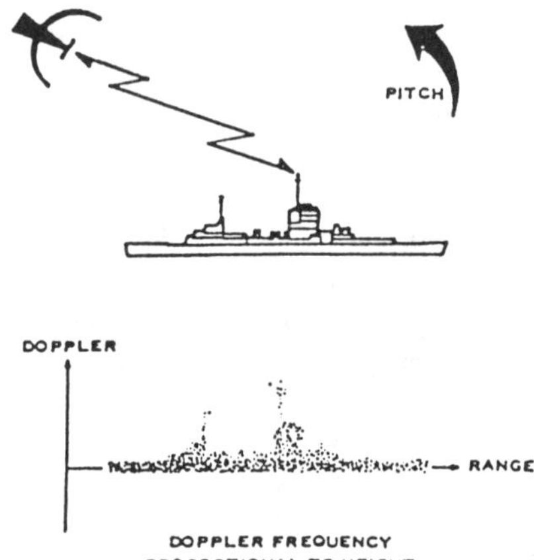

Figure J.11 Pitch motion imaging.

A unique doppler processor has been developed by Texas Instruments that produces an updated output for each time input and requires only N complex multiplications per range cell update. Each range cell, therefore, requires N^2 complex multiplications per time aperture. This technique produces *a real-time continuous presentation* of range-doppler imagery and is called a continuous Fourier transform (CFT).

The processor transforms 256 points for 256 range cells at a rate of 256 operations per second. The doppler filter resolution is equal to the reciprocal of the effective aperture time τ_A. This implies that PRF and number of pulses presummed can be used to control the doppler filter resolution while keeping the number of filters constant. Thus, the complexity of the doppler processor and display circuitry is minimized. Providing the ISAR operator with control of τ_A allows doppler resolution and physical size of the image in the doppler dimension to be adjusted to compensate for ship motion at the time of observation.

The display buffer stores image data and converts them to a form compatible with a multipurpose display system.

J.3.7 AN/APS-134(V) Summary

The AN/APS-134 radar system is an evolutionary improvement of the AN/APS-116 system. It greatly enhances the periscope detection mission and provides the high-altitude capability required of modern maritime surveillance aircraft. An optional multiple-target tracking feature allows for long-term maintenance of the surface plot. Physical characteristics are detailed in Table J.4.

Table J.4 AN/APS-134 (V) Physical Characteristics

Unit	Weight (pounds)	Dimensions (H × W × D inches)	Power Dissipation (watts)
Transmitter	174	13 × 22 × 21	3,495
Synchronizer-exciter	36	10 × 10 × 18	155
Receiver-pulse compressor	48	11 × 13 × 16	149
Power supply	44	9 × 12 × 19	365
Antenna	62	36 × 41 × 27	221
Signal data converter	50	8 × 15 × 20	310
Waveguide pressurizer	15	13 × 12 × 5	
Total	429		4,695
Optional radar control/display	98	16 × 23 × 23	300
TOTAL	527		4,995

When available, addition of ISAR ship imaging will allow real-time target classification to take place at standoff ranges that are beyond the lethality of current and projected surface-to-air weapons.

J.4 References for Appendix J

[J.1] H. R. Ward, "The RAMP PSR, A Solid-State Surveillance Radar," IEE International Radar Conference, RADAR 87, London U.K., October 1987, pp. 150-154.

[J.2] J. C. MacDonald and H. R. Ward, "RAMP - Canada's New ATC System," Journal of ATC, September 1986.

[J.3] J. D. Dyck and H. R. Ward, "RAMP's New Primary Surveillance Radar," Microwave Journal, December 1984.

[J.4] "RAMP Radar Modernization Project," Aerospace Canada International, July/Aug. 1984.

[J.5] D. Wooley, "Canada Leads with Stand-Alone SSR," Interavia, September 1984.

[J.6] K. Milne, "A Survey of Primary Radars for Air Traffic Systems," AGARD 20th Symposium, Cambridge, Mass., May 1975.

[J.7] V. B. McCrae and H. R. Ward, "A Three Pulse Moving Target Detector," IEEE 1980 International Radar Conference Record, Arlington, VA, April 1980, pp. 206-210.

[J.8] D. A. Ethington, "The AN/TPQ-36 and AN/TPQ-37 Firefinder Radar Systems," EASCON, 1977.

[J.9] Eli Broockner, ed., *Aspects of Modern Radar*, Artech House, 1988.

[J.10] J. M. Smith and R. H. Logan, "AN/APS-116 Periscope-Detecting Radar," IEEE Trans. **AES-16**, No. 1, January 1980, pp. 66-73.

Appendix K

RULES OF THUMB AND
SENSITIVITY OF PERFORMANCE
TO RADAR PARAMETERS

K.1 Frequency Selection for Radar Applications

The range of frequencies over which different radar systems have been built
to operate, covers a spectrum nearly 17 octaves wide; from HF radars in the 3
MHz region to millimeter-wave radars operating at 300 GHz or higher. The inverse
relationship between the physical size of RF components (such as waveguide and
antenna) and frequency of operation plays a significant part in the selection of
operating frequency for most airborne and similar applications where small size,
efficient packaging, and limited prime power are the driving considerations, but
the effects of the earth's atmosphere and its weather usually dominate that selection
for most radar systems that must operate in, or look through, that atmosphere.

Three different atmospheric phenomena which influence the propagation of
radar energy (*attenuation, refraction,* and *diffraction*), and the effects of a fourth
influence, *reflection* from the earth's surface, were discussed in Chapter 8. A fifth
effect, that of *backscatter* due to rain or chaff, was described in Chapter 5. Table
K.1 summarizes the predominant applications for radars by operating frequency
band, including a qualitative evaluation of the primary effects of weather in that
band. It should be noted that these relationships hold for the general case, and that
special-purpose radars may fall outside this categorization.

The symbols used here are defined in Appendix B.

Radar Frequency

Typical Applicatons

Band Designation	Frequency Coverage	Weather Effects*	V.Long Range Surveillance	Long Range Surveillance	Short Range Surv/Acq	Long Range Tracking	S.Range Fire Control	Missile Term. Homing
HF	3-30 MHz	almost none	•					
VHF/UHF	30-300 MHz 300-1000MH	very slight		• •	• •			
L	1000-2000 MHz	slight		•	•			
S	2000-4000 MHz	moderate		•	•		•	
C	4000-8000 MHz	moderate			•	•	•	•
X	8000-12000 MHz	mod/strong			•	•	•	•
Ku	12-18 GHz	strong					•	•
Ka	27-40 GHz	severe						•
MMW	40-300 GHz	very severe						•

*rain & atmospheric absorption

Table K.1 Radar Application by Frequency Band

K.2 Rules of Thumb

K.2.1 Frequency and Wavelength

$$f = \frac{c}{\lambda} = \frac{3 \times 10^8 \ (\text{m/s})}{\lambda} \frac{}{(\text{m})} = \frac{30 \ (\text{GHz})}{\lambda \ (\text{cm})}$$

K.2.2 Average Power

$$P_{\text{av}} = \frac{P_t \tau}{t_r} = P_t \tau f_r \quad \text{where } P_t \text{ is peak power}$$

K.2.3 Unambiguous Range

$$R_u = \frac{c t_r}{2} = \frac{c}{2 f_r}$$

K.2.4 Solid Antenna Beam Angle

$$\psi_b = \theta_a \theta_e = \frac{4\pi}{G_t L_n} \quad \text{steradians}$$

K.2.5 Observation Time (Time-on-Target)

$$t_o = \frac{\theta_3}{\omega_a} = \frac{t_o \psi_b}{\psi_s}$$

K.2.6 Horizon Range

$$R_{ht} = \sqrt{2ka}(\sqrt{h_t} + \sqrt{h_r}) \approx 4123(\sqrt{h_t} + \sqrt{h_r}) \ \text{m}$$

K.2.7 Doppler Frequency

$$f_d = \frac{2v_r}{\lambda}$$

K.2.8 Angle Accuracy (Thermal Noise)

K.2.8.1 Sector Scanning

$$\sigma_\theta \approx \frac{0.5\theta_3}{\sqrt{(S/N)_m n}}$$

K.2.8.2 Monopulse Tracking

$$\sigma_\theta \approx \frac{\theta_3}{2\sqrt{(S/N)n}}$$

K.2.8.3 Conical Scan

$$\sigma_\theta \approx \frac{0.7\theta_3}{\sqrt{(S/N)_m n}}$$

K.2.9 Antenna Gain and Beamwidth

Parameter	Uniform Illumination	Tapered Illumination
Rectangular		
Gain	$G_0 = \pi T$	$G_m = \pi T \eta_a$
$G_0 = \dfrac{4\pi A}{\lambda^2}$	$G_0 = \dfrac{9.84}{\theta_a \theta_e}\,(\text{rad})$	$G_m = \dfrac{11.3}{\theta_a \theta_e}\,(\text{rad})$
	$G_0 = \dfrac{32,300}{\theta_a \theta_e}\,(\text{deg})$	$G_m = \dfrac{37,100}{\theta_a \theta_e}\,(\text{deg})$
Number of Elements		
$T = n_a \times n_e$	$T = \dfrac{2A}{\lambda^2}$	$T = \dfrac{10,300}{\theta_a \theta_e}\,(\text{deg})$
Beamwidth		
	$\theta_a = 0.866\lambda/w$ $\theta_e = 0.866\lambda/h$	$\theta_a = 1.19\lambda/w$ $\theta_e = 1.19\lambda/h$
Circular Antennas		
Gain	$G \approx \dfrac{(\pi D)^2}{\lambda^2}\eta_a$	$G \approx \dfrac{6D^2}{\lambda^2}\quad(\eta_a = 0.6)$
Beamwidth	$\theta_3 = 1.02\dfrac{\lambda}{D}$	$\theta_3 \approx 1.25\dfrac{\lambda}{D}$

K.3 Sensitivity of Performance

K.3.1 First-Order Effects

To illustrate the sensitivity of radar performance to various radar parameters, we will make use of the search radar equation (3.2.3) given in Chapter 3:

$$R_m^4 = \frac{P_{av}A_r t_o \sigma}{4\pi\psi_s kT_s D_0(1)L_s}$$

and note that radar frequency does not appear. Note also that nothing appears in the equation relative to the type of waveform or modulation imposed on a carrier frequency. This tells us that given a desired search volume ψ_s, and search frame time t_s, the radar range depends primarily on the power-aperture product and the target radar cross sectional area. The effects of radar frequency are implicit in the receiving antenna aperture area A_r, and are also inherent in the system loss factor L_s, where several of the individual loss components are a function of operating frequency.

The second important point illustrated by this form of the range equation is that the range is proportional to the ratio (t_s/ψ_s). This simply reflects that fact that as the time taken, by an antenna with beamwidth θ_3, to scan the same search volume increases, the observation time (or time-on-target) increases as well, and with it the total energy exchanged with the target. Selection of the radar's frame time is a compromise among such factors as the maximum design target speed, the probability of detection (and false alarm) that is required per scan, and the magnitude of the search volume. If the radar is an integral part of an air defense system, then additional factors must be considered: the detection range required on a given size target, e.g., with a cumulative probability of 0.9, to allow an engagement by some minimum acceptable intercept range.

K.3.2 Controllable Parameters and Their Effects on Radar Range

K.3.2.1 Power vs. Antenna Gain

The range equation shows that in order to double the detection range against a given size target by increasing the average power of the radar, we would need to increase the average power by 12 dB. If, for example, the average power of the radar were 350 watts, it would have to be increased to about 5500 watts to achieve a doubling in detection range. If the radar is a high-prf pulsed doppler radar operating with a 1% duty cycle, this means that the peak power would have to be increased from 35 kW to over one-half of a megawatt. Even if such an increase were technically feasible, the implications for system design modifications to handle such an increase, independent of economics, size, and other operational factors, are formidable.

Increasing the antenna dimensions, and hence gain, offers an alternative that may be acceptable in some situations, e.g., in ground or ship-based applications. Here the 12 dB increase can be divided between the transmit and receive antenna gain. Because antenna area is directly proportional to antenna gain, increasing the antenna size to four times its original area will achieve 6 dB increase in gain for both transmit and receive antennas, and this increase can be achieved by doubling the linear dimensions of a rectangular antenna (or the diameter of a circular antenna). Figure K.1 shows the influence of increasing average power and antenna receiving area on detection range for the S-band search radar described in Chapter 3, where the range in km is described by the equation $R = 15.85 P_{av} A_r$.

The steep slope due to increase in receiving area as compared to average power is clearly seen. The "baseline" radar point is shown in the figure, having an average power of 110 W, a receiving antenna area of 8 m^2, and a free-space detection range of 86 km vs. a 1 m^2 target.

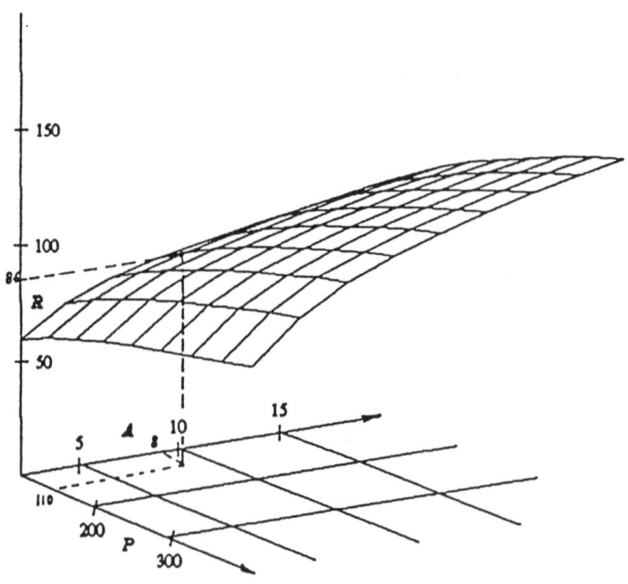

Figure K.1 Effect of power-aperture on radar range.

These two extremes have been discussed to illustrate a point, but real-world requirements seldom offer "either-or" solutions of the nature just described. In doubling the antenna area, for example, the antenna beamwidth is automatically reduced by one-half, and this results in an increase in the time required to cover the search volume. Depending on the specific radar application, this increased scan time may or may not be acceptable.

Most successful radar designs are the result of judicious trade-offs of just such factors as these, and generally include maximizing the power-aperture product in addition to including design features which minimize the controllable sources of loss.

K.3.2.2 System Losses and Noise

In the example used in Chapter 3, it was shown that for a single-frequency search radar, system losses on the order of 20 dB can be typical, with the dominant loss due to target fluctuation when a high probability of detection per scan is required ($L_f \approx 8$ dB for $P_d = 0.9$/scan). At the cost of additional complexity, both time and frequency diversity could be incorporated in a search radar design to reduce fluctuation loss to some fraction of its value for a nominal search radar. If fluctuation loss were decreased by these means to very low levels, e.g., 0.5 dB or so, the detection range would increase by a factor of about 1.5. A closer examination of the remaining loss contributors, which total 12 dB or so, shows little opportunity for any additional reduction in the system loss budget, in that most of these losses are the unavoidable consequence of nature.

The system noise power density kT_s, which also appears in the denominator of the range equation, includes the effects of antenna and RF component thermal and ohmic losses, in addition to receiver thermal noise. Placing the transmitter and receiver close to the antenna terminals (as in an active array radar) will reduce the transmission line loss component of system noise, and the use of low-noise "front-end" RF amplifiers will minimize the receiver noise figure for a given radar. If we again use the S-band radar example given in Chapter 3, where $T_s = 967$ K, and assume that by employing the techniques just described we have reduced the line losses to 0.5 dB, and have also reduced the receiver noise figure to 2.5 dB, then a new $T_s' \approx 400$ K results. Thus, the detection range will be increased by the factor $(T_s/T_s')^{1/4}$, or by approximately 1.25. Two points are noted: (1) we are unlikely to achieve any further reduction in system noise power density, since the other contributors to T_s are beyond our control, and (2) the improvement in detection range achieved by reducing system noise level applies only in a benign environment, for if the radar were required to operate against a background of noise jamming, the advantage just gained would be quickly neutralized.

K.3.2.3 Effect of Target RCS

Typically, radar range performance is referenced to a target having an average radar cross sectional area of 1 m^2 (0 dBm^2). The range for other cross sections can then be easily computed from the relationship:

$$R_{new} = R_{ref}\left(\frac{\sigma_{new}}{\sigma_{ref}}\right)^{1/4} = R_{ref}\,10^{0.025\sigma_{dB}}$$

where the term σ_{dB} is the new target radar cross section expressed in dBm^2. An easily remembered rule-of-thumb states that for every 10 dB change in target RCS (or any other parameter in the range equation), radar range changes by a factor of 1.78. Thus, if the range of our example S-band radar is 86 km vs. 1 m^2, then its range versus a 10 m^2 target will be 1.78 × 86, or 153 km, if all other radar parameters remain the same. Similarly, the range versus a target whose cross section is 0.1 m^2 will be 86/1.78, or 48.3 km.

The Artech House Radar Library

David K. Barton, *Series Editor*

Active Radar Electronic Countermeasures by Edward J. Chrzanowski

Airborne Pulsed Doppler Radar by Guy V. Morris

AIRCOVER: Airborne Radar Vertical Coverage Calculation Software and User's Manual by William A. Skillman

Analog Automatic Control Loops in Radar and EW by Richard S. Hughes

Aspects of Modern Radar, by Eli Brookner, *et al.*

Aspects of Radar Signal Processing by Bernard Lewis, Frank Kretschmer, and Wesley Shelton

Detectability of Spread-Spectrum Signals by Robin A. Dillard and George M. Dillard

Electronic Homing Systems by M.V. Maksimov and G.I. Gorgonov

Electronic Intelligence: The Analysis of Radar Signals by Richard G. Wiley

Electronic Intelligence: The Interception of Radar Signals by Richard G. Wiley

EREPS: Engineer's Refractive Effects Prediction System Software and User's Manual, developed by NOSC

Handbook of Radar Measurement by David K. Barton and Harold R. Ward

High Resolution Radar by Donald R. Wehner

High Resolution Radar Imaging by Dean L. Mensa

Interference Suppression Techniques for Microwave Antennas and Transmitters by Ernest R. Freeman

Introduction to Electronic Warfare by D. Curtis Schleher

Introduction to Sensor Systems by S.A. Hovanessian

Logarithmic Amplification by Richard Smith Hughes

Modern Radar System Analysis by David K. Barton

Monopulse Principles and Techniques by Samuel M. Sherman

Monopulse Radar by A.I. Leonov and K.I. Fomichev

Multifunction Array Radar Design by Dale R. Billetter

Multisensor Data Fusion by Edward L. Waltz and James Llinas

Multiple-Target Tracking with Radar Applications by Samuel S. Blackman

Multitarget-Multisensor Tracking: Advanced Applications, Yaakov Bar-Shalom, ed.

Over-The-Horizon Radar by A.A. Kolosov, et al.

Principles and Applications of Millimeter-Wave Radar, Charles E. Brown and Nicholas C. Currie, eds.

Principles of Modern Radar Systems by Michel H. Carpentier

Pulse Train Analysis Using Personal Computers by Richard G. Wiley and Michael B. Szymanski

Radar and the Atmosphere by Alfred J. Bogush, Jr.

Radar Anti-Jamming Techniques by M.V. Maksimov, *et al.*

Radar Cross Section by Eugene F. Knott, *et al.*

Radar Detection by J.V. DiFranco and W.L. Rubin

Radar Propagation at Low Altitudes by M.L. Meeks

Radar Range-Performance Analysis by Lamont V. Blake

Radar Reflectivity Measurement: Techniques and Applications, Nicholas C. Currie, ed.

Radar Reflectivity of Land and Sea by Maurice W. Long

Radar System Design and Analysis by S.A. Hovanessian

Radar Technology, Eli Brookner, ed.

Receiving Systems Design by Stephen J. Erst

Radar Vulnerability to Jamming by Robert N. Lothes, Michael B. Szymanski, and Richard G. Wiley

RGCALC: Radar Range Detection Software and User's Manual by John E. Fielding and Gary D. Reynolds

SACALC: Signal Analysis Software and User's Guide by William T. Hardy

Secondary Surveillance Radar by Michael C. Stevens

SIGCLUT: Surface and Volumetric Clutter-to-Noise, Jammer and Target Signal-to-Noise Radar Calculation Software and User's Manual by William A. Skillman

Signal Theory and Random Processes by Harry Urkowitz

Solid-State Radar Transmitters by Edward D. Ostroff, *et al.*

Space-Based Radar Handbook, Leopold J. Cantafio, ed.

Statistical Theory of Extended Radar Targets by R.V. Ostrovityanov and F.A. Basalov

The Scattering of Electromagnetic Waves from Rough Surfaces by Petr Beckmann and Andre Spizzichino

VCCALC: Vertical Coverage Plotting Software and User's Manual by John E. Fielding and Gary D. Reynolds